Recent Trends in Physics Research

(A Proceedings of
XIII Biennial National Conference of
Physics Academy of North East, Pane-2022)

Recent Trends in Physics Research

(A Proceedings of XIII Biennial National Conference of Physics Academy of North East, Pane-2022)

Editor in Chief

Prof. N. Nimai Singh
Manipur University

Editor

Dr. Thongam Gomti Devi
Manipur University

Associate Editors

Prof. A. Dilipkumar Singh
Manipur University

Prof. Kushal Kalita
Gauhati University

Assistant Editors

Dr. N. Chandrachani Devi
Manipur University

Dr. M. Shubhakanta Singh
Manipur University

ALLIED PUBLISHERS PVT. LTD.

Noida • Mumbai • Kolkata • Chennai • Bangalore

ALLIED PUBLISHERS PRIVATE LIMITED

D-5, Sector-2, **Noida**–201 301
Ph. Nos.: 0120-4320295/2542557/4352866 • E-mail: delhi.books@alliedpublishers.com

17 Chittaranjan Avenue, **Kolkata**–700072
Ph.: 033-22129618 • E-mail: cal.books@alliedpublishers.com

15 J.N. Heredia Marg, Ballard Estate, **Mumbai**–400001
Ph.: 022-42126969 • E-mail: mumbai.books@alliedpublishers.com

No. 25/10, Commander-in-Chief Road, Ethiraj Lane (Next to Post Office)
Egmore, **Chennai**–600008
Ph.: 044-28223938 • E-mail: chennai.books@alliedpublishers.com

P.B. No. 9932, No. 15, 3rd Floor (Next to Vijaya Bank), 5th Cross, Gandhinagar, Karnataka,
Bangalore–560009, Ph.: 080-41530285 / 22386239
E-mail: bngl.journals@alliedpublishers.com / apsabng@airtelmail.in

Website: www.alliedpublishers.com

ISBN: 978-93-90951-66-6

Published by Sunil Sachdev and printed by Ravi Sachdev at Allied Publishers Pvt. Ltd.,
D-5, Sector-2, Noida–201 301

Editorial

The "Recent Trends in Physics Research" is a book chapter collected from the papers submitted by participants in XIII Biennial National Conference of Physics Academy of North East (PANE-2022). The PANE-22 was organized by Department of Physics, Manipur University under the Convenorship of Dr. Thongam Gomti Devi from 8-10 November, 2022 and several delegates and participants from across the country and abroad have delivered Keynote address, Plenary talks, invited talks, oral and poster presentation during the three days conference.

This book provides a comprehensive work of the advanced field of different areas of Physics such as Condensed Matter Physics and Material Science, Atomic, Molecular Physics and Spectroscopy, Astrophysics, Cosmology and Nuclear Physics, High Energy Physics, Communication and Instrumentation, Interdisciplinary Research and Atmospheric physics, respectively. The contributed papers in this book chapter showcase the state-of-the-art experimental techniques, methodologies, theoretical simulations, molecular modelling and computational methods of different areas of Physics. The contributed chapters in this book focus on their new research and findings done in their research lab or in collaboration with other researchers working in same disciplines or different disciplines across the country.

Lastly, the book "Recent Trends in Physics research" is the mirror of the PANE-2022 conference and highlighting the research works of North-East Institutions and other Institutions across the country working in several branches of Physics.

I hope that the mass readers will get benefitted from this book, in earning knowledge and ideas which will help them in their ongoing research and pursuing of new research area of interest.

Dr. Th. Gomti Devi

Editorial

Contents

SECTION-II: Atomic Molecular Physics and Spectroscopy

SECTION-III: High Energy Physics

SECTION-IV: Astrophysics, Cosmology and Nuclear Physics

SECTION-V: Communication and Instrumentation

SECTION-VI: Interdisciplinary Research and Atmospheric Physics

Section-I

Condensed Matter Physics and Material Science

Effect of Spin-Orbit Coupling on Pt_2MnGa

Bhargab Kakati* and Munima B. Sahariah
Institute of Advanced Study in Science and Technology,
Guwahati-781035, Assam, India
*E-mail: bhargabkakoti@gmail.com

ABSTRACT: Heusler alloys are currently one of the most studied classes of multicomponent systems. In recent years, researchers have become increasingly interested in Heusler alloys due to their wide range of tunable properties. Recent investigations have identified Mn-based Heusler alloys as one of the potential candidates for spintronics and magnetic storage devices. This chapter uses density functional theory to briefly analyse the ground-state magnetic configuration of the full-Heusler alloy Pt_2MnGa both with and without considering the effect of spin-orbit coupling. The nature of exchange coupling parameters of the system was also discussed in this chapter.

Keywords: Heusler Alloy, Spin-Orbit Coupling, Exchange Coupling.

I. Introduction

In the field of spintronics, the devices utilize the electron spin instead of electron charge to carry information. The addition of the spin degree of freedom gives traditional charge-based electronic devices much-desired properties like fast data processing, nonvolatility, high-density recording, and lower power consumption [1]. Some Heusler alloys have shown to be one such material in the ongoing search for new magnetic materials that are suited for spintronic devices. Heusler alloys are magnetic alloys made up of elements that are not inherently magnetic in their isolated form. These ternary intermetallic compounds have the chemical formula X_2YZ (full Heusler alloy) or XYZ (half Hesuler alloy), where X and Y stand for transition metals and Z stands for a main group element. Mn-based Heusler alloys have features including spin transfer torque, spin hall effect, rare earth free hard magnetism, and high curie temperature that make them particularly interesting for use in spintronics and data storage devices [2-4].

One of the Mn-based Heusler compounds, Pt_2MnGa, was recently found to exhibit some desirable structural and electronic properties. The Pt_2MnGa single crystal system was found to be antiferromagnetic (AFM) at high temperature cubic phase in the first experimental report on the material [5].

A recent study on the compound [6] also reported an AFM magnetic configuration. According to Feng *et al.* [7], the ternary Heusler compound Pt_2MnGa possesses a stable tetragonal phase and is ferromagnetic (FM) in its ground state. At the same time, there is a possibility of having large magnetic-field-induced strains (MFIS) in $Pt_{2-x}Mn_{1+x}Ga$ (x = 0, 0.25, 0.75, 1). In another work done by Roy *et al.* [8], it was found that among Ni_2MnGa, Pt_2MnGa, and Mn_2NiGa, Pt_2MnGa has the least brittleness. The same kind of result has also been reported by Yang *et al.* [9], where they found that Pt_2YZ (Y = Cr, Mn, Fe, and Co; Z = Ga, In, Ge, and Sn) are more ductile as compared to widely known Ni_2-based alloys. Three distinct stable surface configurations have recently been anticipated for the material when taking into account the ferromagnetic phase [10].

In this chapter, we aim to settle the debate about its ground-state magnetic configuration with the help of *ab initio* density functional theory. Since Pt_2MnGa contains a heavy element (Pt), we cannot ignore the effect of spin-orbit coupling (SOC) in our calculations. Due to this, we conducted two sets of computations, one without taking SOC into account and the other with it.

II. STRUCTURAL AND MAGNETIC PROPERTIES

Pt_2MnGa assumes a tetragonal structure corresponding to the space group *I4/mmm* [11], with lattice constants $a = b = 4.0174$ Å and $c = 7.2393$ Å. To obtain the optimized geometry, two sets of calculations were done, one for the FM arrangement and the other for the AFM arrangement. Each of these calculations was done with and without considering the effect of SOC. The optimized lattice constants obtained from these simulations were found to be almost similar for FM and AFM configurations. The implementation of SOC does not significantly affect the lattice parameters. On the other hand, the incorporation of SOC clearly has an impact on the ground-state energy. Even though the SOC lowers the ground state energy of the FM and AFM configurations by more than 8 eV, it does not alter the relative stability of the compound.

The ground-state energy obtained from our calculation revealed that the ferromagnetic spin configuration is the most stable one for Pt_2MnGa. Both cases with and without SOC are consistent with this result, as shown in figure 1(a). To further validate our finding, we have plotted the exchange coupling parameters (J_{ij}) of Pt_2MnGa up to the 17th of its nearest neighbours with the help of the SPR-KKR package. The J_{ij} provides information on the magnetic interactions and relative spin orientations of the electrons between two nearby atoms. The J_{ij} is shown as a function of distance from Mn in Figure 1(b), where Mn is positioned at the center.

The oscillating nature of J_{ij} suggests that there is an indirect exchange occurring through RKKY interaction, in which the magnetic exchange is mediated by conduction electrons [12]. Up to roughly the tenth nearest neighbour (NN), we can see competing FM and AFM interactions before the interaction eventually dies out. The first NN interaction is AFM in nature, but the second through fifth are FM. The next nearest neighbour (NNN) interaction is substantially stronger than the NN interaction. The FM phase is more stable than the AFM phase as a result of the cumulative effect of all these interactions.

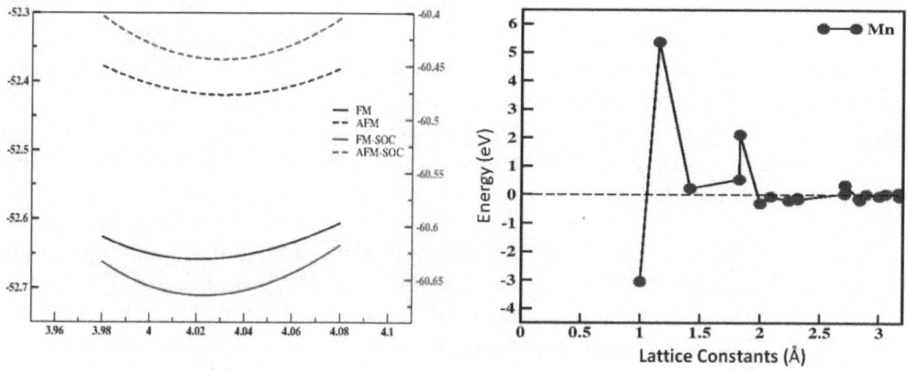

(a) (b)

Figure 1: (a) Ground State Energy *vs* Lattice Parameter Plots for Pt_2MnGa. (b) Exchange Coupling Parameter (J_{ij}) of Pt_2MnGa for Mn at Center

III. CONCLUSION

Density functional theory has been used to determine the ground state magnetic configuration of the full Heusler alloy Pt_2MnGa. Two sets of computations were performed, one for the FM arrangement and the other for the AFM arrangement. These computations were done both with and without taking SOC into account. The SOC has been factored into the computation because of the presence of the heavy element Pt in Pt_2MnGa. The FM spin arrangement is found to be the most stable in both cases. SOC has the effect of making the structures energetically more stable, despite the fact that the relative stability of the magnetic structures remains constant. The plot of exchange coupling parameters helps us further understand the obtained result. In the future, we intend to investigate the non-collinear magnetic behaviour of Pt_2MnGa reported by Sanjay *et al.* [11].

Acknowledgement

The authors, M.B.S. and B.K., would like to thank DST-SERB for their financial assistance with project CRG/2019/005604.

References

[1] Wolf, S., Awschalom, D., Buhrman, R., Daughton, J. von Molnár, V.S., Roukes, M., Chtchelkanova, A.Y. and Treger, D., Spintronics: A spin-based electronics vision for the future, *Science,* 294, 1488–1495 (2001).

[2] Felser, C., Wollmann, L., Chadov, S., Fecher, G.H. and Parkin, S.S. Basics and prospective of magnetic heusler compounds, *APL materials,* 3, 041518 (2015).

[3] Graf, T., Felser, C. and Parkin, S.S., Simple rules for the understanding of heusler compounds, *Progress in solid state chemistry,* 39, 1–50 (2011).

[4] Elphick, K., Frost, W., Samiepour, M., Kubota, T., Takanashi, K., Sukegawa, H., Mitani, S. and Hirohata, A., Heusler alloys for spintronic devices: Review on recent development and future perspectives, *Science and technology of advanced materials,* 2(1) 235–271 (2021).

[5] Hames, F., Crangle, J., Ferromagnetism in Heusler-type alloys based on platinum-group or palladium-group metals, *Journal of Applied Physics*, 42(4), 1336–1338 (1971).

[6] Aguilera-Granja, F., Aguilera-del Toro, R.H. and Morán-López, J.L., A first principles systematic study of the structural, electronic, and magnetic properties of heusler X2MnZ with X = Fe, Co, Ni, Cu, Ru, Rh, Pd, Ag, Pt, Au and Z = Al, Si, Ga, Ge, In and Sn, *Materials Research Express*, 6, 106118 (2019).

[7] Feng, L., Liu, E., Zhang, W., Wang, W. and Wu, G., First-principle investigation of electronic structure, magnetism and phase stability of heusler-type $Pt_{2-x}Mn_{1+x}Ga$ alloys, *Journal of Magnetism and Magnetic Materials*, 377, 40–43 (2015).

[8] Roy, T., Gruner, M. E., Entel, P. and Chakrabarti, A., Effect of substitution on elastic stability, electronic structure and magnetic property of Ni–Mn based heusler alloys: An ab initio comparison, *Journal of Alloys and Compounds*, 632, 822–829 (2015).

[9] Yang, X., Zhang, W., Wang, Y., Du, M. and Feng, S., First-principles investigation of martensitic phase transitions in platinum-rich full-Heusler alloys, *Materials Today Communications,* 24, 101012 (2020).

[10] Fernandez-Escamilla, H., Takeuchi, N., Perez-Tijerina, E. and Guerrero-Sánchez, J., Pt_2MnGa (001) surface stability and its effect on the magnetic and electronic properties: A dft study, *Materials Today Communications*, 33 (2022) 104467.

[11] Singh, S., D'souza, S., Nayak, J., Suard, E., Chapon, L., Senyshyn, A., Petricek, V., Skourski, Y., Nicklas, M., and Felser, C., *et al.*, Room-temperature tetragonal non-collinear heusler antiferromagnet Pt2MnGa, *Nature communications*, 7, 1–6 (2016).

[12] Ashcroft, N.W. and Mermin, N.D., Solid state physics, chapter 32, *Thomson Learning, Inc.*, 715–718 (1976).

Effect of Ion Doping on the Photocatalytic Performance of g-C_3N_4 under Visible Light

M.K. Jaiswal[1,2,*], S. Deka[1,2] and B. Choudhury[1,2]

[1]Materials and Energy Laboratory, Physical Sciences Division,
Institute of Advanced Study in Science and Technology (IASST),
Paschim Boragaon, Vigyan Path, Guwahati, India

[2]Academic of Scientific and Industrial Research (AcSIR),
Ghaziabad, Uttar Pradesh-201002, India

*E-mail: jaiswal.space@gmail.com

ABSTRACT: The photocatalytic (PC) activity of graphitic carbon nitride (g-C_3N_4) doped with potassium (K), magnesium (Mg), and boron (B) is studied under visible light. The modified g-C_3N_4 shows tunable optical bandgap energies and structural distortions as compared to the pristine g-C_3N_4. The doped sample, MgCN shows 83.59% MB degradationunder visible lightwhich was three times quicker than g-C_3N_4. Also, the doped samples' current density (both anodic and cathodic) were found to be almost two times greater as compared to the pristineg-C_3N_4.

Keywords: Photocatalysis, Defects, Bandgap, Doping.

I. Introduction

Graphitic Carbon nitride (C_3N_4) have emerged as promising materials for use in various applications such as photocatalytic water splitting, water decontamination, CO_2 reduction, etc due to its unique properties such as high catalytic activity, good stability, and shows absorption in the visible light region [1]. Still, surface modification is needed in order to overcome the shortcomings of low visible light absorption and fast carrier recombination which impacts the efficiency of the photocatalyst. Several methods such aselement doping, composites, and defect engineering are employed for modifying photocatalyst surface in order to enhance the photocatalytic performance of a photocatalyst [2,3]. Out of these methods doping metals and non-metals results in electronic and structural changes to the g-C_3N_4 framework [4].

In this study, we have shown the photocatalytic and electrochemical activity of g-C_3N_4 doped with metal ions such as K, Mg, and B. The influence of defects and dopants on the photocatalytic performance of the samples with methylene blue (MB), methyl orange (MO), and rhodamine B (RhB) dyes, under visible light. Electrochemical current density measurement shows a high anodic and cathodic current density in ion-doped g-C_3N_4. The Mott-Schottky plot determines the band structure for each composite, which helps us design the photocatalytic mechanism.

II. RESULTS AND DISCUSSION

A. Experimental Section

(a) *Synthesis of K, Mg, and B Doped g-C_3N_4*

Pristine g-C_3N_4 (CN) nanosheets were synthesized at 500°C via calcination of urea [2]. Synthesis of K doped CN follows as. 1 g of K_2SO_4 was dissolved in 10 mL of double-distilled water. Into this solution, 10 mg of dry urea was added. This mixture was oven-dried at 80°C for 6 h. The dried product was than air-calcined at 500°C for 2 h. The final product is named as KCN, was collected, washed, and oven-dried at 80°C. In similar procedure, we synthesized MgCN and BCN using magnesium chloride and boric acid as the precursors for Mg and B, respectively [5].

(b) *Structural and Optical Study*

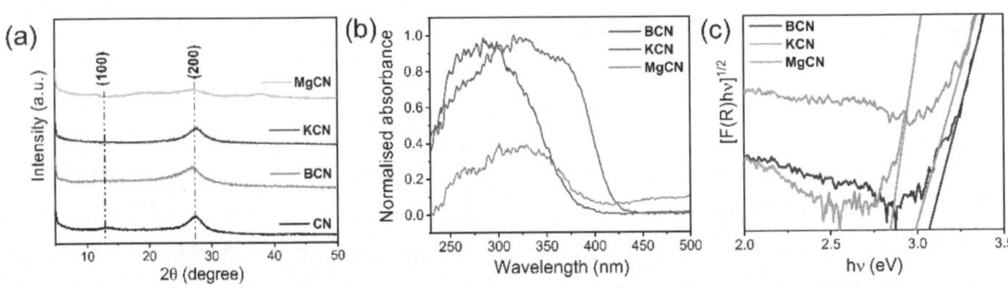

Figure 1: (a) XRD Patternpristine and Doped CN, (b) Kubelka-Munk (K-M) Absorption Plot (c) Taucs Plot of KCN, MgCN and BCN

For structural and optical study XRD pattern and UV- Visible diffuse reflectance spectra of samples were taken. The XRD patterns it is shows the presence of (002) and (100) crystallographic planes of graphitic phase of CN for pure and doped CN samples (Figure 1a). The (100) plane has diminished intensity which implies thatafter doping some distortion of the on-plane ordering of the s-heptazine motifs has occurred [2, 5]. Also, from Figure 1 bit is clear that doped CN shows absorption in the range 250 nm to 420 nm. And the spectral shift in BCN, KCN, and MgCN is a possible result of the binding of K, B, and Mg with the C/N sites of CN. From the Tauc's plot

(Figure 1c), the estimated band gap of the samples are 2.82 eV (KCN), 2.91 eV (MgCN), and 3.15 eV (BCN).

(c) *Photocatalytic Study*

Photocatalytic (PC) degradation of organic dyes like methylene blue (MB), methyl orange (MO), and rhodamine B (RhB) was performed with 300 Watt Xenon lamp with a 390 nm UV cut-off filter. The details of the photocatalytic experiment is as follows. At first, 1 mg of the catalyst are added in a 50 mL of MB (5 mg/L) dye solution. The mixture is stirred for 30 min in the dark for an adsorption/desorption equilibrium. After dark stirring, the solution is kept under illumination. At each 10 min interval, 5 mL aliquot is collected, centrifuged, and the catalyst free solution is taken for absorption measurement. The degradation is measured following the equation:

$$\text{Degradation (\%)} = [(C_0 - C_t)/C_0] \times 100 \qquad \ldots (1)$$

where C_0 is the absorbance of MB solution before illumination, C_t is the absorbance after illumination at time.

The region of "*dark*" measures the adsorption of MB over the catalysts. The maximum MB degradation is found to be 83.59% with MgCN, and the minimum is 38.92% with BCN (figure 2a). Whereas the photodegradation of MO and RhB under visible light are poor for all the samples as compared to that of MB (figure 2 (b,c)). The enhanced photocatalytic activity of MgCN with MB is attributed to the cationic nature of MB which helps in theadsoption of MB firmly to surface of the anionic catalyst (MgCN).Whereas the anionic surface of MO and RhB, however, makes for a poor interaction with the catalyst [5].

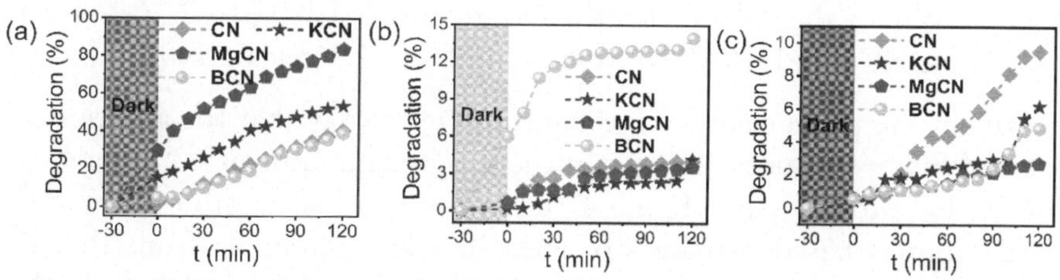

Figure 2: % Degradation *vs.* t (min) Plot of CN, KCN, MgCN and BCN with (a) MB (b) MO and (c) RhB as the Test Dyes under Visible Light

(d) *Electrochemical Study*

The linear sweep voltammetry (LSV) is conducted to measure the current density of the samples. Figure 3(a) shows the LSV curves of the samples in the anodic and cathodic region. MgCN delivers a maximum current density of 1.53 mA $cm^{-2}$and KCN has a maximum current density of 3.8 mA cm^{-2} in the cathodic region [5].

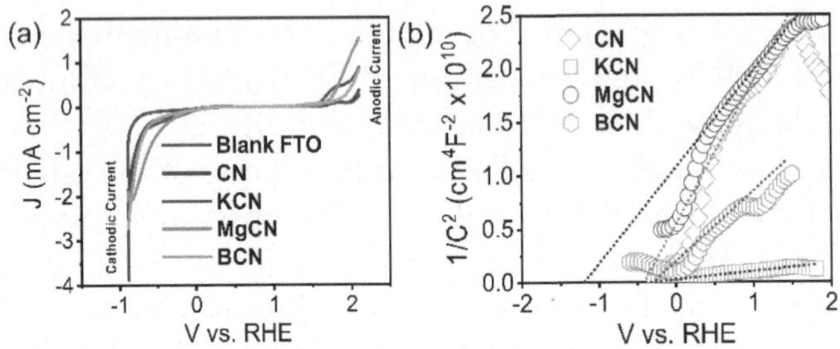

Figure 3: Linear Sweep Voltammetry (LSV) Measurement of Pure and Doped Samples. (b) Mott-Schottky Plot for CN, KCN, MgCN, and BCN in 0.1 M KCl Electrolyte.

The Mott-Schottky (M-S) plot, as shown in Figure 3(b) gives positive slopes for all the samples, indicating an *n*-type semiconductor characteristic. From the M-S plot (C^{-2}*vs.* applied voltage), we have determined the flat band potential:

$$\frac{1}{C^2} = \frac{2}{q\varepsilon\varepsilon_0 N_D}\left(E_{\text{applied}} - E_{FB} - \frac{kT}{q}\right) \quad \ldots (2)$$

C, q, ε, ε_0 and N_D denote the total capacitance, charge of an electron, semiconductor dielectric constant, free space permittivity, and donor density, respectively. $E_{applied}$ and E_{FB} are the applied and flat band potentials, k is the Boltzmann constant (1.38×10^{-23} JK^{-1}), and T is room temperature [5]. The flat band (E_{FB}) or conduction band potential (E_C) is found to be – 0.42 V (CN), – 0.35 V (KCN), – 1.20 V (MgCN) and – 0.29 V (BCN), respectively. Thus, we obtain the valence band edge potentials (E_{VB}) as +2.36 V for CN, +1.97 V for KCN, +1.46 V for MgCN, and +2.77 V for BCN [5].

To explain the photocatalysis mechanism, we can say that the addition of K, Mg, and B ions induces changes in the s-heptazine unit of CN (as observed from XRD pattern). Also, figure 4 shows the band structure of CN and the doped samples based on the VB and CB edge potentials from the Mott-Schottky plot to better understand the mechanism of photocatalysis. As we see, the CB edge potentials of all the samples KCN (-0.35 V), and MgCN (-1.20 V) are more negative than the standard reduction potential of $O_2/O_2^{\bullet-}$ (-0.33 V). Therefore, the photogenerated electrons on the CB of these samples easily generate $O_2^{\bullet-}$ species and thus enhances the photocatalytic activity of the catalyst [5].

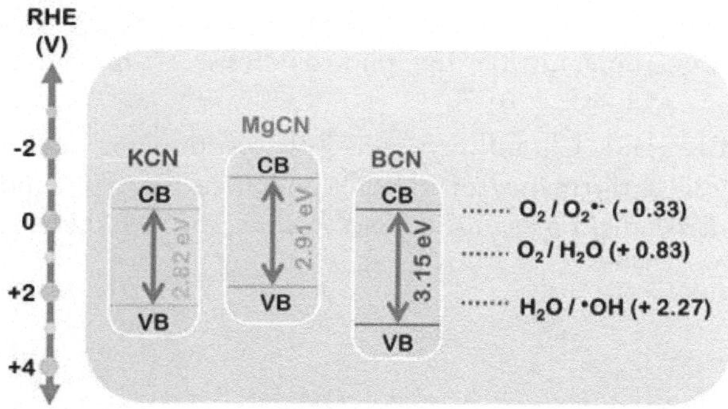

Figure 4: Potential Band Structure for Each Samplebased on the Edge Potential Measurements from the Mott-Schottky Plot for the Valence Band (VB) and Conduction Band (CB)

III. Conclusion

In conclusion, K, Mg, and B are incorporated into the interplanar gap, interlayer space, and lattice site of CN. The excellent visible-light photocatalytic activity of the doped samples is due to: (1) the presence of reactive edges that serve as excellent sites for electron transfer to adsorbed dye molecules (2) a reduced band gap that absorbs a maximum portion of visible light and (3) suitable CB band edge potential for producing $O_2^{\bullet -}$ species to take part in the photocatalytic reaction.

Acknowledgement

Authors acknowledge DST INSPIRE for providing financial support during this study under grant No. DST/INSPIRE Fellowship/2019/IF190107 and SAIC, IASST Guwahati for supporting with the characterization facilities.

References

[1] Wang, X., Maeda, K., Thomas, A., Takanabe, K. Xin G., Carlsson, J.M., Domen, K. and Antonietti, M., A metal-free polymeric photocatalyst for hydrogen production from water under visible light, *Nat. Mater.*, 8, 76-80 (2009).

[2] Choudhury, B., Paul, K.K., Sanyal, D., Hazarika, A. and Giri, P.K., Evolution of nitrogen-related defects in graphitic carbon nitride nanosheets probed by positron annihilation and photoluminescence spectroscopy, *J. Phys. Chem. C*, 122, 9209-921 (2018).

[3] Li, Y., Jin, S., Xu, X., Wang, H., and Zhang, X., Excitonic effects on photophysical processes of polymeric carbon nitride, *J. Appl. Phys.*, 127, 170903 (2020).

[4] Jiang, L., Yuan, X., Pan, Y., Liang, J., Zeng, G., Wu, Z. and Wang, H., Doping of graphitic carbon nitride for photocatalysis: A review, *Appl. Catal. B: Environ,* 217, 388–406 (2017).

[5] Jaiswal, M.K. and Choudhury, B., Role of dopants and defects on the photocatalytic performance of g-C_3N_4 under visible light and sub-band gap excitation, *Journal of Physics D: Applied Physics*, 55, 504002 (2022).

Self-Powered Broadband Photodetector Based on Metal Nitride Nanostructure

Jyotisman Bora[1,2] and Arup R. Pal[1]

[1]Plasma Nanotechnology Laboratory, Institute of Advanced Study in Science and Technology, Guwahati-781035, India

[2]Department of Physics, Gauhati University, Guwahati-781014, India

E-mail: jyotismanb512@gmail.com

ABSTRACT: The application of plasmonics has been increasing day by day, especially in the arena of optoelectronics due to the ability of light confinement in sub-wavelength regime. The charge carriers in plasmonic nanomaterials are primarily generated due to Localized Surface Plasmon Resonance (LSPR) and the interband transition phenomena. In this report, we have briefly discussed the above mentioned charge carrier generation pathways of plasmonic Titanium Nitride (TiN) and their successful extraction by creating a Schottky junction with different semiconductors for the application in broadband photodetection.

Keywords: Photodetector, Metal-nitride, Plasmon Excitation, Interband Transition.

I. Introduction

Commercial photodetectors are traditionally made of semiconductors such as Si, Ge, GaAs, InP, etc. All these semiconductors are suitable for photodetection only when the incident photon energy is larger than the band gap energy. In the case of sub-bandgap excitations, the absence of absorption results in no photon detection. So, the semiconductor can't detect the light signal lower than the band gap wavelength. Moreover, in infrared photodetectors, the major problem of lattice heating is due to the low bandgap of the used semiconductor. Thus, to minimize the heating effect, these detectors need an efficient cooling facility, which increases the cost as well as the volume of the fabricated device. These problems for detecting low energetic photons can be addressed by fabricating a plasmonic photodetector. A plasmonic photodetector can generate bandgap-independent charge carriers. Here, extra energy levels for sub-bandgap absorption are created by placing metallic nanostructures on a high-bandgap semiconductor. This allows us to overcome the limitation arising from the intrinsic band gap of the semiconductor. The hot carriers generated from the metal surface move towards the semiconductor due to a built-inelectric field at the interface, which helps to detect the low energetic photons. The generated carriers can be utilized for various applications such as photocatalysis, photovoltaics, photodetection, etc. [1, 2].

Usually, gold (Au) and silver (Ag) are the most used conventional plasmonic materials in the field of photodetection. But the expensiveness, intrinsic loss mechanism (for Au), and tendency toward oxidation (for Ag) of these conventional plasmonic materials make room for some alternative plasmonic material. Recently, metal nitride-based nanostructures have gained tremendous importance as an alternative plasmonic material. These nanostructures are nowadays preferred over conventional plasmonic materials (Au, Ag, etc.) due to their cost efficiency, low dielectric loss, and high-temperature stability of the metal nitrides. Among the family of plasmonic metal-nitrides, titanium Nitride (TiN) is widely used in optoelectronics due to its high electron conductivity and better CMOS compatibility [3, 4, 6, 7].

In the case of plasmonic materials, the charge carrier generation occurs via two routes: (i) the intraband or LSPR-based charge generation and (ii) the interband transition-assisted carrier generation process. Both these carriers can be extracted using energetically favourable semiconductors [3, 4].

II. Discussion

This report briefly discussed the available literature on plasmonic TiN-based photodetectors, with a special reference to our works on extracting interband-generated charge carriers of plasmonic TiN for UV light harvesting. As mentioned in the introduction part, the Localized Surface Plasmon Resonance (LSPR) and interband excitation phenomena in plasmonic nanomaterial is the main driving mechanism for generating charge carriers in such materials. So far, the LSPR property of plasmonic nanomaterials has been widely used for optoelectronic applications compared to interband-generated carriers.

A. LSPR-based Charge Carrier Generation

In 2016, *Hussain et al.* first reported the fabrication of a plasmonic TiN-based hybrid photodetector. In their work, the hot carriers generated from TiN-LSPR have been confined mainly to the visible region [5]. In 2019, by varying the stoichiometry of the TiN, *Podder et al.* reported that the LSPR of TiN could be tuned from visible to near-infrared (NIR) region. They fabricated two different sets of photodetectors by using the visible and NIR-TiN with a p-type CuO semiconductor. Their study demonstrates a strong prospect of plasmonic TiN nanoparticle-based infrared photodetector, which may overcome the limitations of conventional infrared detectors [6]. In another work, *Podder et al.* fabricated a visible-NIR photodetector using a TiN/TiO_2 Schottky junction. They have observed that a particular stoichiometric TiN shows LSPR in both visible and NIR region. The XPS study reveals that the visible LSPR is mainly due to the titanium nitride phase, while the titanium oxynitride phase contributes to NIR absorption. They recorded maximum photoresponsivity of 230 mA/W at a wavelength of 950 nm, which seems promising for use as a plasmonic NIR photodetector [7].

B. Interband-Assisted Charge Carrier Generation

Theoretically, it has been observed that most of the plasmonic materials show the interband transition in the UV region. This electronic transition occurs from the inner *d* to the *sp* band and can produce efficient charge carriers. So far, most of the low-cost UV (high energy photons) detectors are based on highband gap semiconductors, but in terms of performance, they are insufficient to fulfil all the criteria. Thus, to enhance the performance of such detectors, it becomes necessary to combine interband carriers of metal nanostructure with the existing excitonic carriers of semiconductors so that benefit of both materials can be achieved simultaneously [3,4].

In 2020, Podder *et al.* successfully reported a self-powered UV detector where the interband carriers of plasmonic Au and the excitonic carriers of ZnO simultaneously participated. They have found that the contribution of interband carriers drastically improved the detector performance compared to the pristine semiconductor device [8].

In the case of TiN, the existing theoretical literature reveals that the transition from the N-*p*→Ti-*d* state is mainly responsible for the interband excitations. According to Patsalas *et al.*, the N-*p* state of TiN can spread up to a maximum energy of 5.5 eV below the Fermi position, leading to the origin of various transitions from N-*p* to Ti-*d* state. Theoretical work also reveals that most probable interband transitions occur corresponding to strong absorption of light with energies of 2.3, 3.9, and 5.6 eV. Such literature indicates that illuminating TiN with an energetic UV source will generate charge carriers via interband transition. To the best of our knowledge, we are the first group to experimentally extract these interband transition-generated carriers (electrons) of TiN and use them for practical applications [3].

In one of our works, we have extracted the interband transition generated charge carriers of TiN in a metal nitride-based device geometry with tantalum nitride (TaN_x) as the semiconductor. Among the family of metal nitride semiconductors, $TaN_x$is mainly celebrated for photocatalytic performance, corrosion resistance ability, superior dielectric behavior, etc. But in this report, we have explored the use ability of TaN_x in photodetection. Use of a nitride semiconductor additionally provides better durability of the fabricated detector. The photoactive component, i.e., TiN/TaN_x, shows two strong absorption peaks, one in the UV region and the other in the NIR region. The former indicates the probable interband transition in TiN nanostructures, whereas the later is due to its LSPR or intraband transition. Due to the high band gap, the excitonic effect of TaN_x further enhances the light absorption and charge carrier generation in the UV region. The fabricated device shows a maximum photoresponsivity of 42 mA/W and a specific detectivity of 5.6×10^{10} Jones at the wavelength 310 nm. The LSPR of TiN observed in the NIR region offers a photoresponsivity value of 6.8 mA/W at a wavelength of around 840 nm. The promising UV and NIR photoresponsivity values in a single TiN-based device in the self-powered configuration are highly commendable. This study opens a new avenue for TiN-based broadband photodetector covering the entire range from UV to NIR[3].In another work, we have combined the interband transition generated charge carriers of plasmonic titanium nitride (TiN) with non-centrosymmetric ZnO thin films to fabricate a highly responsive and ultrafast UV photodetectors. Due to the high bandgap (3.73 eV), the band-to-band transition in ZnO occurs in the UV-A region. As mentioned, the interband of TiN is also observed in this region. So, upon proper UV illumination, charge carriers are separately generated in both the semiconducting and plasmonic components of the fabricated device. As a result, the performance of the fabricated TiN/ZnO device drastically enhances compared to the pristine Zn Odevice. The fabricated UV photodetector exhibits a substantial photoresponsivity of 50 mA/W in self-powered condition. The device offers an ultrafast response time of 21 μs due to the non-centrosymmetric wurtzitephase of zinc oxide (ZnO) semiconductor [4].

III. Conclusion

This report discusses the different pathways to extract the charge carriers from the plasmonic TiN. The LSPR of TiN is primarily observed in the visible region of electromagnetic spectrum but can be tuned to the NIR region by simply varying the stoichiometry of the synthesized TiN. The contribution of oxynitride phase is mainly responsible for shifting the LSPR from visible to NIR. The interband carriers of TiN can also be used to detect UV radiation. Combining interband carriers with the excitonic carriers of a high band gap semiconductor can enhance the overall performance of the fabricated detector. From this study, we can infer that a Plasmonic TiN-based photodetector has considerable potential to stand as a highly responsive broadband photodetector covering the entire range from UV-Visible to NIR region.

ACKNOWLEDGMENT

JB acknowledges Institutional Fellowship support from the Institute of Advanced Study in Science and Technology (IASST), Guwahati, India. The authors also acknowledge Deepshikha Gogoi, Santanu Podder, Bablu Basumatary, Khomdram Bijoy Singh, and Shakyadeep Bora for their healthy suggestions while preparing this article.

REFERENCES

[1] Lin, C. and Liu, C.W., Metal-insulator-semiconductor photodetectors, *Sensors* 10, 10 (2010).

[2] Shinde, S.L., Ishii S. and Nagao. T., Sub-band gap photodetection from the titanium nitride/germanium hetero structure, *ACS Appl. Mater. Interfaces* 11, 21965 (2019).

[3] Bora, J., Podder, S., Gogoi, D., Basumatary, B. and Pal, A.R., An all metal nitride nanostructure configuration: Study and exploitation in efficient photodetection, J. Alloys Compd. 879, 160460 (2021).

[4] Podder, S., Bora, J., Thakur, S., Gogoi, D., Basumatary, B., Bora, S.M., Adhikary, N. and Pal, A.R., Interband transition in plasmonic titanium nitride and its contribution towards ZnO based pyro-phototronic application, *Mater. Chem. Phys.,* 275, 125290 (2022).

[5] Hussain, A.A., Sharma, B., Barman, T. and Pal, A.R., Self-powered broadband photodetector using plasmonic titanium nitride, *ACS Appl. Mater. Interfaces,* 8, 4258 (2016).

[6] Podder, S. and Pal, A.R., Hot carrier devices using visible and NIR responsive titanium nitride nanostructures with stoichiometry variation, *Opt. Mater,* 97, 109379 (2019).

[7] Podder, S. and Pal, A.R., Plasmonic visible-NIR photodetector based on hot electrons extracted from nanostructured titanium nitride, *J. Appl. Phys.,* 126, 083108 (2019).

[8] Podder, S., Basumatery, B., Gogoi, D., Bora, J. and Pal, A.R., Pyro-phototronic application in the Au/ZnO interface for the fabrication of a highly responsive ultrafast UV photodetector, *Appl. Surf. Sci.,* 537, 147893 (2021).

Deciphering the Piezo-Photocatalytic and Photocatalytic Behaviour of $BiVO_4$ Nanorods with Anti-Bacterial Activity

S. Deka[1,2,*], M.K. Jaiswal[1,2] and B. Choudhury[1,2]

[1]Materials and Energy Laboratory, Physical Sciences Division
Institute of Advanced Study in Science and Technology (IASST),
Paschim Boragaon, Vigyan Path, Guwahati-781035, India

[2]Academic of Scientific and Industrial Research (AcSIR),
Ghaziabad, Uttar Pradesh-201002, India

*E-mail: suv.deka.rv@gmail.com

ABSTRACT: A facile hydrothermal method was employed to fabricate $BiVO_4$ nanorods. The as-synthesized (B180) and calcined (C300, C500, C700) $BiVO_4$ crystallizes into the monoclinic scheelite phase. The prepared nanorods exhibited extended visible-near-infrared (NIR) absorption.The photo and piezo responsive behaviourwere tested with methylene blue (MB) as a standard organic pollutant. Ultrasonication was used to feed mechanical vibrations into $BiVO_4$.It is found that piezo-photocatalysis is dominant for a short period, but eventually it slows with time. It can be ascribed to sonication's fracturing effect, which further exacerbates the catalytic process due to the damage to the catalytic sites. Additionally, the photocatalytic inactivation of Gram-positive (Staphylococcus aureus) and Gram-*negative (Escherichia coli* K12*) bacteria were investigated under visible light irradiation using C300 and C500 samples. Within 120 min, C500 exhibited superior mortality rates compared to C300.*

Keywords: Piezo-Photocatalysis, Sonication, Photoinduced Bacterial Inactivation.

I. Introduction

Piezo-photocatalysis, which combines the piezoelectric and photo-responsive behaviour of materials, is considered a promising strategy for water treatment. An n-type bimetallic oxide semiconductor, $BiVO_4$, has a low band gap (2.4 eV), suitable band edge position, strong hole mobility, ionic conductivity, ferroelasticity, and comparatively high resilience towards photo-corrosion. The monoclinic scheelite structure is the ideal candidate for the photocatalytic studyof $BiVO_4$. Earlier studies have demonstrated that $BiVO_4$ has both flexoelectric and piezoelectric properties [1, 2].

As an established piezomaterial, herein, we report the photocatalytic and piezo-photocatalytic responses of bismuth vanadate ($BiVO_4$) nanorods under ultrasonic frequency and visible light excitation. We have produced monoclinic scheelite $BiVO_4$ nanorods using the hydrothermal technique. Furthermore, the photo induced inactivation of E. coli K12 and S. aureus were also investigated utilizing the prepared nanorods.

II. Results and Discussions

A. Hydrothermal Synthesis of $BiVO_4$

$BiVO_4$ nanorods were synthesized via a hydrothermal method using $Bi(NO_3)_3 \cdot 5H_2O$ and NH_4VO_3 as the starting precursors. These precursors were first dissolved in 30 mL of 2 M dilute HNO_3 aqueous solution separately. Then the solutions were mixed and stirred at 60°C for 30 min. Meanwhile, ethylene diamine was added until the pH of the solution reached 6–7.

Then the solution was poured into a 100 mL Teflon liner which was sealed in an autoclave and was kept in oven at 180°C for 8 h. The uncalcined sample was designated B180, while the samples calcined at 300°C, 500°C, and 700°C were designated C300, C500, and C700, respectively [3].

B. Structural, Morphological and Optical Study

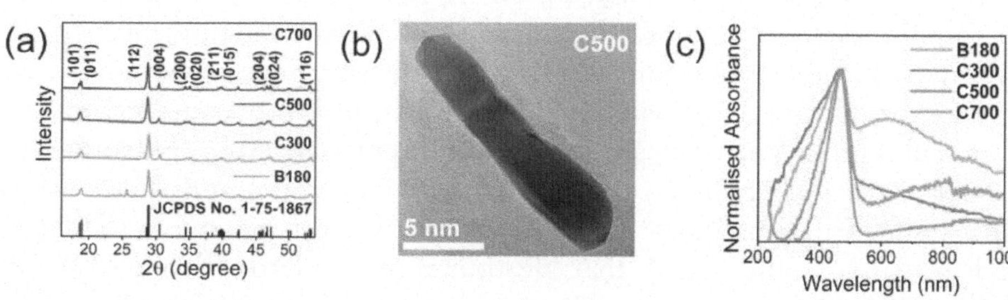

Figure 1: (a) XRD Pattern, (b) TEM Image of C500 (c) Kubelka-Munk (K-M) Absorption Plot of B180, C300, C500 and C700. "Reprinted (adapted) with permission from {DOI: 10.1021/acsanm.2c02072}. Copyright {2022} American Chemical Society."

Figure 1a shows the X-ray diffraction peaks of the synthesized $BiVO_4$. All four samples (B180, C300, C500, and C700) exhibit crystallographic planes that can be indexed to the monoclinic scheelite phase (JCPDS PDF No. 01-075-1867). The TEM images of C500 is displayed in figure 1b which shows nanorodmorphology.Figure1c displays the Kubelka-Munk absorption plot of the samples derived from their diffuse reflectance spectra (DRS). The 467 nm peak undergoes a spectral red-shift in the calcined samples, showing that the monoclinic scheelite system experiences a change in its electronic structure. The hybridized Bi 6p - O sp^2 to V 3d state transition results in the band gap of $BiVO_4$ [4].

C. Piezo-photocatalytic and Photocatalytic Study

We made two arrangements to monitor the degradation of Methylene Blue (MB). It included photocatalysis and piezo-photocatalysis. Piezo-photo-catalytic degradation (LS_0) was investigated by simultaneously applying ultrasonic frequency with visible light irradiation. However, in case of photocatalytic arrangement only light was applied. The procedure has been described as (LS_t). A 20 W LED bulb was used as a visible light source for investigating the degradation profile. In order to provide ultrasonic vibrations, ultrasonic bath was used, delivering a frequency of 35 kHz.

Water was used as the sonication medium which was changed after an interval of every 10 min to avoid further heating of dye solution.Prior to irradiation, the adsorption–desorption equilibrium of MB on the catalyst's surface was observed by stirring the suspension in the dark for a duration of 30 min.Figure 2(a,b) shows the degradation plot of MB degradation. Both photocatalytic and piezo-photocatalytic processes led to 97.63 and 97.13% MB degradation in 40 min [4]. The two process can be categorized into two time domains: (a) 0–10 min and (b) 10–40 min (figure 3 a,b). In 0–10 min, piezo-photocatalysis is more dominant than photocatalytic process (LS_t). When mechanical stress is initiated via ultrasonic vibrations, it constantly leads to creation of acoustic cavitation bubbles in the low-pressure cycle. This eventually grow and collapse in the high-pressure cycle. Subsequently, implosion of such bubbles creates high speed microjets and shock waves. This results in formation of localised hotspot (with temperature (5,000 K) and pressure (1,000 atm)) that dissociates water into $H^{\bullet}$ and $^{\bullet}OH$ by thermocatalytic decomposition [5]. Additionally, the cavitation phenomena generate extremely high pressure which induces polarisation charges in the $BiVO_4$ surface. As a consequence, an internal electric field builds up, because of accumulation of localized positive and negative polarization charges on opposite crystal surfaces. This built in electric field spatially separates the e^- and h^+ pairs, thereby lowering down their recombination rates which eventually takes part in catalytic process as shown in Figure 3a.

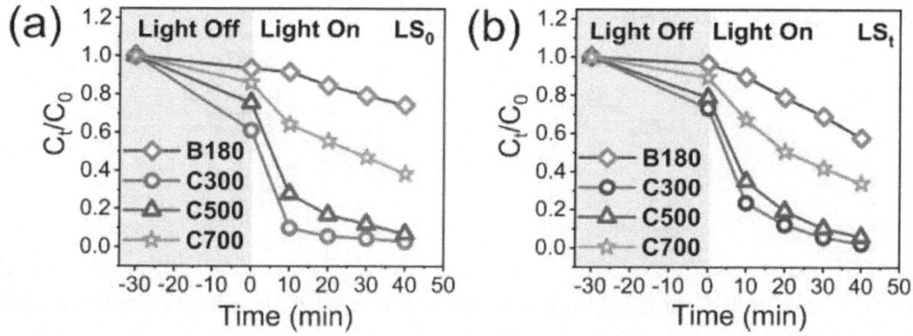

Figure 2: The C_t/C_0 vs. t Plot for MB Degradation of $BiVO_4$ under (a) Piezo-photocatalysis (LS_0) and (b) Photocatalysis (LS_t). "Reprinted (adapted) with permission from {DOI: 10.1021/acsanm.2c02072}. Copyright {2022} American Chemical Society."

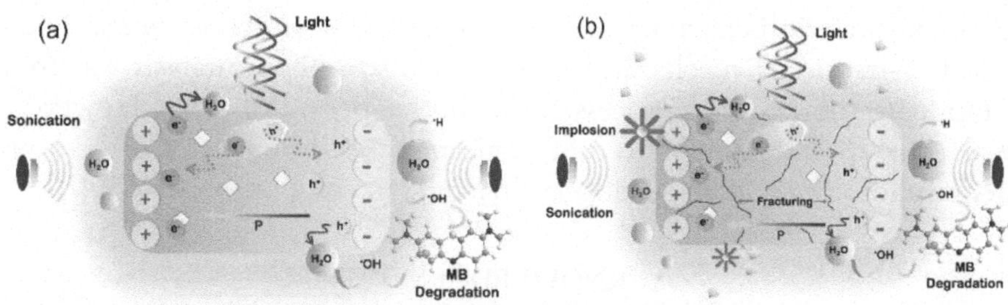

Figure 3: Mechanism of Piezo-photocatalytic Performance at (a) 0–10 min, (b) 10–40 min. "Reprinted (adapted) with permission from {DOI: 10.1021/acsanm.2c02072}. Copyright {2022} American Chemical Society."

However, in 10–40 min domain, it was found that piezo-photocatalysis slows down with time. This can be ascribed to "fracturing effect" of sonication as shown in figure 3b. It was found that for a longer time duration, low frequency sonication has adverse effects which leads to the damage of the catalytic sites which in turn affects the entire catalysis process.Therefore, photocatalysis is more pronounced in this time domain [3].

D. Photocatalytic Bacterial Inactivation

Figure 4(a,b) shows photocatalytic inactivation of *S aureus* and *E coli* K12 using C300 and C500 samples. Within 120 min, C500 exhibited 97.24 and 90.44% mortality rates compared to C300, exhibiting 72.83 and 71.16% against S. aureus and E. coli K12, respectively. When $BiVO_4$ is exposed to light, electron and hole pairs were generated which then reacted with adsorbed oxygen to form various reactive oxygen species (ROS). These ROS trigger enhanced oxidative stress and is primarily responsible for the suppression of bacterial growth [3].

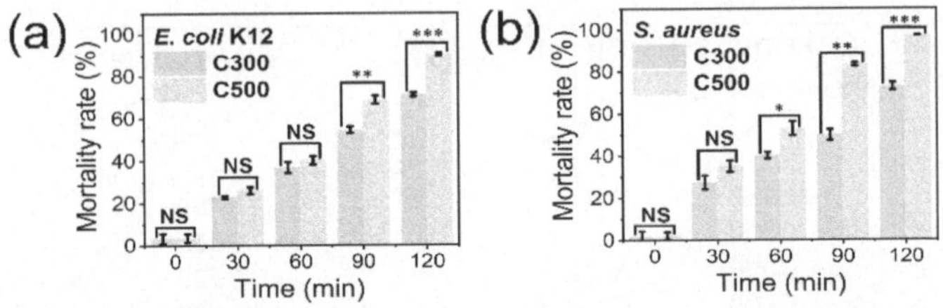

Figure 4: Mortality Rate of (a) *E. coli* K12 and (b) *S. aureus* with C300 and C500 treatment. "Reprinted (adapted) with permission from {DOI: 10.1021/acsanm.2c02072}. Copyright {2022} American Chemical Society."

III. Conclusion

$BiVO_4$ nanorods were synthesized hydrothermally. The piezo-photocatalytic and photocatalytic behaviour of the prepared nanostructures were studied under low-frequency mechanical vibrations. Piezo-photocatalysis was found to dominate the catalytic process for a shorter period of time. However, it retards due to fracturing effect of sonication. Additionally, C500 showed superior photocatalytic inactivation of bacteria in comparison to C300.

Acknowledgement

The authors would like to thank SAIC IASST Guwahati for the TEM and XRD facilities.

References

[1] Shi, L., Zhuo, S., Abulikemu, M., Mettela, G., Palaniselvam, T., Rasul, S. and Wang, P., Annealing temperature effects on photoelectrochemical performance of bismuth vanadate thin film photoelectrodes, *RSC Adv.*, 8, 29179-29188 (2018).

[2] Shao, P.W., Liu, H.J., Sun, Y., Wu, M., Peng, R.C., Wang, M. and Chu, Y.H., Flexoelectric Domain Walls Originated from Structural Phase Transition in Epitaxial $BiVO_4$ Films, *Small*, 18, 2107540 (2022).

[3] Deka, S., Devi, M.B., Khan, M.R., Keerthana, Venimadhav, A. and Choudhury, B., Piezo-photocatalytic and photocatalytic bismuth vanadate nanorods with antibacterial property, *ACS Appl. Nano Mater.*, 5, 10724-10734 (2022).

[4] Tan, H.L., Amal, R. and Ng, Y.H., Alternative strategies in improving the photocatalytic and photoelectrochemical activities of visible light-driven $BiVO_4$: A review, *J. Mater. Chem. A*, 5, 16498-16521 (2017).

[5] Amaniampong, P.N. and Jérôme, F., Catalysis under ultrasonic irradiation: A sound synergy, *Curr. Opin. Green Sustain. Chem.*, 22, 7–12(2020).

Investigation of Electrochemical Properties of Polyaniline-Polyvinyl Alcohol Composite Nanofibers Prepared by Electrospinning

S. Baruah[1,*], D. Mohanta[1] and C.A. Betty[2]

[1]Department of Physics, Tezpur University, Sonitpur-784028, India

[2]Chemistry division, Bhabha Atomic Research Centre, Mumbai-400085, India

*E-mail: susmitabaruahsb@gmail.com

ABSTRACT: Simple electrospinning technique has been employed to prepare conducting polymer Polyaniline and Polyaniline-Polyvinyl alcohol (PANI-PVA) composite nanofibers. Morphological studies of the prepared nanofibers have been carried out using FESEM. The electrochemical studies have been performed by employing a three-electrode system comprising of synthesized electrode as working electrode, platinum as counter electrode and Ag/AgCl as reference electrode at large. To understand the charge transfer kinetics between the electrode and electrolyte interface the synthesized electrodes have been characterized by cyclic voltammetry (CV) and electrochemical impedance spectroscopy (EIS) in a three-electrode setup. Our results indicate that the hybrid electrospun composite nanofibers would provide better charge transfer kinetics as compared with the pristine one. Moreover, effective electroactive area is calculated for both the fabricated electrodes using Randles Sevcik equation.

Keywords: Electrospinning, Nanofibers, Polyaniline, Electrochemical.

I. Introduction

Electrospinning has been considered as one of the simplest and effective method for making large quantities of nanofibers. Afiber spun by electrospinning has the characteristic of being long and uniform in diameter and ranging from a few nanometers up to several micrometers in diameter [1]. The unique properties of nanofibers have made them a promising material for the fabrication of sensors due to their high surface area, high electrical and thermal conductivity, flexibility, porosity, and portability. Nanofibers are ideally suited for large immobilization sites that require large surface areas, resulting in a high degree of interaction with analytes. Biosensors based on nanofibers are therefore anticipated to exhibit a higher level of sensitivity, an extensive shelf life, and faster response times [2]. Researchers are attracted towards electrospun nanofibers because of their wide application in various fields, including biosensors, energy storage, drug delivery systems, biomedical materials, and biomedical devices [3–5].

On the other hand, among all conducting polymer Polyaniline is gaining considerable attention because of its environmental stability, excellent process ability, high electrical conductivity and many more. Despite this, bulk PANI does not perform well in terms of electrochemical performance due to its low electroactive surface area in contact with the electrolyte. Therefore, one-dimensional nanostructures of PANI with various morphologies (e.g., nanofibers, nanotubes, nanowires etc.) have a beneficial effect on improving the electrochemical properties of the material by providing structural integrity, reduced diffusion lengths, high surface areas, and conducting pathways that enhance the electrochemical properties of the material [6]. A wide range of recent developments in electrospinning have been made that is aimed at obtaining composite nanofibers from conductive polymer nanofibers. This includes the introduction of an electrospinnable polymer such as polyacrylonitrile (PAN), poly(vinyl alcohol) (PVA) and poly(vinyl pyrrolidone) (PVP) that can be used to produce composite nanofibers. PVA, with one of the highest densities of functional hydroxyl groups among these polymers, has some excellent properties such as its hydrophilicity, biocompatibility, processability andnontoxicity. Adding PANI to PVA, provides viscosity to solutions, which is essential when electrospinning is being done. In this work, we are presenting morphological and electrochemical studies of prepared PANI and PANI-PVA composite nanofibers.

II. Results and Discussion

PANI and PANI-PVA composite nanofibers were developed by using facile Electrospinning unit. Electrospinning of PANI was performed at 19.2 kV with a 12 cm distance between two electrodes with RPM 960. In order to improve the viscosity, electrospinning of PANI with PVA was carried out at 16 kV with a 10 cm distance between two electrodes with RPM 980.

A. Morphology of PANI and PANI-PVA Composite Nanofibers

The FESEM micrograph of PANI particles covering the aluminium foil and PANI-PVA composite fibers were taken at 1 μm scale range is displayed in Figure 1. Pristine PANI shows some micro to nano particles which got agglomerated and forms some clusters like structure and appeared highly porous. Formation of very less number of fibers probably due to the low viscosity of prepared solution and solubility limitations of PANI in NMP. To improve the viscosity and processibility of PANI, one polymer namely PVA, is chosen as a binder. The morphology of PANI-PVA composites shows a large number of smooth fibers shown in Figure 1(b). However, due to resistance of the jet to the extensional flow some beads are formed.

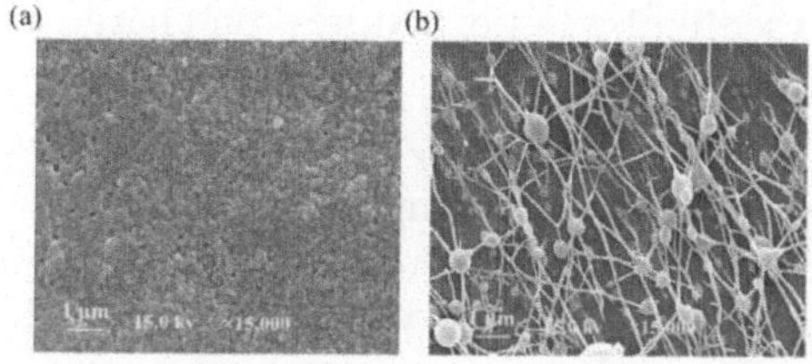

Figure 1: FESEM Micrograph of (a) Polyaniline (PANI) (b) Polyaniline (PANI)-PVA Composites on Aluminium Foil

B. Electrochemical Studies

Cyclic voltammetry has been recorded for prepared PANI/ITO and PANI-PVA/ITO electrodes in presence of 100 mM PBS and 5mM $[FeCN)_6]^{3-/4-}$ as an electrolyte at a scan rate 20 mV/s within a potential range of –0.4 V to 0.8 V. Electrochemical Impedance Spectroscopy (EIS) was performed for all the modified electrodes within frequency window 1 Hz - 1 MHz, with peak-to-peak ac voltage taken as 10 mV and 0 V DC bias.

C. Cyclic Voltammetry of PANI and PANI-PVA Composite Nanofibers

To know the electroactive properties of the synthesized sample, Cyclic Voltametry (CV) was performedand shown in Figure 2(a). CV was run within the potential window –0.4 V to 0.8 V and at a scan rate 20 mV/s in presence of a three-electrode setup comprising of the synthesized electrode as the working electrode, Platinum (Pt) as a counter electrode and Ag/AgCl as a reference electrode using 1 mM PBS with 5 mM $[Fe(CN)_6]^{3-/4-}$ as an electrolyte. The area under the curve of CV pattern increased after incorporation of PVA into PANI as depicted in Figure 2 (a), which is because of better conductivity and capacitance response of the PANI-PVA composite system.

D. Impedance Spectroscopy of PANI and PANI-PVA Composite Nanofibers

To understand the charge transfer kinetics between the interface of the electrode and the electrolyte, Electrochemical Impedence specroscopy (EIS) has been carried out. EIS was performed within frequency range from 1 Hz to 1 MHz for fabricated PANI/ITO, PANI-PVA/ITO as shown in the Figure 2(b). The solid lines illustrates the impedence data whilethe fitted data is represented by the dotted line. The equivalent circuit for fitting the impendence data is displyed in the inset of Figure 2(b). Equivalent circuit parameters; R_1 represents series resistance, as the semicircle portion of the Nyquist plot is represented by R_{ct}, it denotes the charge transfer resistance. C_2 arises due to the double layered capacitor at the electrode and elctrolyte interfaces and W_3 represents the Warburg impedence arises due to diffusion. PANI/ITO shows R_{ct} value 482.2 ohm, after incorporation of PVA into the system the R_{ct} value drops to 471.9 ohm resulting in enhancement in the charge transfer kinetics between electrode and electrolyte interface in the PANI-PVA composite nanofiber system.

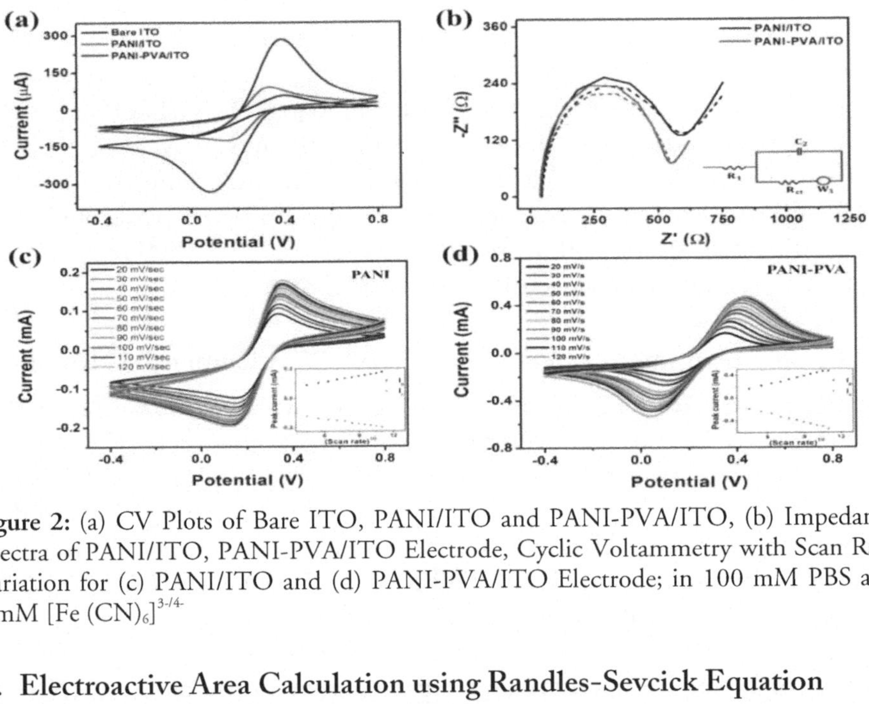

Figure 2: (a) CV Plots of Bare ITO, PANI/ITO and PANI-PVA/ITO, (b) Impedance Spectra of PANI/ITO, PANI-PVA/ITO Electrode, Cyclic Voltammetry with Scan Rate Variation for (c) PANI/ITO and (d) PANI-PVA/ITO Electrode; in 100 mM PBS and 5 mM $[Fe(CN)_6]^{3-/4-}$

E. Electroactive Area Calculation using Randles-Sevcick Equation

The variation of CV patterns at different scan rates from 20 mV/Sec to 120 mV/Sec for PANI/ITO and PANI-PVA/ITO electrode is presented in Figure 2(c) and (d) respectively. With good uniformity the peak current increases steadily with scan rate. The intensity of the peak current is found to be depend on the concentration of the electro-active species and the electroactive surface area of the working electrode, according to Randles Sevcik equation. The linear variation of the cathodic and anodic peak current with square root of the scan rate is displayed in the inset of Figure 2(c) and (d). Further, using Randles Sevcik equation, the effective electro-active surface area of the PANI/ITO and PANI-PVA/ITO electrode is calculated using slope of the current vs. square root of scan rate obtained from the inset of Figure 2(c) and (d). The relationship between peak current and square root of scan rate is given below,

$$I_p = 2.69 \times 10^5 n^{3/2} A D^{1/2} V^{1/2} C \quad \ldots (1)$$

Where all the characters have their standards sence [7]. The calculated active electrochemical surface area for PANI/ITO electrode is 0.307 cm^2 and for PANI-PVA/ITO electrode is 0.368 cm^2.

III. Conclusion

Facile electro spinning technique has been employed to synthesize PANI-PVA composite nanofibers. FESEM micrograph showed the formation of smooth and abundant fibers. Electrochemical characterization namely CV and EIS confirmed better charge transfer kinetics of the PANI-PVA composite system. Using Randles Sevcik equation the effective electrochemical surface area is calculated and enhancement in the active surface area is found in the PANI-PVA composite system. Smooth fibrous membrane of PANI-PVA composite is expected to provide enhanced binding sites for biomolecules and better sensitivity and linearity.

Acknowledgement

Authors are highly grateful to BRNS, Mumbai, for providing financial support (project number- 58/14/09/2019-BRNS). We are thankful to SAIC and UGC SAP DRS II of department of Physics, Tezpur University for providing instrumentation facilities.

References

[1] Senthamizhan, A., Balusamy, B. and Uyar, T., Glucose sensors based on electrospun nanofibers: A review, *Anal. Bioanal. Chem.,* 408, 1285–1306 (2016).

[2] Cetin, M. Z. and Camurlu, P., An amperometric glucose biosensor based on PEDOT nanofibers, *RSC Adv.,* 8, 19724–19731 (2018).

[3] Sun, H., Mei, L., Song, C., Cui, X. and Wang, P., The in vivo degradation, absorption, and excretion of PCL-based implant, *Biomater,* 27, 1735–1740 (2006)

[4] Chandra, R. and Rustgi, R., Biodegradable polymers, *Prog. Polym. Sci.,* 23, 1273–1335 (1998).

[5] Lu, W., Sun, J. and Jiang, X., Recent advances in electrospinning technology and biomedical applications of electrospunfibers, *J. Mater. Chem.* B 2, 2369–2380 (2014).

[6] Chaudhari, S., Sharma, Y., Archana, P. S., Jose, R., Ramakrishna, S., Mhaisalkar, S. and Srinivasan, M., Electrospun Polyaniline Nanofibers Web Electrodes for Supercapacitors, *J. Appl. Polym. Sci.,* 129, 1660–1668 (2013).

[7] Baruah, B. and Kumar, A., Electrocatalytic Acitivity of rGO/PEDOT:PSS Nanocomposite towards Methanol Oxidation in Alkaline Media, *Electroanal*, 30, 1–15 (2018).

Stoichiometry Dependent Property Modifications in Nanosized Spinel Zinc Ferrite

K.U. Aiswarya and K.J. Arun*
Department of Physics, SreeKeralavarma College, Thrissur-680011, Kerala, India
*E-mail: drarunkj@gmail.com

ABSTRACT: In the present work, Zinc ferrite nanoparticles having Zn: Fe in the ratios 1:2, 1:1 and 1:0.5 have been prepared by wet chemical method. Cubic spinel $ZnFe_2O_4$ nanoparticles with average particle size of the order 40 nm have been obtained. The effect on structural, morphological, optical and electrical properties by changing the stoichiometric composition has been investigated using powder XRD analysis, SEM, FTIR, UV-VIS-NIR spectroscopy and Dielectric measurements. An attempt has been made to correlate the observed property variations with the variations in stoichiometric composition of the material.

Keywords: Zinc Ferrite, Nanoparticles, Stoichiometry.

I. Introduction

Ferrites are class of compounds composed of oxidised iron and other metals. There is a widespread usage of ferrites due to their unique properties involving applications in the fields of biomedicine, ferrofluids, magnetic refrigeration systems, sensors, catalysts, anode materials and so on [1].

Zinc ferrite has got significant attention from researchers owing to some of its unique behaviours like high electrical resistivity, low eddy current loss, high radiation damage resistance, and high temperature stability etc. [4]. Recently they have been used for hyperthermia treatment of cancer, semiconductor photo catalysis, absorbent material for hot gas desulphurization, photochemical hydrogen production etc. [3]. There are many methods available for the synthesis of $ZnFe_2O_4$ nanoparticles such as hydrothermal, co precipitation, sol gel, solvothermal, ball milling, self propagating low temperature combustion method etc. Because of simplicity, reduced cost, time saving and lesser number of precursors, wet chemical synthesis is performed in this work. Structural, Optical, Morphological, dielectric studies are performed and presented in this work.

II. Experimental Methods and Results

Zinc chloride and Ferric chloride purchased from Sigma Aldrich co. are used as the starting materials without any further purification. Samples with Zinc to Fe ratio 1:0.5 is termed as ZF1. ZF1 is prepared by adding 20 ml of 0.1 M Zinc chloride solution and 20 ml of 0.05 M Ferric chloride solution and stirred continuously at 80°C for 3 hours. 0.1N sodium hydroxide solution is used to adjust the pH value to 10. Brown coloured precipitate obtained is washed with distilled water several times and filtered and dried at 100°C. Samples with Zinc to Fe ratio1:1 is termed as ZF2 is synthesised by reacting 20 ml of 0.1 M aqueous solutions of Zinc chloride and Ferric chloride at a pH of 10 at 80°C for 3 hours. Samples with Zinc to Fe ratio 1:2 is termed as ZF3is synthesised by reacting 20 ml of 0.1 M aqueous solutions of Zinc chloride and Ferric chloride at a pH of 10 at 80°C for 3 hours. Three samples of Zinc ferrite are prepared and studied.

Structure of the Zinc ferrite nanoparticle is studied using X-ray diffraction analysis. The powder X-ray diffraction pattern is recorded (Figure 1) for three different stoichiometric ratios of zinc ferrite by using Cu Kα radiation (λ = 1.541Å) over a 2θ range of 0 to 90°. Intensity and interplanar d spacing values of entire observed diffraction peaks perfectly matches with the cubic spinel form of zinc ferrite having lattice constants a = 8.443 A° with space group Fd3m(227). (hkl) planes are identified with the data using Match 3 software and compared with the JCPDS card No: 22-1012. The average particle size of the nano product is evaluated using the Debye- Scherrer's equation [2, 6].

$$D = K\lambda/\beta \cos\theta$$

where, *D* is the mean size of the crystallites (nm). The value of crystallite shape factor K is taken as 0.9, β is the full width half maximum (FWHM) in radians and θ is the Braggs angle. Average particle size of ZF1, ZF2, ZF3 are found to be 42.62 nm, 40.37,39.54 nm. It is observed that the peaks corresponding to various (hkl) values of the material for the three different stoichiometric ratios has displaced slightly in angle. It is because as stoichiometry changes, electronic arrangement varies without affecting the structure causing lattice strain and the net result will be a minor shift in the peaks.

In order to study the lattice strain, Williamson-Hall (W-H) plot [3, 5] is drawn using the equation,

$$\beta \cos\theta = \frac{K\lambda}{D} + 4\varepsilon \sin\theta$$

Figure 2 is plotted with 4sinθ along the x-axis and β cos θ along y axis and is linear fitted for the three different ratios. The points are scattered from the straight line because of the lattice strain that may arise from the excess volume of grain boundaries due to dislocation. Slope and Y intercept of the fitted line represents strain and the particle size. For all samples negative strain is observed due to lattice shrinkage. Particle size and strain of ZF1, ZF2, ZF3 are 43.34 nm, 39.96 nm, 38.84 nm and -6.82×10^{-4}, -5.406×10^{-4}, -9.217×10^{-4}. The size of nano particles obtained from the W-H plot are in close agreement with XRD analysis.

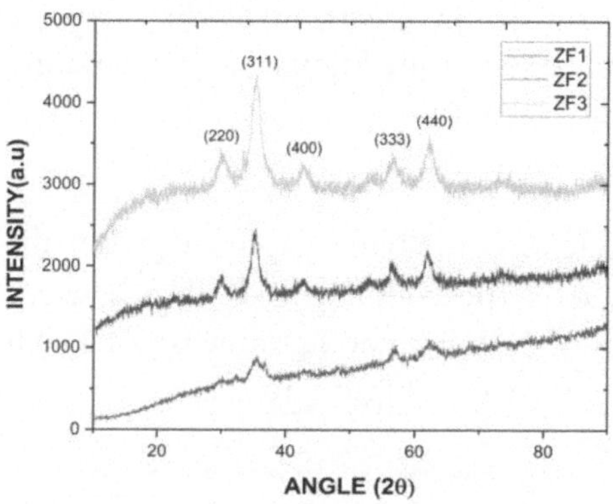

Figure 1: XRD Pattern

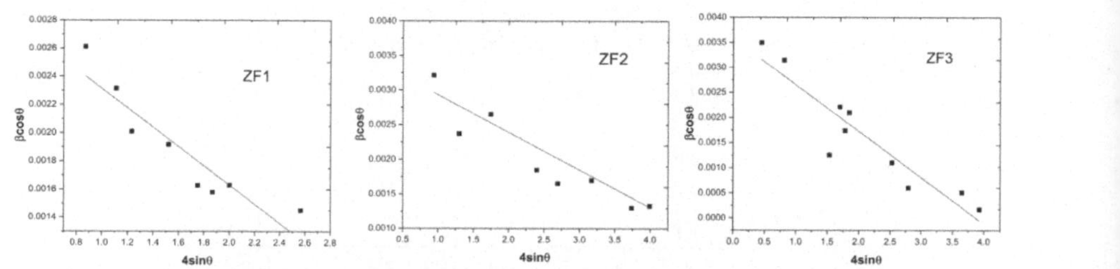

Figure 2: W-H

Figure 3: SEM Images

Scanning electron microscopic images of the nanoparticles are shown in the figure 3, which clearly shows the morphology of randomly dispersed nano flakes of $ZnFe_2O_4$, the size of which is in close agreement with the XRD results.

In order to confirm the formation of Zinc ferrite from its precursors and to ascertain the various functional groups present FT-IR spectrum is recorded in the range 400–4000 cm^{-1}. The position of the various peaks in the three samples are shifted slightly due to the change in stoichiometric ratios. Assignment of vibrations are noted in Table 1.

The UV-VIS-NIR absorption spectrum of the sample is taken in the range 200 nm to 2500 nm is shown in Figure 5a. The energy gap of the sample is found using the Tauc relation,

$$(\alpha h\nu)^n = A(h\nu - E_g)$$

where, A is a constant, E_g is the band gap energy, h is the Plank constant and ν is the frequency. α is the coefficient of absorbance. n is a constant that depends on the band gap type. The E_g value of ZF1, ZF2, ZF3 are 3.10 eV, 3.31 eV, 3.45 eV are shown in Figure 5b.

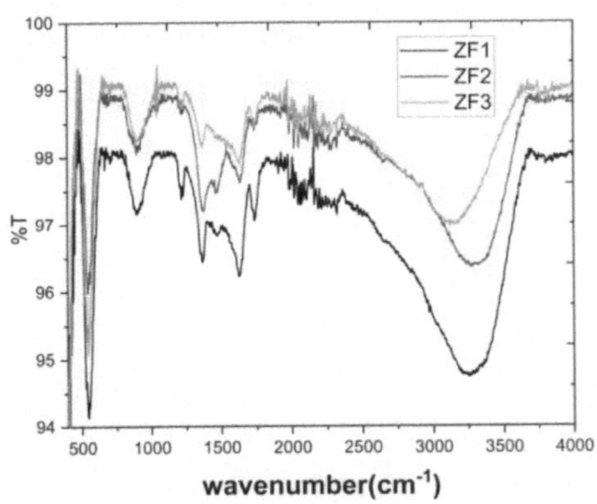

Figure 4: FT-IR Spectrum

Table 1

Wave Number (cm^{-1})	*Assignment of Vibrations*
485	Zn-O Vibration
540	Fe-O Vibration
910,1365	Overtones
3260,1620	Peaks due to moisture absorption

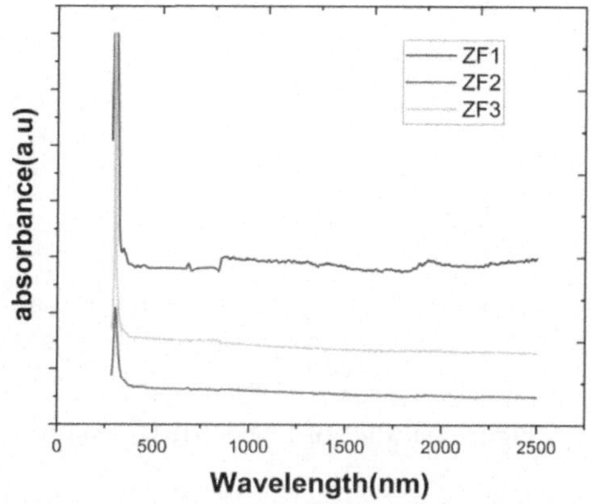

Figure 5a: UV-VIS-NIR Absorbance Spectra

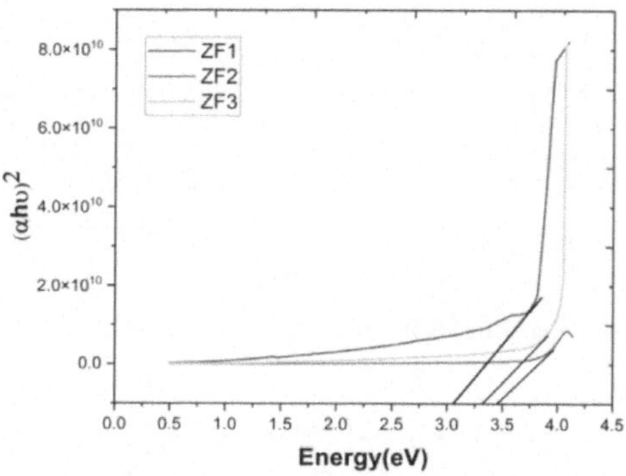

Figure 5b: Plot of Optical Band Gap Energy

Dielectric constant, ε of zinc ferrite pellet is calculated using the following equation,

$$\epsilon = \frac{t \times Cp}{A \times \epsilon 0}$$

where, *t* is the thickness, C_p = Equivalent parallel capacitance which is obtained from the data of measurement, ε_0 = permittivity of vacuum = 8.854 × 10^{-12} (F/m), A = Area of the pellet. Dielectric dissipation factor, tan δ or loss tangent of test material is obtained is shown in figure (6a). The relative permittivity ε of the pellets versus log f is drawn and is shown Figure 6b. The relative permittivity of the pellets decrease with increasing of frequency in all the three samples [3]. For all the ratio dielectric constant has the value between 100–200 but exact value is different for different stoichiometric ratios of the material.

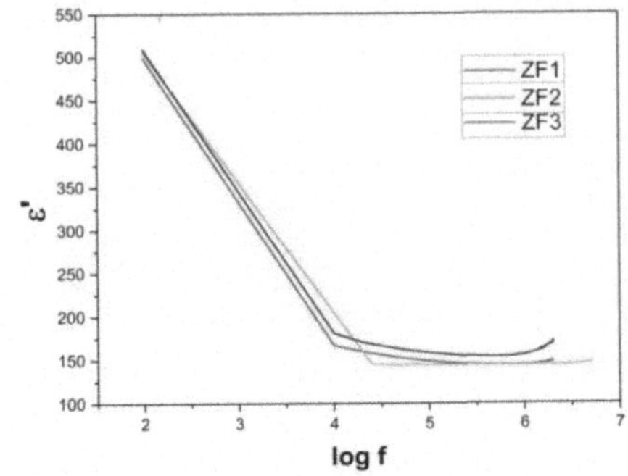

Figure 6a: Plot of Dielectric Constant

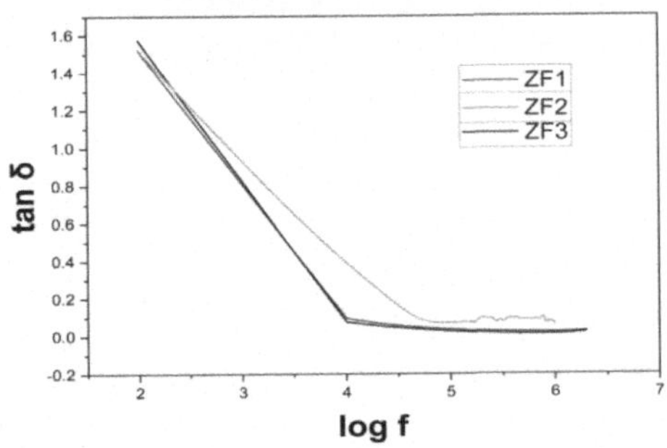

Figure 6b: Variation of Loss Tangent with Frequency

III. CONCLUSION

$ZnFe_2O_4$ nano particles of three different stoichiometric compositions have synthesised by wet chemical method. The effect on structural, morphological, optical and electrical properties by varying stoichiometry are studied using XRD analysis, SEM, FTIR, UV-VIS-NIR spectroscopy analysis and Dielectric analysis and expalined.

ACKNOWLEDGEMENT

Authors acknowledge Sophisticated Test and Instrumentation Centre (STIC), Cochin for providing SEM characterization facilities.

REFERENCES

[1] Yang, J.M. and Yen, F.S., Evolution of intermediate phases in the synthesis of zinc ferrite nanopowders prepared by the tartrate precurssor method, *Journal of alloys and compounds*, 450, 387–394 (2008).

[2] Naseri, M.G., Elias Saion, B., Hashim, M., Abdul Halim Shaari and Hossein Abasstabar Ahangar, Synthesis and characterization of zinc ferrite nanoparticles

by a thermal treatment method, *Solid State Communications*, Vol. 151(14–15), 1031 1035 (2011).

[3] Bardhan, A., Ghosh, C.K., Mitra, M.K., Das, G.C., Mukherjee, S. and Chattopadhyay, K.K., Low temperature synthesis of zinc ferrite nanoparticles, *Solid State Sciences*, Vol. 12(5), 839–844 (2010).

[4] Daliya, S. Mathew and Ruey-Shin Juang, An overview of the structure and magnetism of spinel ferrite nanoparticles and their synthesis in microemulsions, *Chemical engineering Journal*, 129, 51–65 (2007).

[5] Khorsand Zak, A., Majid, W.H. Abd., Abrishami, M.E. and Ramin Yousefi, X-ray analysis of ZnO nanoparticles by Williamson–Hall and size–strain plot methods., *Solid State Sciences*, 13(1), 251–256 (2011).

[6] Sinthiya and M. Maria Angelin, Synthesis of zinc ferrite ($ZnFe_2O_4$) nanoparticles with different capping agents, *International Journal of ChemTech Research*, Vol. 7, 2144–2149 (2015).

Magnetron Sputtered Functional Gradient Coating as Impediment to Crack Propagation

Hemen Kakati
Department of Physics, Nalbari College, Nalbari-781335, India
E-mail: hemenkakati@gmail.com

ABSTRACT: Magnetron sputtered coating is a novel candidature in thin film deposition because it uses the technique of atom by atom deposition that helps in getting very thin film with enhanced protective properties such as hardness, adhesion, corrosion etc. In this article we have discussed the various surface properties on which the behaviour of crack propagation through the interfaces of the films depends on. Depending on the theory and experimental results of various works it has been established that a functional gradient coating can have more advantages over general conventional coating which can also act as a resistant material to crack propagation. It is also proposed that the relative surface properties of material degradation can also be enhanced by means of the multi-layered gradient coating deposited by magnetron sputtering technique.

Keywords: Gradient Coating, Crack Propagation, Magnetron Sputtering, Corrosion, Hardness.

I. Introduction

The ever increasing demand for design and production engineering led the scientists to search for new material with significant improvement of the properties for economic benefits. Magnetron sputtering is a PVD coating used for deposition of thin films using atom by atom deposition [1-2]. It is observed that Depositions of hard coating are found to be accompanied by high internal stress. This in turn leads to lattice distortion and thermal mismatch. Due to these reasons it is very difficult to produce hard coating with good adhesion properties [3]. Often the thin ceramic PVD coating contains columnar structure, micro-cracks, pinholes and high porosity. This leads to very fast surface degradation of thin film in hostile environment. Considering all this facts including adhesion, hardness, corrosiveness the functional gradient coating has been proposed which has superior properties than conventional single layered coating. The Science of crack propagation through thin films is the associated problem which has been studied for better understanding of degradation of the surface [4-5].

Functional gradient coatings are the coatings where the surface properties are changing gradually with thickness/dimension. The gradual change in surface characteristic is achieved by different mechanism like gradual change in condition of the system like pressure, temperature, power of target materials, and gradient change in stoichiometry by varying plasma parameters etc. can also be produced.

Recently Functional gradient materials are supposed to have great prospects in biomedical application and are considered to be promising method for stress-shielding effect [6]. In this article, various surface characteristics along with crack propagation behaviour have been discussed in order to have a film with superior system integrity.

II. Deposition Process by Magnetron Sputtering

The magnetron sputtering is a well-known PVD technique for deposition of thin films [7]. In a magnetron sputtering the target material is screwed to the magnetron which is powered either by DC or RF negative voltage that in turn attracts the positive ions of inert gas from the plasma. The bombardment of the high energetic positive ions on the target leads to sputtering of the material atom by atom that get deposited on the substrate placed on the side or in the line. The deposition characteristics of the material can be controlled by varying the plasma characteristics such as temperature, pressure and power. The stoichiometry of the deposited material can also be controlled by changing the pressure of reactive gas introduced to the plasma with which the sputtered atom get combines to form compound thin film.

Instead of a single magnetron, double magnetron or triple magnetron system can be utilized for depositing required compound film. The distance, power of the various magnetrons can be varied for getting the desired functional gradient coating. In case of insulating material the power supply we can use of RF frequency which produces self-bias voltage on the target that can be utilize for the desired sputtering. In generally argon gas is used as the sputtering gas. Oxygen or nitrogen can be utilized as the reactive gas for compound thin film deposition. The magnetron sputtering technique can be preferred over other techniques as here the deposition process is elementary i.e. atom by atom. Therefore the graded materials can be achieved with continuous variation/minute variation by changing the system and plasma parameters gradually.

For the deposition process, the system need to be evacuated to a very low pressure of 10^{-6} mbar and then inert gas is injected through mass flow controller and the pressure may be maintained at 10^{-3} mbar for production of plasma. The require DC/RF power supply is switch on with slowly increasing the voltage to get the plasma. The discharge power is set at the desired value for getting the gradient coating.

III. Advantage of Functional Gradient Coating over Other Coating

We can have mainly three different types of coatings like single layered, multilayered and functional gradient coating. Considering the limitations of other coating a functional gradient coating is preferred which is based on various characteristics of surface as discussed below.

Due to vast difference of the properties of the substrate and coating material only a few combination are found to achieve the required level of system integrity. To achieve the better performance the characteristics that are considered are:

(a) Elastic moduli (stress discontinuity)
(b) Thermal expansion coefficients
(c) Crystal structure and
(d) Chemical compatibility.

The cracks produced in the film may be result of various phenomenons produced at:

(a) **Surface:** The surface of the film is most exposed to the surrounding environments. The surface gets the first interaction with work piece and the durability of the film depends mainly on the characteristics such as adhesive or abrasive wear properties. The degradation of surface also results due to interaction with the environment owing the properties like corrosion, oxidation.
(b) **Film:** Not only the surface the characteristics of the film are also an important criterion for its strength. The film characteristics that are important to achieve a good integrity are hardness, strength (Fatigue), internal stress, fracture toughness.
(c) **Substrate/Film:** One of the main important for a good quality thin film over a substrate depends on the interaction of the molecules between the surface and the film. The adhesive force is counted in this case, which should be of high order value to get the required work done by instrument. The adherence of the film on the substrate depends on various characteristics like interaction of the substrate and layer, the thermal expansion misfit.

The thin films are always accompanied by intrinsic stresses. Thermal and intrinsic stresses fall in the category of residual stress. The combination of residual and mechanical stress leads to failure of the coating. The tensile stress leads to surface cracks that propagate through the film.

For a brittle surface crack may penetrate the substrate even it is unstressed. As a result the interface get debond which leads to surface delamination. Compressive stress leads to interface or edge delamination. These failures deteriorate thin film performance.

For tool cutting or abrasive and erosive wear resistance, the coating thickness plays vital role. For this thin film have got the advantage however various studies reveals that thin film has columnar structure accompanied with micro cracks, pinholes, and transient grain boundaries. The thin films often have a high degree of through-coating porosity; which leads to accelerated pitting corrosion. Thus failure at the coating/substrate interface in the case of hostile environment conditions has been observed.

To rectify these drawbacks in thin films, different types of coatings such as multicomponent, hybrid, multiphase, heterogeneous, multilayer, micro laminate, gradient and composite coatings are studied to get the superior quality. In a multilayer coating, one or more intermediate layers are applied between the actual coating and the substrate. This in turn improves the compatibility providing matching of chemical bonding and thermal expansion coefficient. The difference between multilayer and gradient coating is that gradient coating refers to a system where composition gradually varies from one layer to another. Thus there is no sharp transition between the layers exists unlike the multilayer where the transition is abrupt.

A. Crack Propagation

Crack propagation is an important features of a coating on which the surface behavior of a coated material depends such as:

(a) Adhesion
(b) Corrosion
(c) Hardness

Hardness is one of the important factor on which the crack propagation depends. At an interface between material having similar elastic properties but different hardness, the propagation of the cracks depends on the direction of approach of the crack to the interface. It is observed that the crack coming from softer side to hard material stops or diverted. However in the opposite case i.e. when the crack approach from hard side to softer side, the crack may cross the interface and enter the softer material. Thus depending on the utility of the material the gradient characteristics of the film may be varied.

Even a relatively thin layer of lower hardness may be utilized as impediment to crack propagation. Therefore it is postulated that the coating fracture resistance can be maximized by depositing multilayer coating of alternating thin layers of hard and soft material (ductile).

Wear in multilayer thin film is produced by removal of material layer by layer. As a multilayer thin film has a number of interfaces, it is postulated the cracks gets deflected by the interface and thus the deterioration of the film may be prohibited. It is henceforth concluded that multilayer thin films are tougher than single layer thin film.

However, multilayer thin film has drawbacks due to the way they are produced. As the interface accompanied by sharp change in the characteristics of the film, the layers give rise to film failure due to growth of cracks at the interface. This is the main reason why the functional gradient coating is preferred. In a functional gradient coating unlike the sharp variation of characteristics in the interface, the variation is gradual and thus the crack growth as well as propagation of cracks can be overcome effectively.

IV. Conclusion

In this article, we have seen a comparison between a single layered thin film, multi-layered thin film and gradient coating. It is explained that a magnetron sputtering technique is more feasible for production of gradient coating as here the film is deposited atom by atom. A multilayer thin film is found to have relatively low compressive residual stress than single layered film. In a gradient coating the porosity get lowered because the open structure from surface to substrate is interrupted by repeated nucleation at interfaces between sub layers. As a result gradient coatings have more wear and corrosion resistance than single layered film. Any Crack produced in a gradient coating suffers deflection due to weak interfaces and thus crack propagation can be prohibited. By magnetron sputtering technique, favourable gradients to have low residual stress can be produced. Due to difference of elastic moduli of various surfaces the crack gets deflected. The gradient coating also provides the reduced bending stress.

References

[1] Kelly, P.J. and Arnell, R.D., Magnetron sputtering: a review of recent developments and applications, *Vacuum,* 56, 159 (2000).

[2] Gudmundsson, J.T. and Lundin, D., 1-Introduction to magnetron sputtering, *High Power Impulse Magnetron Sputtering,* 1 (2020).

[3] Durand-Drouhin, O., Santana, A.E., Karimi, A., Derflinger, V.H., and Schutze, A., Mechanical properties and failure modes of TiAl(Si)N single and multilayer thin films, *Surface and Coatings Technology,* 163, 260 (2003).

[4] Xia Cedric, Z. and Hutchinson, W.J, Crack patterns in thin films, *Journal of Mechanics and Physics of Solids,* 48, 1107 (2000).

[5] Hattar, K., Misra, A., Dosanjh, M.R.F., Dickerson, P., Robertson, M. and Hoagland, R G., Direct observation of crack propagation in Copper-Niobium multilayers, *J. Eng. Mater. Technol.* 134, 021014 (2012).

[6] Shi, H., Zhou, P., Li, J., Liu, C. and Wang, L., Functional gradient metallic biomaterials: Techniques, Current Scenery, and Future Prospects in the Biomedical Field, Front Bioeng. *Biotechnol,* 8 (2021).

[7] Kakati, H., Pal, A R., Bailung, H. and Chutia, J., Deposition of nanostructured crystalline and corrosion resistant alumina film on bell metal at low temperature by RF magnetron Sputtering, *Applied Surface Science,* 255, 7403 (2009).

Surface Topography and Surface Roughness Study of CoTbNi Alloy Thin Films Using 3D Optical Profilometer

R.K. Basumatary[1,2]*, R. Brahma[2] and S.K. Srivastava[3]

[1]Department of Physics, Darrang College, Tezpur-784001, India

[2]Department of Physics, Bodoland University, Kokrajhar-783370, India

[3]Department of Physics, Central Institute of Technology Kokrajhar, Kokrajhar-783370, India

*E-mail: rajibkrbasumatary@gmail.com

ABSTRACT: The surface topography and roughness of ternary alloy thin films of $Co_xTb_yNi_z$ fabricated on Si<100> substrate by co-sputtering of $Co_{85}Tb_{15}$ target with Ni target using DC magnetron sputtering has been reported. The thin film thickness and the film surface roughness play an important role in tuning various properties like magnetic property, conductivity and other properties of thin film materials. Thus the measurement of film thickness and surface roughness is one of the primary step to optimize thin film preparation. The thickness of the films are kept fixed at 120 nm. The roughness parameters Sa and Ra of both as-deposited and post-annealed films measured in ISO 25178 Height and ISO 4287 Amplitude standard are found to increase significantly with sputtering power of Ni target. The post annealing of the films results larger film surface roughness.

Keywords: Thin Films, Surface Roughness, CoTbNi Alloy, Magnetron Sputtering.

I. Introduction

Surface roughness influences a number of material properties such as magnetic properties of thin films [1,2], conductivity of thin films [3], scattering of light from surfaces [4,5], wear of mechanical surfaces [6] etc. One of the commonly used surface profilometers like AFM uses stylus or tip that measures vertical deflection of tip while scanning over the surface and gives the measure of surface roughness. A significant drawback of stylus type surface profilometers like AFMs is that the data are distorted by the finite size of the tip or stylus [7,8]. Other non-stylus profilometers like ellipsometer and profilm 3D uses polarization and interference of light to measure the surface profile. In the present work we report the measurement of surface roughness of thin films deposited using dc magnetron sputtering and variation of surface roughness as a function of sputtering power. The surface roughness indicator parameters Sa and Ra are defined in ISO 25178 Height and ISO 4287 Amplitude standard respectively. The equation by which Sa is measured is $S_a = \frac{1}{A}\int\int Z(x,y)dxdy$[9] and Ra is measured by $R_a = \frac{1}{l}\int_0^l Z(x)dx$ [10], A and *l* being the scanned area and sampling length of the sample.

II. Experimental Details

Ternary alloy thin films of RE-TM $Co_xTb_yNi_z$ have been prepared using dc magnetron sputtering technique. The alloy thin films were deposited on Si<100> substrate by co-sputtering of $Co_{85}Tb_{15}$ target and Ni (99.99% pure) target at room temperature under vacuum pressure ~5×10^{-5} mbar (base pressure) and Ar as sputtering gas. The $Co_{85}Tb_{15}$ target was sputtered at constant power 261 watt and Ni target was sputtered at four different powers 285, 308, 332 and 356 watt to vary the Ni content of the films. The deposition parameters such as base pressure, Ar gas flow, target-to-substrate distance and magnetron angle were kept same to ensure similar deposition condition. A set of samples deposited under the same conditions were annealed at 400° for 60 minutes to study the influence of annealing on the thin films surface profile. The deposition rate of the targets were estimated by repeated deposition of thin films for different time under same deposition conditions. The film thickness, surface topography and surface roughness were measured using profilm 3D optical profilometer. The composition of the films were estimated using EDX spectroscopy.

III. Results and Discussion

The composition of deposited thin films are estimated using EDX spectroscopy. One representative EDX spectra and elemental colour mapping of EDX spectra of as-deposited film at 285 watt is presented in Figure 1. The obtained composition (presented in Table 1) of the films shows that the Ni content of films increases with increase in Ni target sputtering power. The elemental colour mapping of EDX spectra shows that the films are uniform and homogeneous. To estimate the deposition rate of the film for each sputtering power, the film was deposited on partially masked substrates repeatedly for three times for different deposition time maintaining the same deposition parameters. The masks were removed after deposition and step height of films were measured using profilm 3D optical profilometer. One representative 3D coloured image of step height obtained for 24 minute deposition time is presented in the Figure 2. The step height was determined from the difference of the average height of the film and substrate as indicated in Figure 2. The step height was found to be 577.7 nm for 24 minute deposition time and hence the deposition rate per minute is 24.1 nm. The average deposition rate per minute was then calculated for each deposition. For each sample the same process has been followed to estimate deposition rate. The actual thin films were then deposited to keep the overall film thickness of 120 nm. The film surface roughness were estimated using 3D optical profilometer and expressed in Sa (ISO 25178 Height) and Ra (ISO 4287 Amplitude) respectively. The 3D image of surface topography of film, colour contrast z-scale and line profile of as-deposited films are presented in Figure 3. The estimated Sa and Ra for both as-deposited and post annealed thin films are presented in Table 1 and the variation of Sa and Ra as a function of Ni target sputtering power is plotted in Figure 4. The Ra was measured along the middle line of the scanned area of sample. The value of Sa increases from 0.691 nm for film sputtered with Ni target at 285 watt to 1.195 nm for film sputtered with Ni target at 356 watt. On the other hand, the Ra values are found to be smaller than Sa and increase with sputtering power of Ni target from 0.353 nm to 0.512 nm. Similar variation of Sa and Ra are observed for post annealed samples. The value of Sa increases from 0.548 nm to 1.285 nm and Ra from 0.451 nm to 0.528 nm with increase in sputtering power of Ni target from 285 watt to 356 watt. It is also observed that the value of both Sa and Ra of annealed thin films are higher than the as-deposited thin films.

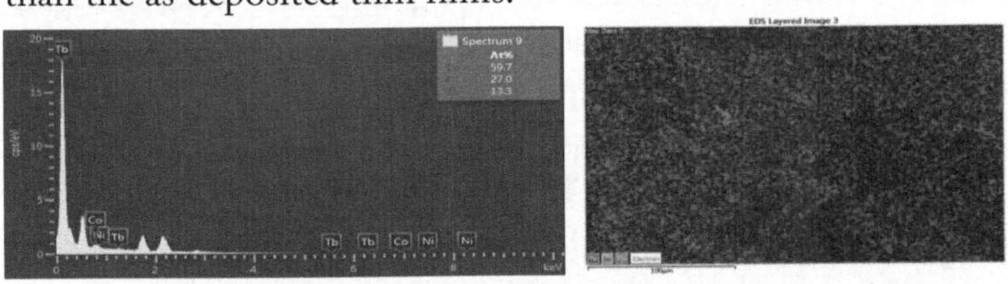

Figure 1: EDX Spectrum (left) and Elemental Colour Mapping (right) of CoTbNi thin Film with Ni Target Sputtered at 285 Watt

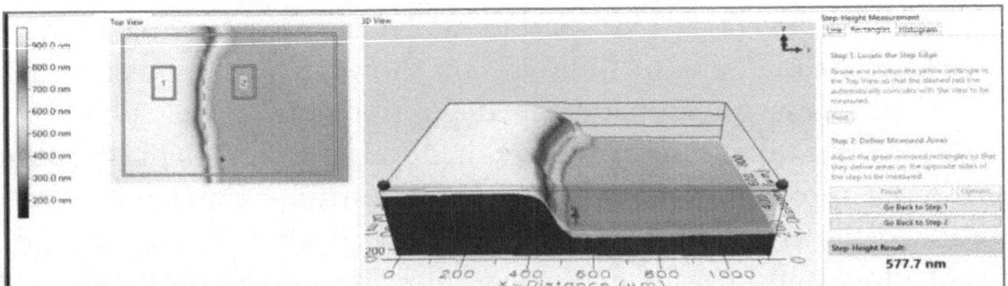

Figure 2: 3D image of step height measurement with colour contrast z-scale of CoTbNi film with Ni target sputtered at 285 watt. The figure at the right bottom of the figure is the average height of the film surface from substrate.

Table 1: Composition of Deposited thin Films and Value of Roughness Parameters Sa and Ra of as-Deposited and Post-Annealed Thin Films

$Co_{85}Tb_{15}$ Sputtering Power (watt)	*Ni target Sputtering Power (watt)*	*Composition*	*Surface Roughness nm*			
			As-deposited		*Annealed*	
			Sa (nm)	*Ra (nm)*	*Sa (nm)*	*Ra (nm)*
261	285	$Co_{60}Tb_{13}Ni_{27}$	0.691	0.353	0.548	0.451
	308	$Co_{36}Tb_{18}Ni_{46}$	0.802	0.411	0.947	0.470
	332	$Co_{38}Tb_{8}Ni_{54}$	0.783	0.492	0.912	0.511
	356	$Co_{32}Tb_{7}Ni_{61}$	1.195	0.512	1.285	0.528

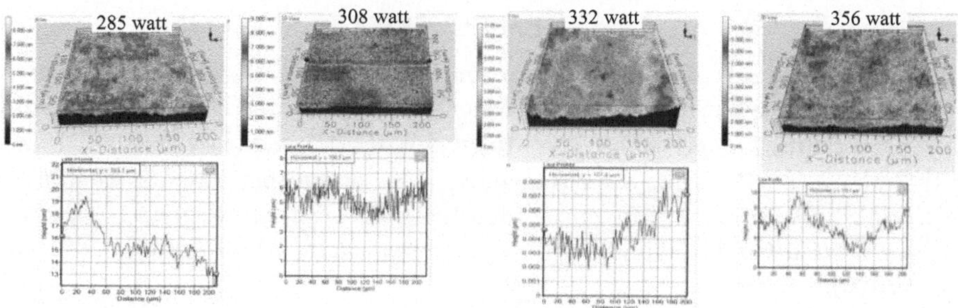

Figure 3: 3D Picture of Surface Topography, Colour Contrast z-scale and Line Profile of As-deposited CoTbNi thin Films of Ni Target Sputtered at Four Different Power

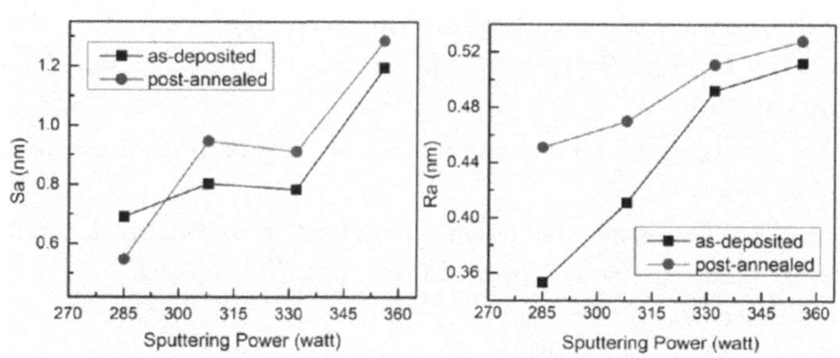

Figure 4: Variation of Sa and Ra of Thin Films as a Function of Ni Target Sputtering Power

IV. CONCLUSION

In conclusion we have deposited CoTbNi ternary alloy thin film by co-sputtering of $C_{85}Tb_{15}$ target and pure Ni target using dc magnetron sputtering system. The deposition rate of the samples were determined by creating step height and measured using 3D optical profilometer. The thickness of the thin films were kept 120 nm for all the films. The deposition rate is found to increase with increase in Ni target power as ejection of Ni from the target increases with sputtering power. The surface roughness parameters Sa and Ra for both as-deposited and annealed films increases with sputtering power of Ni target. The annealing of films leads to increase in Sa and Ra which is likely due to the migration of atoms during the annealing of films.

ACKNOWLEDGEMENT

We acknowledge the financial support for a project through "Early Career Research Award" from DST-SERB India, with sanction order number (ECR/2016/000713).

REFERENCES

[1] Basumatary, R.K., Aakansha, Basumatary, B., Brahma, B., Hussain, R., Ravi, S., Brahma, R. and Srivastava, S.K., Magnetic Property of CoTbNi Ternary Alloy Thin Films, *J. Supercond. Novel. Magn.* 33(10), 3165–3170 (2020).

[2] Brahma, B., Hussain, R., Basumatary, R.K., Aakansha, Ravi, S., Brahma, R., and Srivastava, S.K., Influence of Cu Insertion Layer on Magnetic Properties of Co-Tb/Cu/Co-Tb Thin Films, *J. Supercond. Novel Magn.*, 33(9), 2891–2897 (2020).

[3] Luo, E.Z., Heun, S., Kennedy, Wollschläger, M.J. and Henzler, M., Surface roughness and conductivity of thin Ag films, *Phys. Rev. B* 49, 4858 (1994).

[4] Bennett, J.M. and Dancy, J.H., Stylus profiling instrument for measuring statistical properties of smooth optical surfaces, *Appl. Opt.* 20, 1785 (1981).

[5] Stover, J.C., Optical Scattering, Measurement and Analysis -McGraw- Hill, New York (1990).

[6] Nowicki, Multiparameter representation of surface roughness, *Wear 102,* 161 (1985).

[7] McCool, J.I., Assessing the effect of stylus tip radius and flight on surface topography measurements, *Trans. ASME* 106, 202 (1984).

[8] Song, J.F. and Vorburger, Stylus profiling at high resolution and low force, T.V., *Appl. Opt.* 30, 42 (1991).

[9] Rafieian, F., Mousavi, Dufresne, M.A., Yu, Q., Polyethersulfone membrane embedded with amine functionalized microcrystalline cellulose, *Int. J. Biol. Macromol.,* 164, 4444–4454 (2020).

[10] Sin, K. and Sivertsen, J.M., Surface roughness and magnetic properties of in situ heated and postannealed thin films of perpendicular barium ferrite, *J. Appl. Phys.* 75, 5972(1994).

Synthesis, Properties and Photo-Catalytic behavior of Silver-Polypyrrole Nanocomposites

Jamuna Mazumdar*and Sulochana Deb

Department of Physics, Gauhati University, Guwahati-781014, Assam, India

*E-mail: mazumdarjamuna@gmail.com

ABSTRACT: Polypyrrole-Silver (PPY-Ag) nanocomposite samples have been synthesized by interfacial polymerization method. The samples are prepared for three different concentrations of silver and characterized by Field Emission Scanning electron microscopy (FE-SEM), X-Ray diffraction (XRD), UV-Visible spectroscopy and Fourier transform infrared spectroscopy (FT-IR) techniques. The photocatalytic activity of the samples are studied by photocatalytic degradation of aqueous solution of methylene blue (MB) dye under direct sunlight. The photocatalytic performance increases with increasing the concentration of silver in the samples.

Keywords: Silver-Polypyrrole, Nanocomposite, Photocatalytic Activity, Degradation, Dye.

I. INTRODUCTION

Conducting polymers have been the subject of extensive recent research in recent years due to their distinctive chemical characteristics. Polypyrrole (PPy) is the most widely investigated conducting polymer amongst the known conducting polymers because of its outstanding thermal stability, good electrical conductivity, environmental stability, mechanical qualities and highflexibility in its synthesis [1, 2]. It has vast applicability as electrical gadgets, sensors, biosensors, supercapacitors, polymeric batteries etc. [2, 3].

Metal nanoparticles especially gold, silver, platinum, and copper have attracted a great deal of attention due to their distinctive electrical, optical, catalytic and thermal properties as well their significant applications in numerous disciplines [4, 5]. Conducting polymer-based nanocomposites loaded with metallic nanoparticles provide fascinating systems to study the possibilities of constructing device functionality [6]. As compared to pure PPy, Ag-PPy nanocomposites perform better in the areas of sensing, optoelectronics, catalysis and photocatalysis [7-9]. Therefore, the preparation of Ag-PPy nanocomposites becomes an innovative challenge for researchers.

As a photocatalyst, nanocomposites are crucial in the process of destroying harmful substances under direct sunlight. In order to degrade and decompose organic stains, eliminate bed odors, and convert toxic and non biodegradable organic materials into CO_2 and H_2O, photocatalyst is used. There are various industries like printing, paper, food along with textile industry use wide variety of dyes. These dyes are one of the responsible elements for soil and water pollution. Environment pollution is the greatest danger for humanity. So in order to reduce its toxicity we should look for the approaches to normalize its effect in order to reduce its impact on human and environment. The addition of Silver (Ag) in Polypyrrole (PPy) enhances its photocatalytic activity as noble metals like Au, Ag, Ptprevente-hole pair recombination. The thermal and chemical stability of PPy makes it a stable photosensitizer. Nanostructure of PPy conjugated polymer perform significant photocatalytic activity under UV and Visible light [10, 11]. Here, in this article we report the synthesis, properties and the photocatalytic activity of Silver-Polypyrrole nanocomposite with the degradation of the methylene blue (MB) dye under direct sunlight.

II. Experimental

Synthesis of Ag-PPy nanaocomposite Silver-Polypyrrole nanocomposite is prepared by interfacial polymerization method. For this, silver nitrate ($AgNO_3$) is dissolved in 1M hydrochloric acid (HCl) which is then added to the organic phase containing 0.2M Pyrrole in 10 ml chloroform ($CHCl_3$). While 0.1 M Ammonium Persulphate is dissolved in 1 M HCl which is then slowly added to the above mixture of solution. After 24 hours, the entire solution was filtered and washed several times with ethanol and distilled water. The residue of Ag-PPy nanocomposites thus obtained after purification is then dried and used for further characterization. Three different samples are prepared for three concentrations (3 mM, 5 mM and 8 mM) of $AgNO_3$.

III. Results and Discussion

A. FE-SEM Analysis

The morphological characteristics of the samples are examined by field emission scanning electron microscope (FE-SEM). FE-SEM images show (Figure 1) that composite clusters are mostly of spherical in shape. Size and densityof particles increase with increasing the concentration of silver in the nanocomposites. The average diameter of spherical nanocomposites lies in the range of 450–550 nm.

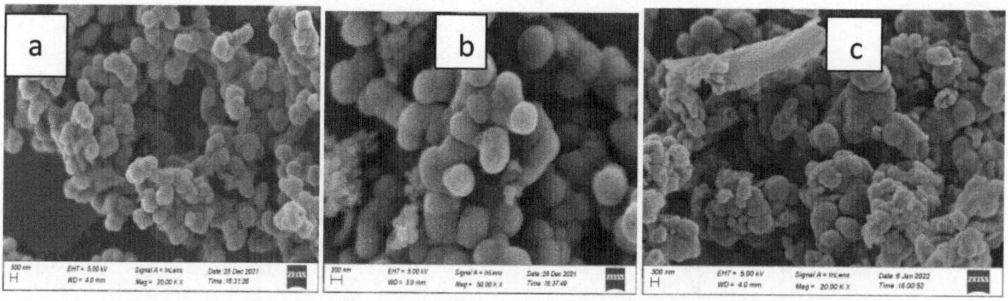

Figure 1: (a, b and c) are FE-SEM Images of Ag-PPy Nanocomposite Samples with Silver Concentration 3 mM, 5 mM and 8 mM.

B. X-Ray Diffraction (XRD) Analysis

Figure 2 represents the XRD pattern of synthesized Ag-PPy nanocomposite samples. The board peak around 2Θ = 24°is due to amorphous nature of PPy[12]. The diffraction peaks at 2θ value 27.2° and 32.1°, 46.1° and 76.3° corresponds to the diffraction lines produced by crystalline planes (226), (264), (200) and (311) indicate the face centered cubic (FCC) nature of silver nanoparticles. The intensity increases with the increase in concentration of silver in the nanocomposite. Allare in agreement with the reported literatures[13, 14].

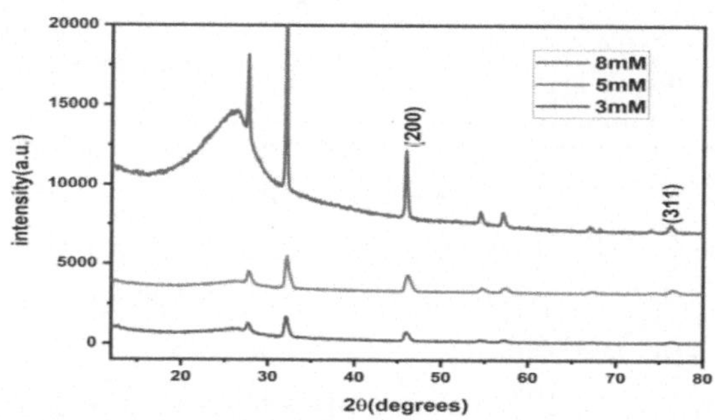

Figure 2: X-Ray Diffraction Patterns of Ag-PPy Nanocomposite Samples for Three Different Silver Concentration

C. UV-Visible Spectroscopy Analysis

Figure 3 shows the UV-Visible absorption spectra of the two samples. There are two board observable peaks at 310 nm and 308 nm (blue shifted) are due to π-π* transition band of PPy. The surface Plasmon resonance (SPR) peak of silver is absent in the spectra due to the reason that PPy coates around silver nanoparticles and the strong scattering from PPy shell possibly screened the surface of silver [15].

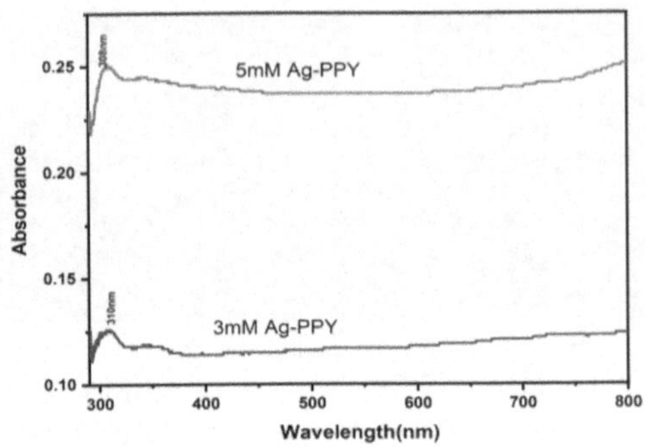

Figure 3: UV-Visible Spectroscopy of Two Samples

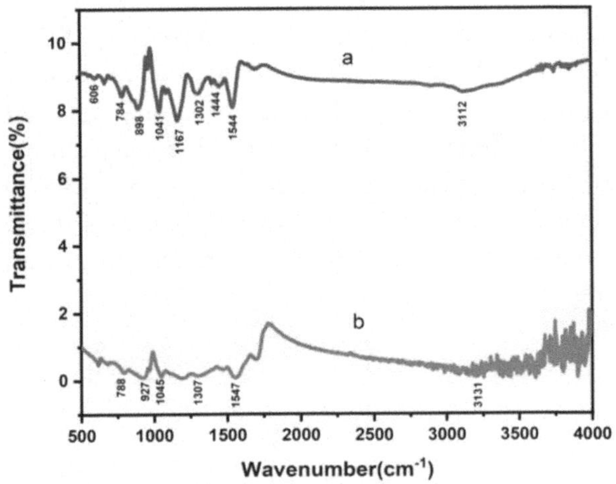

Figure 4: FTIR Spectra of a) 3 mM Ag-PPy b) 5 mM Ag-PPy

D. FT-IR Spectra Analysis

The FT-IR spectra of Ag-PPy nanocomposites shows (Figure 4) variouspeaks such as C-H outer banding vibration (606 cm^{-1}), C-H and N-H in plane deformation vibration (784 cm^{-1}), C-C out of plane deformation stretching (898 cm^{-1}), C-N in plane deformation mode (1041 cm^{-1}), peaks at 1167 cm^{-1} (C-N stretching)indicates the oxidized and doped PPy. The band at 1302 cm^{-1} is related to C-N stretching mode, peaks at 1444 cm^{-1} represent C = C in plane deformation mode and peak at 1544 cm^{-1} representing symmetric Pyrrole ring. There is littile blue shift in the peaks with increasing silver content may be due to interfacial interaction of silver with PPy causing delocalization of charge in the polymeric backbone. All the FT-IR peaks are well matched with previously reported literatures [16, 17].

E. Photocatalytic Activity Analysis

Photoctalytic activity of the prepared samples are examined for degradation of Methylene blue under direct sunlight. The photocatalysts coming in contact with light, electron-hole pair generated. It then react with water and oxygen to create hydroxyl radical (OH) and reactive oxygen. These strong radical species degrade the MB dye into strong molecules of CO_2 and H_2O. As shown below the absorbance of MB decreased with time. The percentage of dye degradation is calculated using the formula,

$$\%\text{Degradation} = \frac{(Co-Ct)}{Co} \times 100$$

Where C_o and C_t are the initial concentration of dye and concentration at time t after irradiation. The degradation after three hours is more for samples with more silver content as shown below.

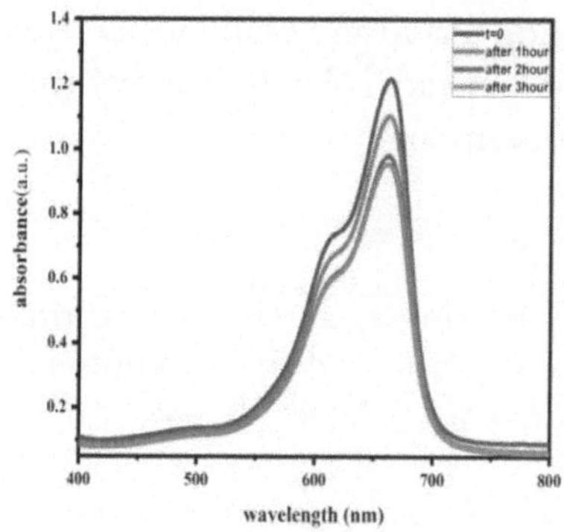

a. 3 mM Ag-PPy

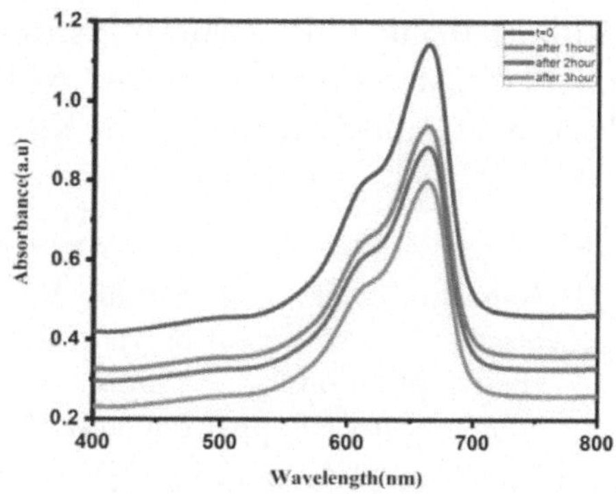

b. 5 mM Ag-PPy

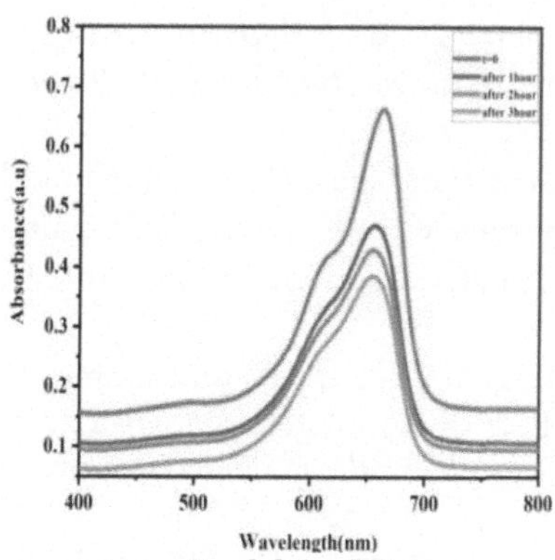

c. 8 mM Ag-PPy

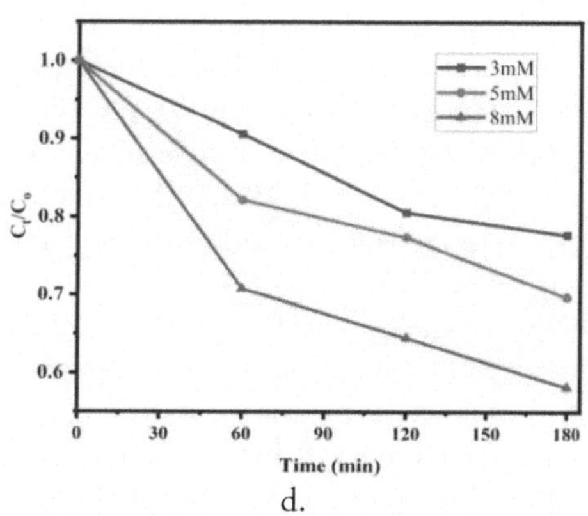

d.

Figure 5: (a, b, c) Absorbance Spectra of MB for Various Samples at Different Time Intervals and Figure 5 (d) Shows Degradation of MB with Irradiation Time for all the Three Samples.

IV. Conclusions

We have successfully synthesized Ag-PPy nanocomposites using simple interfacial polymerization method. FE-SEM images shows that the composite particles are mostly spherical in shape and size of which increases with $AgNO_3$ content in the samples. XRD pattern indicates the presence of silver along with PPy. UV-VIS spectra confirms the formation of PPy. FTIR spectra shows the crosslinking of silver and polypyrrole. Photocatalytic activity of the prepared samples showed degradation of Methylene blue under direct sunlight and % degradation increases with increasing silver concentration in the nano-composite. The synthesized material could be used as a potential material for degradation of various pollutant with high activity under direct sunlight.

Acknowledgements

Authors thank CIF (Central-Instrumentation-Facility), Gauhati University for FE-SEM measurements and SAIF (Sophisticated Analytical Instrument Facility), Gauhati University for XRD measurements.

References

[1] Ramesan, M.T. and Santhi, V., 2017. In situ synthesis, characterization, conductivity studies of polypyrrole/silver doped zinc oxide nanocomposites and their application for ammonia gas sensing. *Journal of Materials Science: Materials in Electronics,* 28, pp. 18804–18814.

[2] Ramesan, M.T. and Santhi, V., 2018. Synthesis, characterization, conductivity and sensor application study of polypyrrole/silver doped nickel oxide nanocomposites. *Composite Interfaces,* 25(8), pp. 725–741.

[3] Yavuz, A., Ozdemir, N. and Zengin, H., 2020. Polypyrrole-coated tape electrode for flexible supercapacitor applications. *International Journal of Hydrogen Energy,* 45(38), pp. 18876–18887.

[4] Wang, N., Guan, B., Zhao, Y., Zou, Y., Geng, G., Chen, P., Wang, F. and Liu, M., 2019. Sub-10 nm Ag nanoparticles/graphene oxide: controllable synthesis, size-dependent and extremely ultrahigh catalytic activity. *Small,* 15(23), p. 1901701.

[5] Yang, M., Lu, F., Zhou, T., Zhao, J., Ding, C., Fakhri, A. and Gupta, V.K., 2020. Biosynthesis of nano bimetallic Ag/Pt alloy from Crocus sativus L. extract: Biological efficacy and catalytic activity. *Journal of Photochemistry and Photobiology B: Biology,* 212, p. 112025.

[6] Camargo, P.H.C., Satyanarayana, K.G. and Wypych, F., Nanocomposites: Synthesis, Structure, *Properties and New Application Opportunities, Materials Research,* Vol. 12, Issue 1, 2009, pp. 1–39.

[7] Yadav, A., Kumar, H., Kumari, R. and Sharma, R., 2022. Progress in the development of metal nanoparticles encapsulated with Polypyrrole plastic nanocomposites: Antibacterial and photocatalytic properties, *Materials Science and Engineering,* B, 286, p. 116085.

[8] Kate, K.H., Damkale, S.R., Khanna, P.K. and Jain, G.H., Nanosilver mediated polymerization of pyrrole: synthesis and gas sensing properties of PPy/Ag nanocomposite, *J. of Nanoscience and Nanotechnology,* Vol. 11, Issue 9, 2011, pp. 7863–7869.

[9] Punnakkal, V.S., Jos, B. and Anila, E.I., 2021. Polypyrrole-silver nanocomposite for enhanced photocatalytic degradation of methylene blue under sunlight irradiation, *Materials Letters,* 298, p. 130014.

[10] Podasca, V.E., Buruiana, T. and Buruiana, E.C., 2019. Photocatalytic degradation of Rhodamine B dye by polymeric films containing ZnO, Ag nanoparticles and polypyrrole, *Journal of Photochemistry and Photobiology A: Chemistry,* 371, pp. 188–195.

[11] Yuan, X., Floresyona, D., Aubert, P.H., Bui, T.T., Remita, S., Ghosh, S., Brisset, F., Goubard, F. and Remita, H., 2019. Photocatalytic degradation of organic pollutant with polypyrrole nanostructures under UV and visible light, *Applied Catalysis B: Environmental,* 242, pp. 284–292.

[12] Folorunso, O., Hamam, Y., Sadiku, R., Ray, S.S. and Adekoya, G.J., 2021. Electrical Resistance control model for polypyrrole-graphene nanocomposite: Energy storage applications, *Materials Today Communications,* 26, p. 101699.

[13] Senthilkumar, P., Yaswant, G., Kavitha, S., Chandramohan, E., Kowsalya, G., Vijay, R., Sudhagar, B. and Kumar, D.R.S., 2019. Preparation and characterization of hybrid chitosan-silver nanoparticles (Chi-Ag NPs); A potential antibacterial agent, *International journal of biological macromolecules,* 141, pp. 290–298.

[14] Mallikarjuna, K., Narasimha, G., Dillip, G.R., Praveen, B., Shreedhar, B., Lakshmi, C.S., Reddy, B.V.S. and Raju, B.D.P., 2011. Green synthesis of silver nanoparticles using Ocimum leaf extract and their characterization, *Digest Journal of Nanomaterials and Biostructures,* 6(1), pp. 181–186.

[15] Chitte, H.K., Bhat, N.V., Karmakar, N.S., Kothari, D.C. and Shinde, G.N., 2012. Synthesis and characterization of polymeric composites embeded with silver nanoparticles, *World Journal of Nano Science and Engineering,* 2(1), pp. 19–24.

[16] Yuksel, R., Alpugan, E. and Unalan, H.E., 2018. Coaxial silver nanowire/polypyrrole nanocomposite supercapacitors, *Organic Electronics,* 52, pp. 272–280.

[17] Moosvi, S.K., Majid, K. and Ara, T., 2016. Studying the electrical, thermal, and photocatalytic activity of nanocomposite of polypyrrole with the photoadduct of K 3 [Fe (CN) 6] and diethylenetriamine. *Materials Research,* 19, pp. 983–990.

Investigation of Nanocrystalline Brookite in the Synthesis, Structural Characterization with the Use of XRD and Raman Spectroscopy

Th. Nando Singh[1,*], **Th. Gomti Devi**[2] **and Sh. Dorendrajit Singh**[2]

[1]Department of Physics, C.I. College, Bishnupur, Manipur-795126, India

[2]Department of Physics, Manipur University, Canchipur, Manipur-795003, India

*E-mail: thoudam.nando@gmail.com

ABSTRACT: Brookite is one of the polymorphs of titanium dioxide (TiO_2) in nature. It is the less stable as compare to the anatase and rutile phase of TiO_2. It is prepared at low temperature non-aqueous sol-gel technique. From X-ray diffraction (XRD) study, it was found that its crystal structure was found to be orthorhombic structure with space group Pcab, crystal parameters a = 5.4558 Å, b = 9.1819 Å, c = 5.1429 Å. Crystal lattice distortion ratios are found to decrease with increase of temperature whereas crystallites sizes are found to increase with the raise of temperature. The anatase-to-rutile phase transition was observed at 550°C which is supported both from the study of XRD and Raman spectroscopy. To the best of our knowledge, there is limited study of this nanoparticle. The study shall bring new insights of its properties applicable to different fields.

Keywords: XRD, Raman Spectroscopy, Nanoparticle.

I. INTRODUCTION

The Titanium dioxide (TiO_2) is a well known highly ionic white metal oxide n-type semiconductor material with diverse range of applications ranging from medicine, chemical to electronic industries. It has been widely used in photocatalysis [1], photovoltaic cells, humidity sensors, Li ion batteries, solar cells [2], hydrogen storage, bio-medical engineering, drug delivery systems, antibacterial materials, cosmetics, sunscreens, and electronics, nontoxic, etc. It has four polymorphs in nature, for the rutile structure including rutile (tetragonal), anatase (tetragonal), brookite (rhombohedral), and TiO_2 (B) (monoclinic). Other high pressure polymorphs have been also reported. Rutile is more stable than anatase and brookite. Anatase and rutile adopted tetragonal symmetry with space group $I4_1/amd$ and $p4_2/mnm$ respectively, p_{bca} and c_2/m for brookite and TiO_2 (B) respectively. TiO_2 is a wide band gap semiconductor with band gap energy of 3.2 eV for the anatase crystal structure, and 3.0 eV for the rutile structure. This wide band gap of TiO_2 is adequate for doping with lanthanides so that the 4f state lies within the gap which decreases the probability of nonradiativede-excitation mechanism. From the above polymorphs of TiO_2 brookite is the least studied candidate.

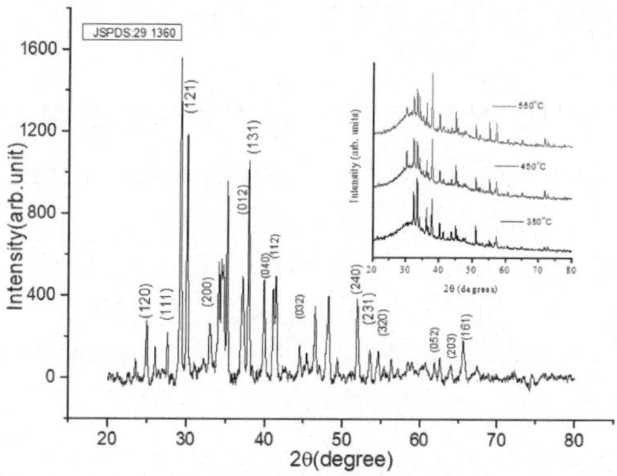

Figure 1: XRD pattern of as prepared brookite sample heated at temperature 550°C and inset is the XRD pattern of as prepared brookite sample heated at temperature 350°C, 450°C and 750°C respectively.

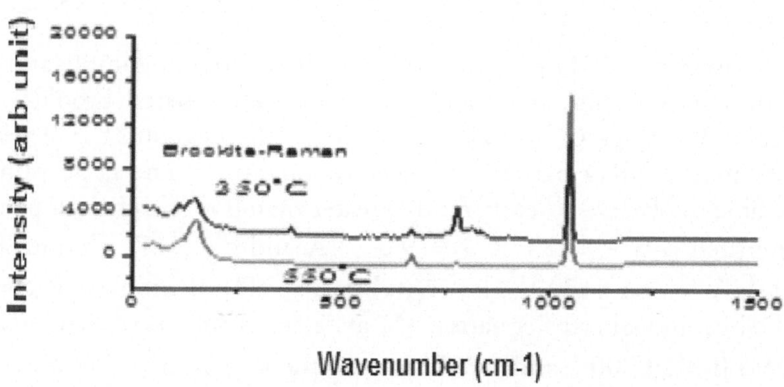

Figure 2: Raman Spectra of as Prepared Brookite Sample Heated at Temperature 350°C and 550°C Respectively.

II. EXPERIMENTAL DETAILS

A. Synthesis of Brookite

The starting materials used in this study were titanium (IV) isopropoxide (TIP) $[Ti(OCH(CH_3)_2)_4$, with purity 98+%, Acros Organics, Newjersey,

U.S.A, europium (III) oxide[Eu_2O_3, with purity 99.99%, HIMEDIA], absolute ethanol (EtOH) [C_2H_5OH, with purity 99.8%, HIMEDIA], sodium hydroxide [NaOH], hydrochloric acid [HCl with purity 35%, HIMEDIA] and Tetraethyl orthosilicate (TEOS) [$Si(OC_2H_5)_4$, with purity 98%, Sigma Aldrich], Acetone[C_3H_6O,with purity 99.8%, HIMEDIA]. All the chemicals were used in a reagent grade and used without further purification.The details of the synthesis of TiO_2 are reported elsewhere [3]. The temperature and pH value are maintained at 10°C and 8-9 respectively.

III. Results and Discussion

Powder X-diffraction (XRD) data are recorded in Ultima IV X-ray diffractometer of Rikagu with nickel filtered CuK_α radiation (λ = 1.54056 Å) having tube voltage 30 kV and current ~20 mA. XRD patterns are recorded in the angular range $20° \leq 2\theta \leq 80°$ with a step size of $\Delta 2\theta = 0.005°$. The crystallite size (D) was computed using Debye- Scherrer relation (DS)[4].

$$D = \frac{K\lambda}{\beta cos\theta} \quad \ldots (1)$$

Where **β** is the full width at half maximum (FWHM) of the diffraction peak in radian; λ (= 0.15406 nm) is the wavelength of X- ray for CuK_α; θ is the Bragg's angle measured in radian; K is a constant taken as 0.89 and known as structural constant.

Figure 1 shows the XRD patterns of as prepared brookite sample annealed at 550°C and inset is also the XRD patterns of as prepared brookite sample annealed at 350°C, 450°C and 750°C respectively. Less number of peaks was observed in the XRD pattern at lower temperature. The large number of peaks is able to observe at temperature greater than 350°C. The cell parameters are computed as a = 10.832Å, b = 9.0172Å and c = 2.6723Å and unit cell volume as V = 261.017$Å^3$. The crystallite size (D) of brookite is computed using Debye and Sherrer equation (1) at selected full width half maximum (FWHM) β = 19.7008° and 1.25583° then their respective sizes are found to be 4.288nm and 66.41 nm. The average crystallite size of brookite as D = 35.349 nm. The absence of the characteristic peak of anatase at 2θ = 62.57° is the confirmation for the occurrence of brookite phase [5] which is also consistent with the result of JCSPD No. 29 1360.

Intensity ratio of brookite was calculated from the XRD data as $\frac{I^{(121)}}{I^{(120)}}$ = 4.29 which is the contrast report of A.D. Paola [6]. This might be due to factors like impurities, particle size, surface area [7].

Raman spectroscopy data were recorded by using 785 nm Xplora1 micro Raman systems. Figure 2 shows the Raman spectra of the as prepared brookite sample annealed at 350°C and 550°C. Five number of bands are able to observed at the Ramman spectra when the sampled is heated at temperature 350°C including one intense band at wave number 1051.18 cm^{-1} and whereas at higher temperature i.e. 550°C three less number of bands are observed. It might be due to the presence of impurities or other phases of TiO_2.The presence of minor impurities cannot detect by the XRD but Raman can detect easily. The presence of intense band at wave number 153 cm^{-1} is the attribute of presence of brookite [8]. And absence of characteristic band of anatase at wave number 516 cm^{-1} in the Raman spectra of as prepared sample heated at temperature 350°C allows for occurrence of brookite.

IV. Conclusion

Brookite nanoparticle has bensynthesied by using non-aqueous sol-gel technique at low temperature and characterized with the used of (XRD) and Laser Raman spectroscopy respectively. The presence of peak at brookite is confirm both from the study of XRD and Raman spectroscopy. We are able to obseve larger size of the crystallite at narrow FWHM and red shift Raman mode at Eg(1) with the increase in heating temperature.

Acknowledgement

The Authors express their most sincere gratitude to FIST-DST, Delhi for instrumentation facilities in the Department of Physics, NERIST, Arunachal Pradesh, India.

References

[1] Beltram, A., Romero-Ocaña, I., Delgado Jaen, J., Montini, T. and Fornasiero, P., Photocatalytic valorization of ethanol and glycerol over TiO_2 polymorphs sustainable hydrogen production, *Appl.Catl. A Gen*, 518, 167 (2016).

[2] B.O'Regan and M. Gratzel, A low cost, high efficiency solar cell based on dye-sensitized, *Nature,* 353 737 (1991).

[3] Singh, T.N., Devi, T.G. and Singh, S.D., Synthesise, characterization and optical studies of gallium activated titania nanoparticle, *Nano-Structure & Nano-Objects,* 11, 65 (2017).

[4] Hu, W., Li, L., Li, G. and Tang, C., Sun L., High-quality brookite TiO_2 flowers: Synthesis, characterization, and dielectric performance. *Cryst. Growth Des.*, 9, 3676 (2009).

[5] Cullity, B.D., Elements of X-ray diffraction, Addison-Wesley Pub. Co.

[6] Paola, A.D., Bellardita, M. and Palmisano, L., Brookite, the Least known TiO_2 Photocatalyst, 3, 36 (2013).

[7] Zhang, H., Banfield, J.F., Structural characteristics and Mechanical and Thermodynamic properties of nanocrystalline TiO_2, *Chem. Rev.* 114, 9613 (2014).

[8] J.G., Ishigaki, T. and Sun, X., Anatase, brookite, and rutile nanocrystals via redox reactions conditions: Phase-selective synthesis and physicochemical properties, *J. hys. Chem.* C111, 4969 (2007).

Structural Properties of Undoped and Ce Doped Y_2O_3 Nano Powders Synthesized by Solution Combustion Method

M. Dutta* and G. Wary[1]

Department of Physics, Cotton University, Guwahati-781001, India

*E-mail: manashreed2111@gmail.com

ABSTRACT: In this work, we report structural, morphological and optical properties of undoped and cerium (Ce) doped yttrium oxide (Y_2O_3) nanoparticles synthesised by a solution combustion method. The incorporation of Ce in the Y_2O_3 host was found to influence the structural, morphological and optical properties of Y_2O_3 nanoparticles. Although the synthesized samples exhibit body-centred-cubic crystal structure with (222) preferential orientation, the doping of Ce shifted the (222) diffraction peak towards lower diffraction angle. The average crystallite size of the nanoparticles were found to decrease from 29.27 nm to 28.93 nm with the increase of Ce doping. Further, the size of the nanoparticles were found to vary from 23 nm to 50.78 nm with increase of the dopant from 1 to 5%.

Keywords: Nanoparticle, Doping, Y_2O_3, Ce.

I. Introduction

The rare-earth Cerium (Ce) ions doped crystals constitute an attractive class of materials that have been extensively used in many kinds of fields including scintillation phosphor, laser medium and white light emitting diode phosphors. The outstanding optical behaviors of trivalent Ce ion has drawn considerable interest due to its popular optical 5d-4f transition. Among various host materials, Yttrium Oxide Y_2O_3 crystals is considered to be the most promising host because of its unique chemical and thermal stability. Ce doped Y_2O_3 crystal is a multifunctional materials that can give rise to applications of energy storage phosphor [1-2].

II. Experimental Details

A. Sample Synthesis

In this study, undoped and Ce doped Y_2O_3 phosphors were synthesized by solution combustion method. We prepared Y_2O_3:Ce samples with five different Ce doping concentration from 1 to 5% of mole.

Experimental steps for synthesis of pure Y_2O_3:

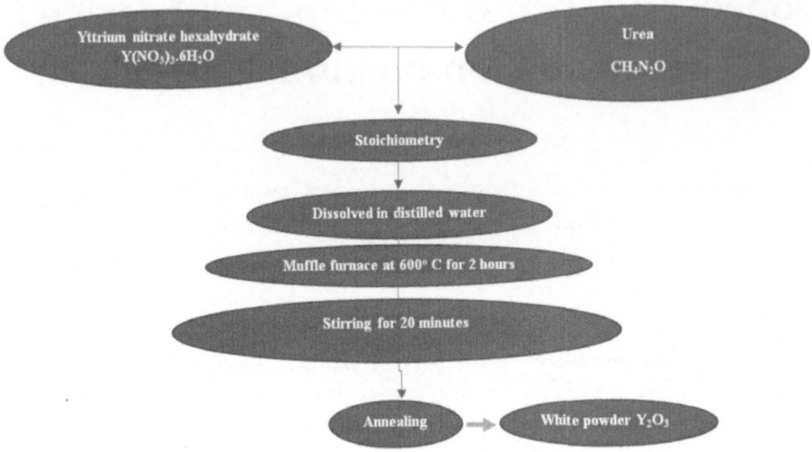

Experimental steps for synthesis of Ce doped Y_2O_3:

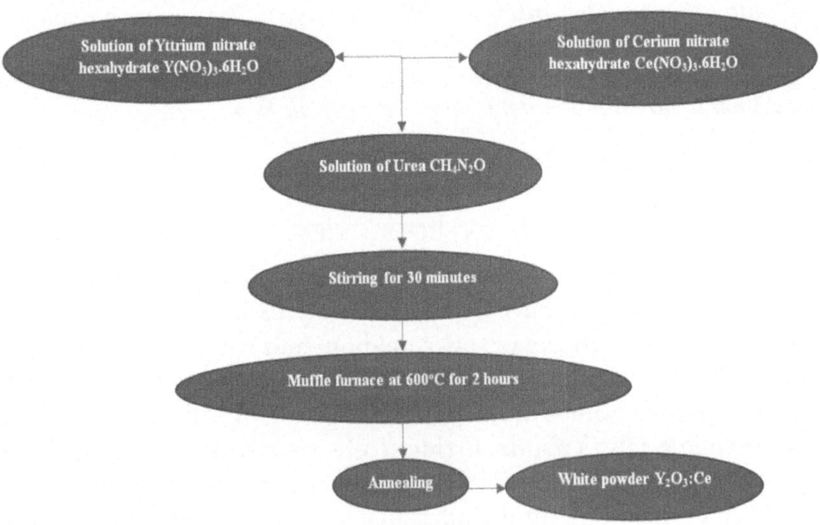

III. RESULTS AND DISCUSSION

A. Structural Analysis

The XRD patterns of the synthesized undoped and Ce doped Y_2O_3 powders are shown in Figure 1. It shows sharp diffraction peak indicating that the synthesized powders are crystalline. A comparison with the Joint Committee Powder Diffraction Standard (JCPDS) data of Y_2O_3 indicates that the XRD patterns are consistent with JCPDS data 89–5591. All of the diffraction peaks indexed to a body centred cubic structure of Y_2O_3, with a dominant (222) peak around 2θ = 29°. The Figure reveals that as doping concentration of Ce increases, all the XRD peaks of Y_2O_3 and Y_2O_3:Ce powders were found to shift towards the lower angle. The Figure 2 shows gradual shifting of (222) peak to the lower angle with increasing the concentration of Ce from 1 to 5 mol%. This indicates the expansion of the lattice parameters [3]. Since the ionic radii of Ce^{3+} ions are larger than that of Y^{3+} ions[4], the increase of lattice parameter with increasing Ce concentration is expected due to the larger ionic radii of Ce^{3+} ions in comparison to that of Y^{3+}ions. The average crystallite size (*D*) of the samples was estimated from XRD data using Scherrer's formula [5]:

$$D = \frac{0.9\,\lambda}{\beta cos\theta} \qquad \dots (1)$$

Where β is the broadening of diffraction line measured at half of its maximum intensity (radian) i.e. FWHM and λ is the wavelength of the X-ray used (1.5406 Å). The estimated values of the crystallite size (*D*) are presented in Table 1 and it shows a steady decrease with increasing Ce concentration from 0 to 5 mol%. In this case, the gradual increase of Ce doping progressively reduces the concentration of Y in the system which results a suppressed grain growth of Ce doped Y_2O_3 samples [6]. The strain induced in the samples were calculated using William-Hall son formula [7]:

$$\varepsilon = \frac{\beta}{\tan\theta} \qquad \dots (2)$$

The calculated values of strain are presented in Table 1. It is found that the strain values show an increasing trend with increasing the concentration of Ce, which confirms the deterioration of the crystalline quality of the synthesized samples with Ce doping [8, 9].

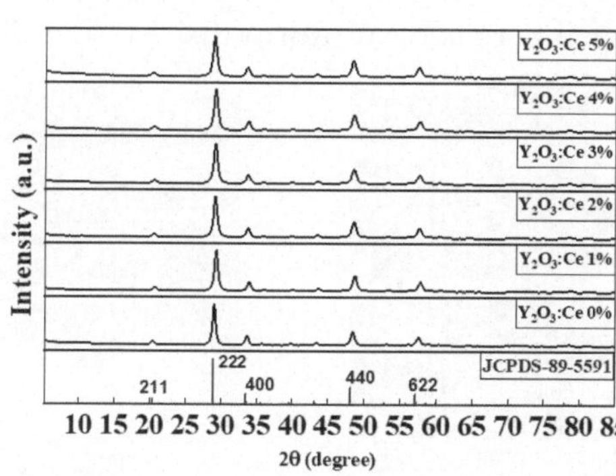

Figure 1: XRD Patterns of Y_2O_3:Ce with the Ce Samples Concentrations of 0%, 1%, 2%, 3%, 4% and 5%

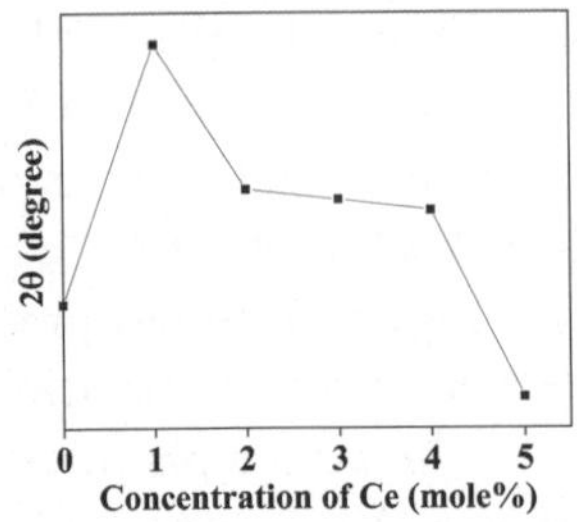

Figure 2: Shifting of (222) Peak of with Ce Concentrations

The JCPDS data 89–5591 has been shown for comparison.

Table 1: Estimated Values of Average Crystallite Size, Microstrain and Particle Size of the Samples

Sample Name	*Average Crystallite Size, D (nm)*	*Strain ($x10^{-3}$)*	*Estimated Particle Size (nm)*
Y_2O_3:Ce (0%)	29.02	189.00	23.00
Y_2O_3:Ce (1%)	29.27	166.00	33.20
Y_2O_3:Ce (2%)	29.13	156.00	37.11
Y_2O_3:Ce (3%)	29.12	150.00	44.00
Y_2O_3:Ce (4%)	29.11	149.00	46.80
Y_2O_3:Ce (5%)	28.93	147.00	50.78

B. Morphological Analysis

The surface morphologies of the synthesized samples were analysed using field-emission scanning electron microscopy (FESEM). Figure 3 shows FESEM images of 0–5% Ce doped Y_2O_3 samples under the same magnification. From FESEM images, it is observed that the samples have irregular spherical particle-like structure. There is no uniform distribution of particle size.

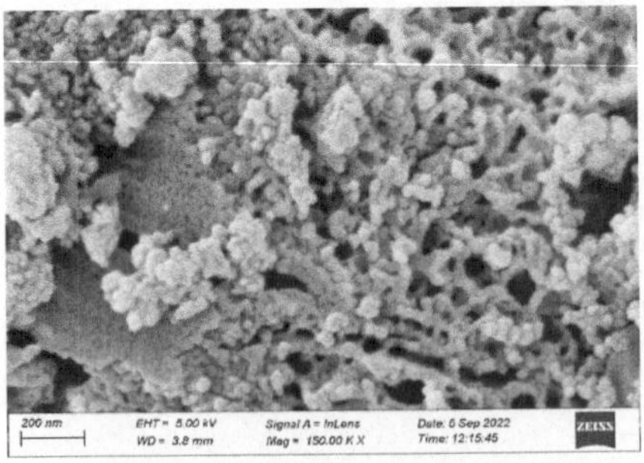

(a) Y_2O_3:Ce0%

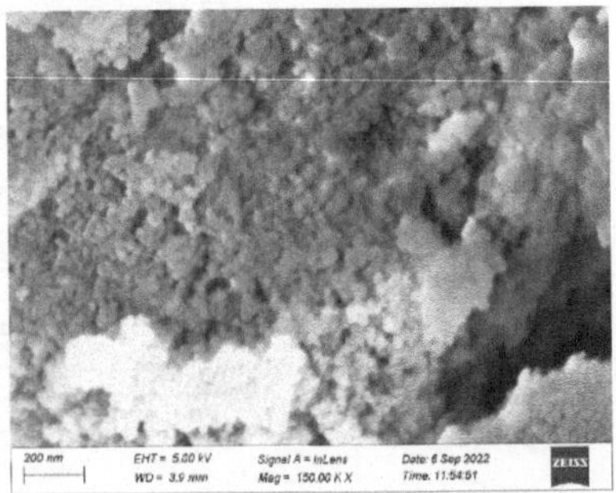

(b) Y_2O_3:Ce1%

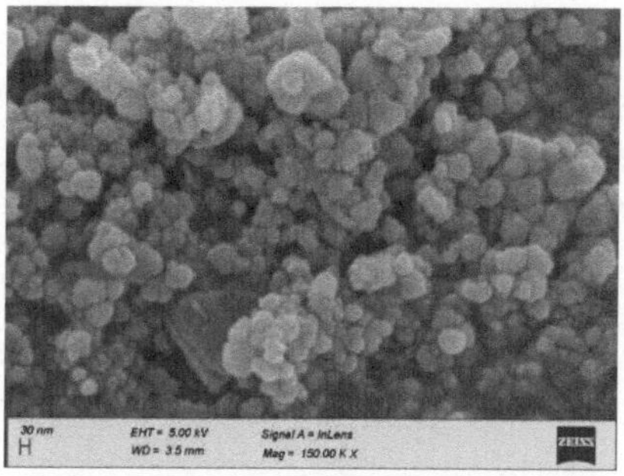

(c) Y_2O_3:Ce2%

(d) Y_2O_3:Ce3%

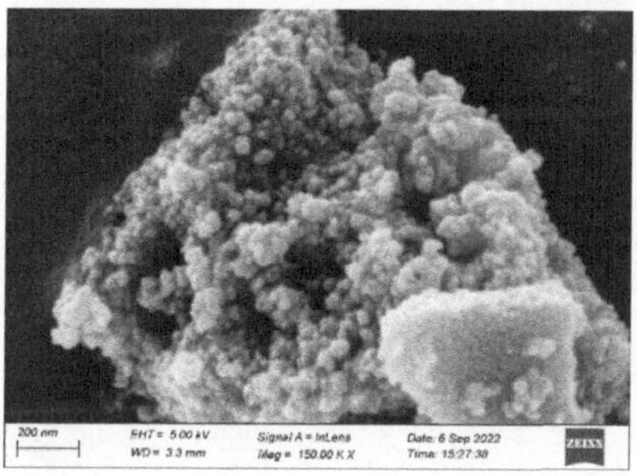

(e) Y_2O_3:Ce4%

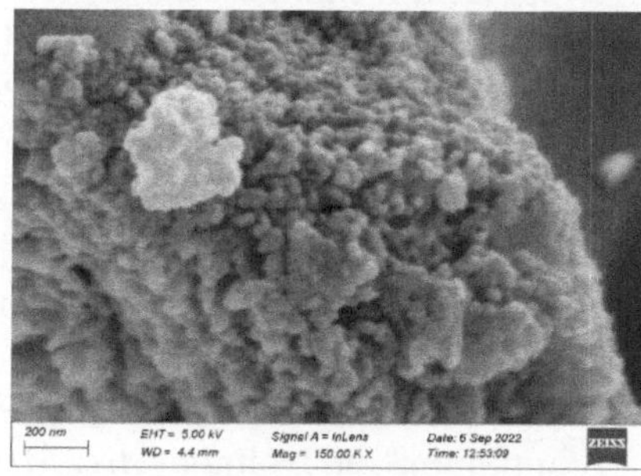

(f) Y_2O_3:Ce5%

Figure 3: FESEM Images of (a) Y_2O_3:Ce(0%), (b) Y_2O_3:Ce(1%), (c) Y_2O_3:Ce(2%), (d) Y_2O_3:Ce(3%), (e) Y_2O_3:Ce(4%) and (f) Y_2O_3:Ce(5%) under 150 kX Magnification.

IV. Conclusion

Within this work the nanopowders of undoped and Ce (1–5%) doped Y_2O_3 were synthesized by solution combustion method. The structural and surface morphological properties of the samples have been studied. The X-ray diffraction analysis reveal that the samples exhibit body centred cubic structure. The crystallite size of the undoped sample was estimated as 29.03 nm but it is found to increase for 1% Ce doped Y_2O_3 sample (29.27 nm) after that the crystallite size is found to decrease up to 28.93 nm for a sample doped with 5% Ce. The samples have spherical structure of diameter ~23.00–50.78 nm.

Acknowledgment

We are thankful to SAIF, Gauhati University, India and IASST, India for the XRD and FESEM reports respectively.

References

[1] Verhiest, K., Almazouzi, A., Wispelaere, N.D., Petrov, R. and Claessens, S., Development of oxides dispersion strengthened steels for high temperature nuclear reactor applications, *Journal of Nuclear Materials,* Vol. 385, No. 2, pp. 308–311, 2009.

[2] Laversenne, L., Kairouani, S., Guyot, Y., Goutaudier, C., Boulon, G. and Cohen-Adad, M.T, Correlation between do pant content and excited-state dynamics properties in Er^{3+}-Yb^{3+}-codoped Y_2O_3 by using a new combinatorial method, *Opt. Mater,* 19, 59–66 (2002).

[3] Suryanarayan, C.M. and Grant Norton, X-Ray Diffraction: A Practical Approach, Plenum Press, New York (1998).

[4] Hassanzadeh-Tabrizi, S.A., Synthesis and luminescence properties of YAG: Cenanopowder prepared by the Pechini method. *Adv. Powder Technol.,* 23(3), 324–327 (2012).

[5] Patterson, A., The Scherrer Formula for X-ray particle size determination, *Phys. Rev.,* 56, 978–982 (1939).

[6] Fujihara, S., Sasaki, C. and Kimura, T., *J European Ceramic Society,* 21, 2109 (2001).

[7] Williamson, G.K. and Hall, W.H., X-ray line broadening from filed aluminium and wolfram, *Acta Metall,* 1 22–31 (1953).

[8] Dapiaggi, M., Geiger, C.A. and Artioli, G., *Am. Miner.,* 90, 506 (2005).

[9] Qin, W. and Szpunar, J.A., *Phil. Mag. Lett.,* 85, 653 (2005).

Structural, Optical, and Dielectric Properties of Magnesium-Doped Barium Iron Niobate Solid Solution

Rimpi Chakravarty and B.N. Parida
Department of Physics, Central Institute of Technology Kokrajhar
(Deemed-to-be-University under MoE, Govt. of India), BTR, Assam, India

ABSTRACT: In this present work, an attempt is made to study the structural, optical, and dielectric properties of $Ba_{0.92}Mg_{0.08}Fe_{0.5}Nb_{0.5}O_3$ solid solution. Preliminary structural investigation through X-ray diffraction analysis delineates the lattice formation and monoclinic crystal structure of the sample. UV-Visible spectroscopy reflects that the sample strongly absorbs visible light and the energy bandgap is estimated to be 1.73 eV. The sample is found to exhibit the universal dielectric response phenomena within the frequency range 100 Hz – 500 kHz for temperatures 25°C–300°C.

Keywords: Perovskite, Dielectric, UV-Visible Spectroscopy.

I. Introduction

Complex perovskite with mixed cations (A'A''BO_3, AB'B''O_3, or A'A''B'B''O_3) is one of the standard modified perovskite system that has been extensively studied over the years. In these compounds generally cations like K^+, Na^+, Ba^{2+}, Pb^{2+}, Bi^{3+}, La^{3+}, etc. at A-site, and Mg^{2+}, Sc^{3+}, Zn^{2+}, Fe^{3+}, Ti^{4+}, Nb^{5+} etc. at B-site are present. Among the various materials synthesized based on this prototype, Barium iron niobate or $BaFe_{0.5}Nb_{0.5}O_3$ (abbreviated as BFN) is a promising lead-free complex perovskite with mixed B-site cations which exhibits intriguing properties. It has been efficiently prepared by techniques like the sol-gel method, solid solution route, co-precipitation, and molten salt method as per reports. Literature reveals that BFN was first introduced as a potential dielectric material by Yokosuka in the year 1995[1]. Subsequently, Saha and Sinha [2], and Intatha *et al.*[3] reported it as a relaxor ferroelectric with a colossal dielectric constant (ε_r) value. In these reports, the observed characteristics of BFN are ascribed to the space charge effect, grain-grain boundary effect, and Maxwell-Wagner polarization. Huang *et al.* [4] divulged that the ε_r value of BFN increases along with temperature, and the crest noted at higher temperatures signifies the presence of oxygen vacancies in the material. Moreover, BFN tends to exhibit its ideal electrophysical properties only when it is sintered at a very high temperature. This, however, augments the oxygen vacancies in the material, and as a result the loss tangent (tanδ) value increases [5]. Several researchers have tried to overcome this issue that hinders the applicability of pure BFN by introducing suitable dopants like Na, La, Bi, and Ca at its A-site. In this particular work, Mg is chosen as the dopant element and a detailed analysis of structural, optical, and dielectric properties of $Ba_{0.92}Mg_{0.08}Fe_{0.5}Nb_{0.5}O_3$ is carried out.

2. Experimental Section

Considering factors like reliability and simplicity the solid solution route is followed for the preparation of the desired sample. Barium carbonate ($BaCO_3$, 99.8% purity), Magnesium carbonate ($MgCO_3$, 95% purity), Ferric oxide (Fe_2O_3, 95% purity), and Niobium pentoxide (Nb_2O_5, 99.9%) purchased from LobaChemie Pvt. Ltd. India are used as the starting raw chemicals and weighed out as per the following chemical reaction:

$$0.92\ BaCO_3 + 0.08\ MgCO_3 + 0.25\ Fe_2O_3 + 0.25\ Nb_2O_5 \rightarrow Ba_{0.92}Mg_{0.08}Fe_{0.5}Nb_{0.5}O_3 + CO_2$$

First, these raw chemicals are mixed in an agate mortar-pestle system and then grounded meticulously in dry and wet conditions (solvent: methanol) for 4.5 hours. This chemical mixture is shifted to an alumina crucible and heated in a limited supply of oxygen at 1275°C in a muffle furnace. The obtained powder is once again grounded to achieve its non-coarse state and is used in various characterizations.

The Rigaku ULTIMA IV diffract to meter with Cu K_α source is used to collect the diffraction data of the sample in Bragg's angle ranging from 20° to 80° scanned at 2°/min. The Shimadzu UV-2600 spectrophotometer is used to monitor the response of the sample to UV-Vis light in the wavelength ranging from 200 to 800 nm. The N4L PSM 1735 Impedance analyzer is utilized to record the dielectric-related parameters across the sample in the frequency range 100 Hz to 500 kHz at temperature intervals between 25°C to 300°C.

III. Results and Discussion

In order to confirm the formation of the sample as well as to carry out its preliminary structural investigation the X-ray diffraction (XRD) technique is used. Sharp and precise diffraction peaks are noticed in the XRD profile (Figure 1) which varies from the initial raw chemicals implying the formation of a new sample. The profile is refined using the computer programs POWD and MAUD and based on the obtained results it can be stated that the sample has a monoclinic crystal structure. Additionally, the basic specifications obtained from refinement are shown in Figure itself.

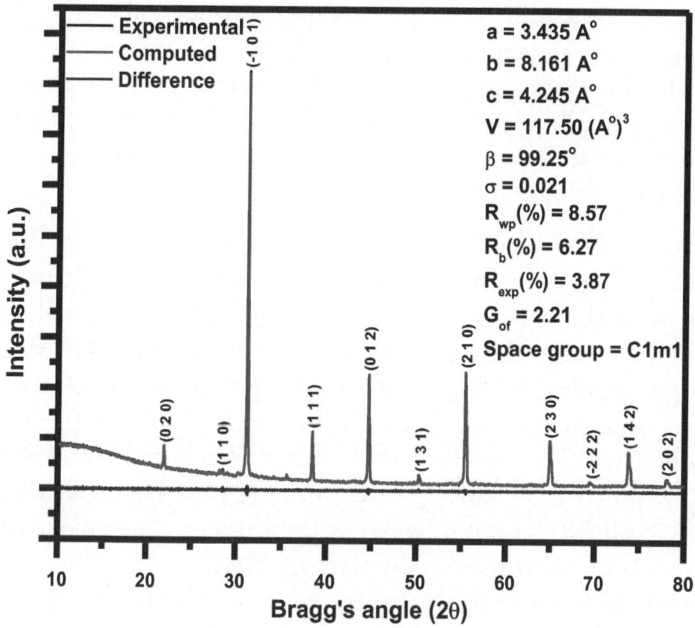

Figure 1: XRD Profile of the Sample

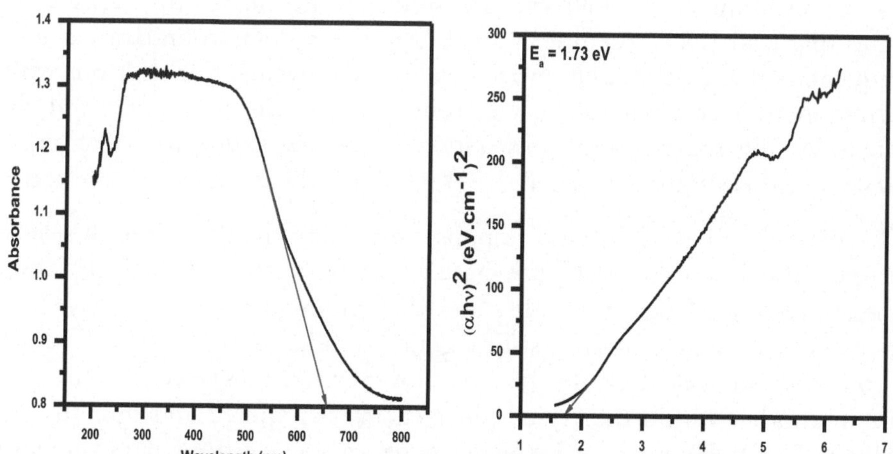

Figure 2: (a) UV-Visible Absorbance Spectrum (b) Tauc's Plot

To determine the functionality of the sample on basis of its response to an externally applied electromagnetic field whose spectral region ranges from ultraviolet to visible light, the UV-Visible spectroscopy analysis is performed. From the UV-Visible absorbance spectrum (Figure 2(a)) it is found that the value of absorbance is maximum at 339.5 nm and the cut-off wavelength is 656.2 nm. This reveals that the sample strongly absorbs visible light. Additionally, Tauc's equation $(\alpha h\nu)^{1/n} = A(h\nu - E_g)$ (where n = ½ for direct allowed transition considered in this study) is applied to the absorbance data and the calculated bandgap energy (E_g) is 1.73 eV (Figure 2(b)). Based on the optical response and resultant E_g value it can be stated that the sample is suitable for fabricating photocatalytic devices.

The dielectric behavior of the sample is analyzed by examining the frequency response of tanδ and ε_r. A decreasing trend is observed (in Figure 3(a)) in the value of ε_r with the enhancement of frequency for all selected temperature intervals. This reveals the influence of several polarization effects such as space charge, dipolar, ionic, and electronic polarizations. Out of them the space charge and dipolar polarization effects make a huge contribution in the lower frequencies. The involved multivalent elements Fe and Nb augment space charge polarization through the hopping electron mechanism between $Fe^{3+} \leftrightarrow Fe^{2+}$ and $Nb^{5+} \leftrightarrow Nb^{4+}$whereas the dipoles generated Ba-O, Mg-O, Fe-O, and Nb-O intensify the dipolar polarization [6]. However, with the enhancement of frequency, this effect becomes ineffective as the hopping electrons and dipoles are unable to cope with fluctuating ac field and eventually get diverged resulting in the drop of ε_r value. This behavior can also be expounded through the Maxwell-Wagner model [6]. This model states that a dielectric ceramic comprises resisting grain boundaries at lower frequencies that extend high impedance to the movement of electrons which results in the accumulation of electrons at the grain boundaries and thus intensifies the space charge polarization effect. As frequency enhances the accumulated electrons get scattered and subsequently the value of ε_r decreases.

The anomaly observed in tanδ response to varying frequency for all selected temperature intervals (in Figure 3 (b)) can be explained through Koop's phenomenological theory [7]. It states that the high value of tanδ at lower frequencies is because of the resisting grain boundaries that cause a hindrance in the movement of electrons. For the movement of these electrons from one site to another within the crystal more energy is required and thus the tanδ value increases. However, as frequency enhances the electrons can easily move through the active conducting/semiconducting grains without consuming much energy and hence tanδ is significantly low. Moreover, the room temperature ε_r and tanδ responses to frequency variation (shown in the inset of Figure 3 (a) & (b)) reveal that the sample depicts high ε_r and low tanδ values at lower frequencies. This suggests that the sample is suitable for capacitive applications.

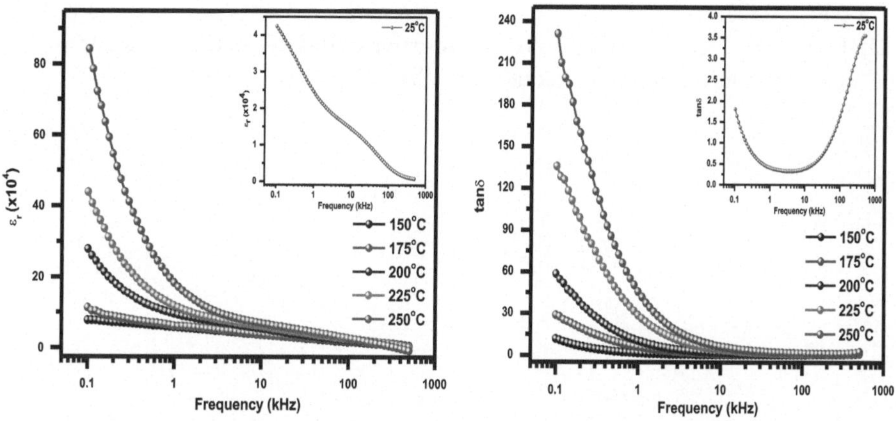

(a) (b)

Figure 3: Variation of (a) ε_r and (b) tanδ with Frequency at Selected Temperatures

IV. Conclusion

The sample is prepared effectively by the solid solution route at optimized experimental conditions. Structural investigation reveals that the sample has a monoclinic crystal structure and belongs to the C1m1 space group. From the strong response of the sample to visible light and the resultant E_g value, it can be stated that the sample is suitable for fabricating photo catalytic devices. The observed dielectric behavior delineates that the sample may be used for designing capacitive parts.

References

[1] Yokosuka, M., Dielectric dispersion of the complex perovskite oxide $Ba(Fe_{1/2}Nb_{1/2})O_3$ at low frequencies, *Jpn. J. Appl. Phys.*, 34, 5338–5340 (1995).

[2] Saha, S. and Sinha, T.P., Structural and dielectric studies of $BaFe_{0.5}Nb_{0.5}O_3$, *J. Phys. Condens. Matter,* 1, 249–258 (2002).

[3] Intatha, U., Eiessayeam, S., Wang, J. and Tunkasiri, T., Impedance study of giant dielectric permittivity in $BaFe_{0.5}Nb_{0.5}O_3$ perovskite ceramic, *Current Applied Physics,* 10(1), 21–25 (2010).

[4] Ke, S. and Huang, H., Giant low-frequency dielectric tunability in high-k $Ba(Fe_{1/2}Nb_{1/2})O_3$ ceramics at room temperature, *J. Appl. Phys.*, 108 (6), 064104 (2010).

[5] Raj, A., Gupta, P., Choudhary, R.N.P., Structural and electrical properties of La^{3+} modified $Ba(Fe_{0.5}Nb_{0.5})O_3$ ceramics, *Journal of Physics and Chemistry of Solids*, 148,109676 (2021).

[6] Chakravarty, R., Parida, R.K. and Parida, B.N., Dielectric and impedance study of 'Ca'-doped $BaFe_{0.5}Nb_{0.5}O_3$ complex perovskite, *Applied Physics A*, 128(8), 647 (2022).

[7] Koops, G., On the dispersion of resistivity and dielectric constant of some semiconductors at audio frequencies, *Phys. Rev.*, 83, 121 (1951).

Synthesis of Crystalline and Defect-Enriched Anatase TiO_2 Using Plasma – Liquid Interaction

P. Kalita[1,2,*], P.J. Boruah[1] and H. Bailung[1,2,3]

[1]Plasma Application Laboratory, Physical Sciences Division, Institute of Advanced Study in Science and Technology (IASST), Guwahati-781035, Assam, India

[2]Academy of Scientific and Innovative Research (AcSIR), Ghaziabad, Uttar Pradesh–201002, India

[3]Department of Physics, Bodoland University, Kokrajhar, Assam–783370, India

*E-mail: 2015parismitakalita@gmail.com

ABSTRACT: Plasma – Liquid Interaction (PLI) offers a non-equilibrium reactive environment for nanofabrication. Although this novel technique is widely used for material fabrication, Plasma- Induced Crystallization (PIC) is one aspect of this plasma process that is still poorly understood. The main concern of this chapter is to understand the PIC of amorphous material in a liquid medium. In the present work we studied the crystal growth of amorphous titanium oxide in correlation with plasma treatment time and the initial pH of the precursor. Here, we have observed that a longer treatment time can introduce defects or dislocations into the material.

Keywords: Plasma – Liquid Interaction (PLI), Reactive Species, Plasma – Induced Crystallization (PIC), Ti and O-vacancy-related Defects.

I. INTRODUCTION

The plasma generation in contact with a liquid medium or Plasma – Liquid Interaction (PLI) at atmospheric pressure provides a non-equilibrium reactive environment, leading to a novel technique for fabricating advanced functional materials[1].In the plasma-liquid interface, a wide range of reactive species is formed, facilitating a chemically reactive zone. This high chemical reactivity leads this process to be quite rapid. PLI is well-known as a rapid process for nanofabrication, where both electrode material and a metallic precursor serve as the raw material. Various reported articles have already explained the possible reaction pathways for the formation of those nanomaterials. However, the crystal growth and defect incorporation into the crystal lattice in a plasma environment is not investigated in detail. We should understand the correlation between the plasma parameters and the synthesized material to develop a clear insight into the synthesis process. In the last decades, this technique has been widely used for the synthesis of noble metal nanoparticles (NPs) (Au, Ag, and Au_xAg_{1-x}, etc.) at different discharge conditions[2–4]. The plasma-generated solvated electrons (e^-_{aq}), H_2O_2, and nitrogen species (NO_2^-/NO_3^-) play the role of reducing agent in the synthesis process. Besides the synthesis of metal NPs, a few reports also describe the synthesis of metal oxide nanomaterials using this technique. Jung *et al.* have synthesized crystalline tungsten oxide decorated TiO_2 photocatalyst through the liquid phase plasma (LPP) interaction[5]. Here, the tungsten ion from the WCl_4 liquid precursor is reduced by the plasma-generated reactive species on the surface of pristine TiO_2, and this composite material is further used for the photo-degradation of diethyle phthalate (DEP) molecules. The pristine TiO_2 is also modified using the LPP method by decorating crystalline Fe and cobalt oxide NPs by the same research group[6,7]. Despite this plasma-assisted material fabrication being a well-known process, the Plasma – Induced Crystallization (PIC) of materials in a liquid medium remains a poorly understood aspect of plasma synthesis. Ongoing research aims to gain a deeper understanding of this process and identify ways to optimize it for specific applications.

The present work demonstrates the rapid crystal growth of amorphous titanium oxide due to the interaction of plasma-generated reactive species and the titanium precursor. During the crystal growth of the material, some Ti and O-vacancy-related defects are incorporated. These findings indicate that plasma treatment time is an effective parameter for controlling the defect density in the crystal structure.

II. Experimental Details

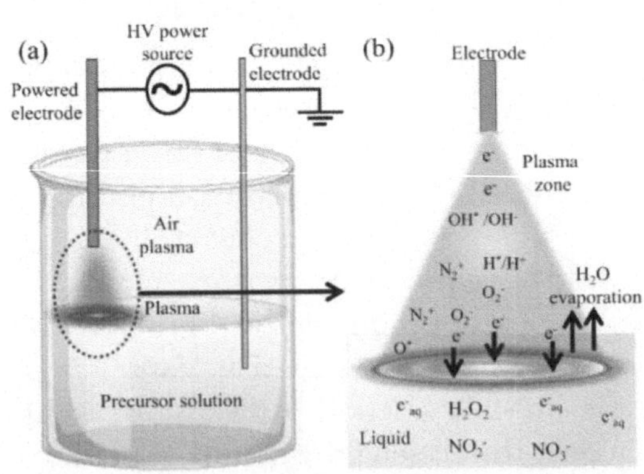

Figure 1: Schematic Diagram of (a) Plasma-Liquid Reactor and (b) The Formation of Various Reactive Species in the Discharge Zone and the Liquid Medium

Figure 1(a) represents the schematic of the plasma reactor where two metal rods have been used as powered and grounded electrodes. The powered electrode (Cu rod) is positioned 1–2 mm above the liquid surface, while the grounded electrode (stainless steel) is submerged within the liquid medium. A high-voltage AC power source (5 kV, 50 Hz) ignites the plasma discharge between the tip of the powered electrode and the liquid medium. Figure 1(b) represents the generation and interaction of plasma-generated reactive species with the liquid medium.

The titanium precursor solution is prepared by hydrolyzing titanium but oxide, which forms amorphous titanium oxide by removing the butoxy group. The precursor solution is plasma treated for different treatment times (X: 0–120 mins). The plasma-treated samples are coded as TX, where X represents plasma treatment time. The non-treated sample is termed as T0.

III. Plasma – Induced Crystallization (PIC) and Defect Incorporation

Figure 2(a) shows XRD patterns of the plasma-treated and non-treated samples. The non-treated (T0) and 5 mins of the plasma-treated sample (T5) show the amorphous nature of the material. However, after 10 mins of plasma treatment, a highly crystalline phase of the material is achieved. It is observed that increasing treatment time reduces the material's crystallinity due to the formation of defects or dislocations in the crystal lattice during the crystal growth process [8]. Recently, our group has also investigated the formation of electron-rich O-vacancy and grain boundary defects during the synthesis of tungsten oxide using the PLI technique [9]. Kong *et al.* and Xu *et al.* used the gas phase plasma engraving method to introduce O-vacancy defects into the TiO_2 and Co_3O_4 nanosheets [10, 11].

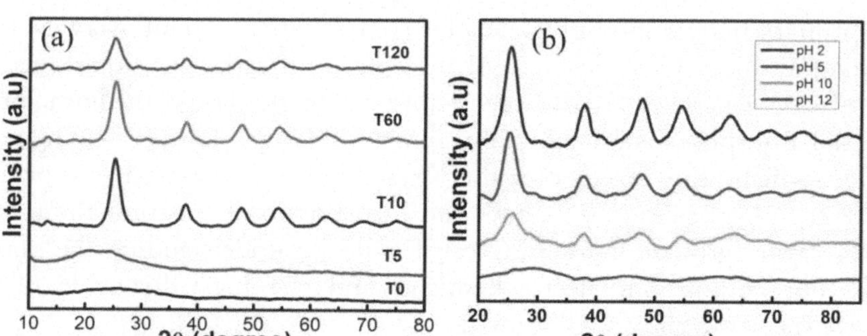

Figure 2: (a) XRD Patterns of (a) Air Plasma Treated Sample at Different Treatment Times and (b) the Samples Prepared at Different Initial pH

Again, due to nitrogen species (NO_3^-/NO_2^-) formation, the plasma-treated solution becomes acidic. Therefore, to understand the role of pH, the same experiment is conducted at different initial pH of the solution. The XRD patterns (Figure 2(b)) reveal that the crystallinity of the material is high at lower initial pH compared to a higher initial pH.

Mariotti *et al.* observed the formation of crystalline Si NPs in a gas-phase non-thermal atmospheric pressure RF microplasma environment using a gas mixture of argon and silane[12]. They observed that the gas temperature of the plasma is below the crystallization threshold of Si. Using two heating models, they explained that NPs could go beyond crystallization threshold temperature, which depends on the heat flux transferred to the nanosystem by the ion fluxes. Kramer *et al.* also studied the crystallization mechanism of Si NPs in atmospheric pressure RF microplasmas using a tandem plasma device[13]. Two separate plasma is produced where upstream plasma is produced using low power, and downstream plasma is produced using variable RF power. In the upstream plasma, amorphous Si seeds are grown from the gas mixture of argon and silane, after which it enters into the secondary plasma, where the crystallization takes place as a function of RF power used. Benčina *et al.* used low-pressure non-thermal oxygen plasma to crystallize TiO_2 nanotubes within seconds, which is synthesized via electrochemical anodization [14]. The key advantage of this PIC is no structural disruption during the treatment due to the short contact time between plasma and the material. However, the PIC of amorphous material in a liquid medium is a bit complex to understand because a part of the heat is transferred to the liquid molecules to initiate a lot of physicochemical phenomena simultaneously.

IV. Conclusion

The primary motivation of this chapter is to deliver a general overview of the Plasma – Induced Crystallization of amorphous material in a liquid medium. The plasma-generated reactive species present in the liquid medium interact with the amorphous material and generate a localized heat on the material's surface which can overcome the crystallization threshold temperature. However, longer plasma treatment can incorporate defects into the material during crystal growth. This study can extend by understanding the plasma–liquid non-equilibrium electrochemistry and the individual role of each reactive species in the crystal growth.

ACKNOWLEDGMENT

The authors would like to acknowledge SAIC-IASST for providing instrumentation facilities.

REFERENCES

[1] Chiang, W.H., Mariotti, D., Sankaran, R.M., Eden, J.G. and Ostrikov, K., Microplasmas for Advanced Materials and Devices, *Adv. Mater.*, 32, 1905508 (2020).

[2] Richmonds, C. and Sankaran, R.M., Plasma-Liquid Electrochemistry: Rapid Synthesis of Colloidal Metal Nanoparticles by Microplasma Reduction of Aqueous Cations, *Appl. Phys. Lett.*, 93, 131501 (2008).

[3] Huang, X.Z., Zhong, X.X., Lu, Y., Li, Y.S., Rider, A.E., Furman, S.A. and Ostrikov, K., Plasmonic Ag Nanoparticles via Environment-benign Atmospheric Microplasma Electrochemistry, *Nanotechnology*, 24, 095604 (2013).

[4] Yan, T., Zhong, X., Evelyn Rider, A., Lu, Y., Furman, S.A. andOstrikov, K., Microplasma-chemical Synthesis and Tunable Real-time Plasmonic Responses of Alloyed AuxAg1−x Nanoparticles, *Chem. Commun.*, 50, 3144 (2014).

[5] Ki, S.J., Park, Y.K., Kim, J.S., Lee, W.J., Lee, H. and Jung, S.C., Facile Preparation of Tungsten Oxide Doped TiO2 Photocatalysts using Liquid Phase Plasma Process for Enhanced Degradation of Diethyl Phthalate, *Chem. Eng. J.*, 377, 120087 (2019).

[6] Lee, H., Park, Y.K., Kim, S.J., Kim, B.H. and Jung, S.C., Fe-decorated TiO_2 Powder Photocatalysts with Enhanced Visible-light-driven Degradation Activities, *Surf. Coatings Technol.*, 307, 1018 (2016).

[7] Lee, H., Park, Y.K., Kim, S.J., Kim, B.H. and Jung, S.C., Titanium Dioxide Modification with Cobalt Oxide Nanoparticles for Photocatalysis, *J. Ind. Eng. Chem.*, 32, 259 (2015).

[8] Kalita, P., Boruah, P. J., Khanikar, R. R. andBailung, H., Plasma–induced Rapid Crystallization and Surface Engraving of Amorphous TiOx(OH)y to Enhance Adsorption and Photocatalytic Activity, *J. Photochem. Photobiol. A Chem.*, 434, 114251 (2023).

[9] Boruah, P.J., Khanikar, R.R. and Bailung, H., Synthesis and Characterization of Oxygen Vacancy Induced Narrow Bandgap Tungsten Oxide (WO3−x) Nanoparticles by Plasma Discharge in Liquid and its Photocatalytic Activity, *Plasma Chem. Plasma Process.*, 401019 (2020).

[10] Kong, X., Xu, Y., Cui, Z., Li, Z., Liang, Y., Gao, Z., Zhu, S. and Yang, X., Defect Enhances Photocatalytic Activity of Ultrathin TiO2 (B) Nanosheets for Hydrogen Production by Plasma Engraving Method, *Appl. Catal. B Environ.*, 230, 11 (2018).

[11] Xu, L., Jiang, Q., Xiao, Z., Li, Xi., Huo, J., Wang, S. and Dai, L., Plasma-Engraved CO_3O_4 Nanosheets with Oxygen Vacancies and High Surface Area for the Oxygen Evolution Reaction, *Angew. Chemie.*, 128, 5363 (2016).

[12] Askari, S., Levchenko, I., Ostrikov, K., Maguire, P. and Mariotti, D., Crystalline Si Nanoparticles below Crystallization Threshold: Effects of Collisional Heating in Non-thermal Atmospheric-Pressure Microplasmas, Appl. Phys. Lett., 104, 163103 (2014).

[13] Kramer, N.J., Anthony, R., Mamunuru, J.M., Aydil, E.S. andKortshagen, U.R., Plasma-induced Crystallization of Silicon Nanoparticles, *J. Phys. D. Appl. Phys.*, 47, 075202 (2014).

[14] Benčina, M., Junkar, I., Zaplotnik, R., Valant, M., Iglič, A. andMozetič, M., Plasma-induced Crystallization of TiO_2 Nanotubes, *Materials (Basel).*, 12, 626 (2019).

Comparative Dielectric Studies on Two Composites of 5O.6+$BaTiO_3$

Ragini Tripathi[1*], P.R. Alapati[2] and Ravindra Dhar[3]

[1]Department of Physics, Sunbeam Women's College Varuna, Varanasi 221001 (UP), India

[2]GKCIET, Maligram 732141, West Bengal, India

[3]Centre of Materials Science, University of Allahabad, Prayagraj-211002 (UP), India

*E-mail: vrtripathi@gmail.com

ABSTRACT: The presence of ferroelectric NPs (FNPs) like barium titanate ($BaTiO_3$) enhances the properties of LC composites without distorting the global nematic director. The thermal and dielectric studies were carried out by doping two different concentrations of $BaTiO_3$ in 5O.6 viz. 0.4 and 0.8 wt.%. It was observed that the findings for these two concentrations are very different. With the doping of 0.4 wt.% the magnitude of dielectric anisotropy slightly decreases in negative order, whereas for 0.8 wt.% the value increases. This reflects the interaction of $BaTiO_3$ NPs with the permittivity components. Furthermore, the presence of $BaTiO_3$ NPs in the LC matrix led to a shift of relaxation frequency to higher region (for 0.4 wt.%) and to lower region (for 0.8 wt.%). The variation of dielectric strength was also observed. The ionic conductivity is almost similar to pure for 0.4 wt.% doped sample, but the value decreases for 0.8 wt.% doped sample. This suggests the probability of aggregation with the increase of concentration.

I. Introduction

The phase sequence exhibited by 5O.6 is N-SmA-SmC-SmB-SmF-SmG [1-2]. Different studies & investigations of this compound were reported, which serves it to be as a model system [3-5]. In this study, our focus is on the modification of dielectric properties of 5O.6 in all mesophases upon the dispersion of $BaTiO_3$ nanoparticles as well as carbon nanotubes (MWCNTs).

The effects of NPs dispersion of LC compounds yields distinct results [6-7] and dielectric properties are found to be concentration dependent [8-11]. On the basis of different reported results [12-13], it could be presumed that the dielectric parameters, such as permittivity, loss, dielectric anisotropy, relaxation frequency and dielectric strength show diverse behavior. Fundamentally they depend upon the host LC material and the dispersion of NPs, also the concentration of dispersed NPs plays very important role. But as such no theory was developed which can completely explain the behavior of dielectric parameters for LC-NP composites. Accordingly, the effect of variation of $BaTiO_3$ NPs concentration in 5O.6 (in two compositions: 0.4 wt. % and 0.8 wt. %) on dielectric properties are discussed.

II. Results and Discussion

The liquid crystal-nanoparticle composites were prepared by adding two different concentrations of $BaTiO_3$ NPs in 5O.6 viz. 0.4 wt. % and 0.8 wt. %. The size of dispersed $BaTiO_3$ NPs is 100 nm. The morphology of the $BaTiO_3$ NPs was investigated by transmission electron microscopy (TEM), which confirmed the uniform distribution of the NPs, in 5O.6 without any agglomerations (shown in Plate 5.1).DSC scans were recorded after five thermal cycles of the sample to ensure the stabilized thermal data of the system. The phase sequence, transition temperatures and enthalpy of pure and dispersed samples determined by DSC measurements in cooling process are shown in Table 5.1 and scan in cooling cycles at scan rate 5°C/min are shown in Figure 5.2 (a). Dielectric measurements as a function of temperature and frequency were made by following the same experimental techniques [14]. The variation of perpendicular ($\varepsilon_{\perp}'$) and parallel ($\varepsilon_{\parallel}'$) components of relative permittivity with temperature for the pure and $BaTiO_3$ NPs dispersed 5O.6 samples are shown in Figure 5.3. According to Maier and Maier theory, the dielectric anisotropy is given as [15],

$$\Delta\varepsilon = \frac{NHF}{\varepsilon_{\circ}}\left[\Delta\alpha - \frac{F}{2kT}\mu^2(1 - 3cos^2\beta)\right]S$$

From the above equation it is obvious that the dielectric anisotropy ($\Delta\varepsilon$) effectively depends upon β, *N*and *S*. In the present study, the dispersion of 0.4 wt. % of $BaTiO_3$ NPs results in a decrease in the perpendicular component of the permittivity and an increase in the value of parallel component of the permittivity. Whereas, when the concentration of NPs was increased to 0.8 wt. %, the perpendicular component slightly increases while parallel component decreases. This leads to a decrease (0.4 wt. %) and increase (0.8 wt. %) in the dielectric anisotropy (in magnitude). It seems that the dispersion of 0.4 wt. % of $BaTiO_3$ NPs in 5O.6 supports the effective parallel component of the dipole moment rather than the effective perpendicular component and the dispersion of 0.8 wt. % of $BaTiO_3$ NPs supports the effective perpendicular component of the dipole moment rather than the effective parallel component. Therefore, the ratio of effective dipole moment changes and results in a decrease (0.4 wt. %) and increase (0.8 wt. %) of the effective βof the dispersed system. Furthermore, the number density (*N*) is expected to decrease with the dispersion of $BaTiO_3$ NPs and due to this $\Delta\varepsilon$ should decrease. Since, ferroelectric NPs possess large dipole moment and high polarizability, therefore with the dispersion of small concentration (< 1.0 wt. %) of ferroelectric NPs, the order parameter is rarely expected to be decreased. Hence, β and *N* appears to be responsible for the decrease in $\Delta\varepsilon$ with the dispersion of 0.4 wt. % of $BaTiO_3$ NPs. Moreover, the presence of ferroelectric NPs in LC matrix increases the order parameter of the dispersed system by the coupling of NPs with macroscopic orientational order of LC [16]. Consequently, the order parameter (*S*) increases that overcompensates the decreasing effect of *N*. Hence, it may be concluded that angle β and order parameter increases after the dispersion of 0.8 wt.% of $BaTiO_3$ NPs which results in the increase of dielectric anisotropy. Furthermore, as per previous studies[17] the increase in the concentration of NPs increases the chances of aggregation. Since, NPs are much larger than LC molecules; therefore due to the ordering of LC they would couple to the surface of NP through surface-energy anchoring due to which it decreases the disorder in the system and hence stabilizes the liquid crystalline phases.

Table 5.1: Phase Polymorphism in Pure and Dispersed 5O.6 as Studied with DSC in Cooling Process

Phase transition	Pure 5O.6		5O.6+0.4 wt.% $BaTiO_3$ NPs		5O.6+0.8 wt.% $BaTiO_3$ NPs	
	T_P(°C)	ΔH(J/g)	T_P(°C)	ΔH(J/g)	T_P(°C)	ΔH(J/g)
Iso-N	71.3	3.368	71.4	2.99	71.3	2.65
N-SmA	59.4	2.479	59.6	2.289	59.5	1.92
SmA-SmC						
SmC-SmB	49.2	6.337	49.2	5.818	49.1	5.603
SmB-SmF						
SmF-SmG	36.5	0.6245	36.6	0.7417	36.5	0.6755
SmG-Cr	-13.1	20.36	-16.0	17.89	-12.6	17.77

Table 5.1: Relaxation Frequency (f_R), Dielectric Strength ($\delta\varepsilon$) and Distribution Parameters (α) of Dielectric Relaxation Modes in Various Phases of Pure and Dispersed Samples

Phases	Pure 5O.6			5O.6+0.4wt.% $BaTiO_3$ NPs			5O.6+0.8wt.% $BaTiO_3$ NPs		
	f_R	$\delta\varepsilon$	α	f_R	$\delta\varepsilon$	α	f_R	$\delta\varepsilon$	α
N (63.0 °C)	1.62 MHz	0.36	0.02	1.87 MHz	0.35	0.04	1.30 MHz	0.36	0.01
SmA (54.0 °C)	435.09 kHz	0.47	0.04	480.45 kHz	0.51	0.02	359.08 kHz	0.37	0.01
SmC(52.0 °C)	326.95 kHz	0.47	0.03	377.41 kHz	0.49	0.02	292.41 kHz	0.36	0.02
SmB(46.0 °C)	16.91 kHz	0.51	0.03	19.38 kHz	0.49	0.01	13.79 kHz	0.35	0.02
SmF(41.0 °C)	9.87 kHz	0.52	0.01	11.51 kHz	0.51	0.02	8.55 kHz	0.38	0.02
SmG(36.0 °C)	5.79 kHz	0.46	0.04	6.81 kHz	0.52	0.03	4.37 kHz	0.39	0.03

Now, with the study at molecular level, the perpendicular component ($\varepsilon_{\perp}^{'}$) was found to be almost invariant in the frequency range of 1 Hz to 10 MHz. This suggests that the dielectric relaxations corresponding to the molecular rotations along their long axes are well above 10 MHz. However, a relaxation mechanism has been observed in parallel component ($\varepsilon_{\parallel}^{'}$). The ranges of relaxation frequencies are found to be different from N to SmG phases (Figure 5.2). The relaxation frequency (Table 5.2) of the observed mechanism has been calculated by the simulation of the experimental data using Cole-Cole equation [18] as described earlier. From the value of α (mentioned in Table 5.2), it is apparent that the observed relaxations are a typical Debye process. From Figure 5.4 it can be observed that the relaxation frequencies are well described by an Arrhenius law.

$$f_R = f_0 \exp\left(\frac{-W_A}{N_A kT}\right)$$

With the help of slopes of the straight lines (see Figure 5.4), activation energies (W_A) were obtained and are shown in Table 5.3. Thus, on the basis of the magnitudes of the f_R and their Arrhenius behavior, the relaxation mechanism could be attributed to the flip-flop motion of molecules about their short axes [19]. One can notice from Table 5.2 that in case of 0.4 wt. % dispersed sample, the relaxation mechanism shifts towards high-frequency side. Thus, the usable frequency range (below relaxation frequency) has been found to be increased, which is important from application point of view. However, the addition of 0.8 wt. % of $BaTiO_3$ NPs to 5O.6 led to a shift of relaxation frequency to lower frequencies. This shifts in the f_R indicate an increase in the relaxation times ($\tau = 1/2\pi f$) for the reorientation of LC molecules. From Table 5.3 it can be observed that except SmA and SmB phase, the activation energy for 0.8 wt. % dispersed sample is comparatively greater than pure 5O.6. Therefore, the increase in activation energy together

Table 5.1: Values of Activation Energy (W_A) in Different Phases of Pure and Dispersed O.6 Samples

Phase	Activation Energy (W_A), kJ/mol		
	Pure 5O.6	5O.6+0.4 wt.% $BaTiO_3$ NPs	5O.6+0.8 wt.% $BaTiO_3$ NPs
N	33.11	30.86	48.02
SmA	40.62	47.89	32.96
SmC	32.07	37.89	55.65
SmB	44.85	37.02	23.87
SmF	27.62	28.27	31.03
SmG	32.87	30.61	72.93

with the increase in relaxation times suggests that a strong interaction exists between the LC molecules and NPs [20]. Since, LC molecules interact with permanent dipole moments of NPs as well as with the surface of NPs (anchoring effect). These interactions hinder the rotation of the LC molecules in the vicinity of the NPs, due to which relaxation frequencies shift to lower side.

The breaks or discontinuity at N-SmA and SmA-SmC phase transitions and a pronounced jump at SmC-SmB in Arrhenius plot (Figure 5.4) is because the ordering of molecules is changed inside the layer [21]. It can be observed that for pure and 0.4 wt. % dispersed sample the dielectric strength (Figure 5.5(a)) is almost temperature independent and falls in the vicinity of SmA-N phase transition temperature. However, the strength of 0.8 wt. % dispersed sample is less than that of pure; also the drop around SmA-N is not as prominent as in rest two samples. The frequency dependence of the total conductivity parallel to the director of pure and dispersed 5O.6 is shown in Figure 5.5(b) in N phase at 63.0°C. It can be seen that for 0.4 wt. % dispersed sample, $\sigma_{t\parallel}$ is almost same with respect to that of pure 5O.6. This indicates that there is no significant movement of the $BaTiO_3$ NPs along the transverse direction of the nematic director. Whereas in case of 0.8 wt. % dispersed sample, $\sigma_{t\parallel}$ decreases in comparison to pure sample. It is most likely to be due to the formation of aggregates by NPs, since higher concentrations of NPs in LC matrix show strong tendency of aggregation [22].

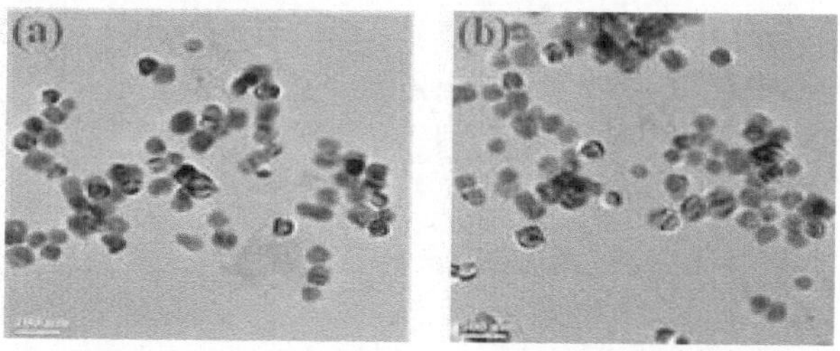

Plate 5.1: TEM Morphology of (a) 0.4 wt. % and (b) 0.8 wt. % $BaTiO_3$ NPs Dispersed in 5O.6

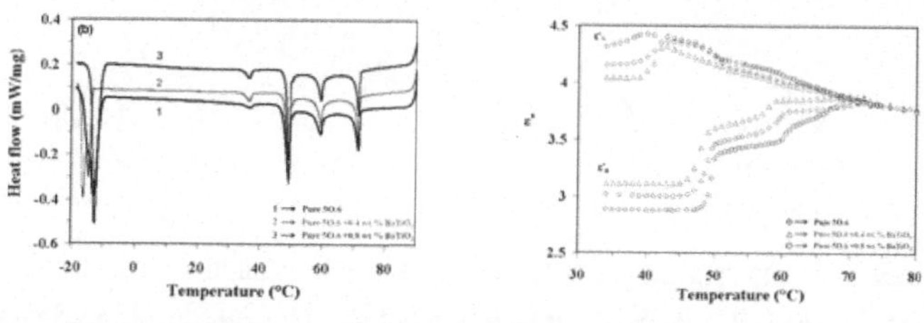

Figure 5.1: **(a)** DSC thermograph of pure and dispersed 5O.6 in cooling cycle. **(b)** Temperature dependence of parallel ($\varepsilon'_{\parallel}$) and perpendicular ($\varepsilon'_{\perp}$) components of the relative permittivity for the pure and dispersed 5O.6.

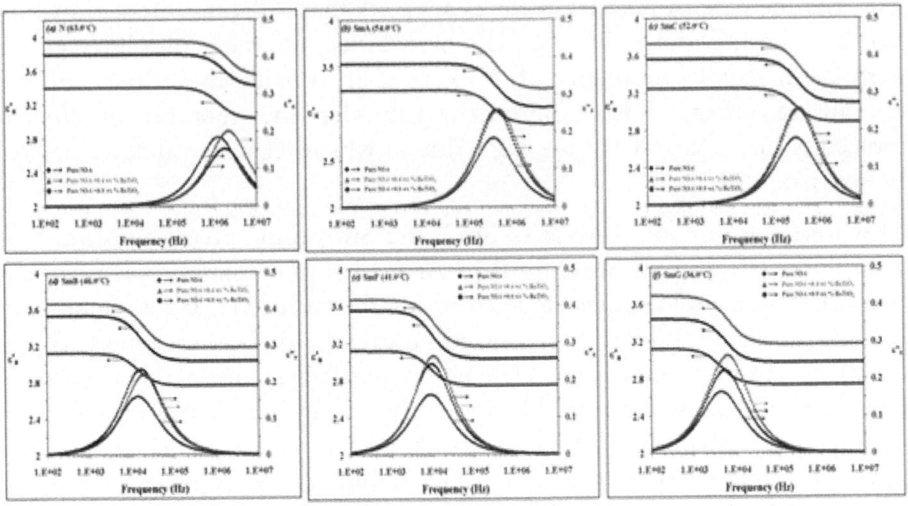

Figure 5.2: Frequency dependence of the parallel component of the permittivity ($\varepsilon_{\parallel}'$) and the loss ($\varepsilon_{\parallel}''$) in (a) N (at 63.0 °C), (b) SmA (at 54.0 °C), (c) SmC (at 52.0 °C), (d) SmB (at 46.0 °C), (e) SmF (at 41.0 °C) and (f) SmG (at 36.0 °C) phase. Values are taken after subtracting the low and high frequency corrections from the measured data.

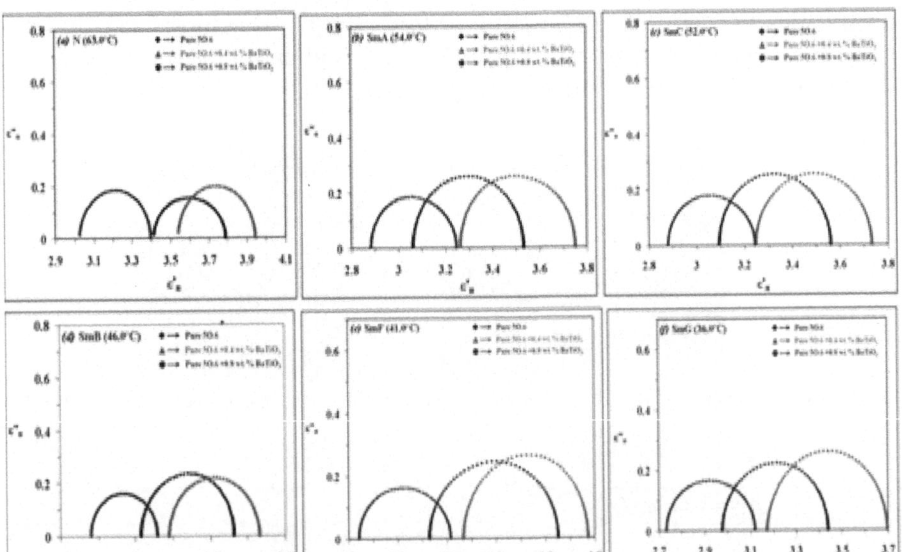

Figure 5.3: Cole-Cole plots showing the variation of loss ($\varepsilon_{\parallel}''$) with permittivity ($\varepsilon_{\parallel}'$) for the pure and dispersed 50.6 in (a) N (at 63.0 °C), (b) SmA (at 54.0 °C), (c) SmC (at 52.0 °C), (d) SmB (at 46.0 °C), (e) SmF (at 41.0 °C) and (f) SmG (at 36.0 °C) phase.

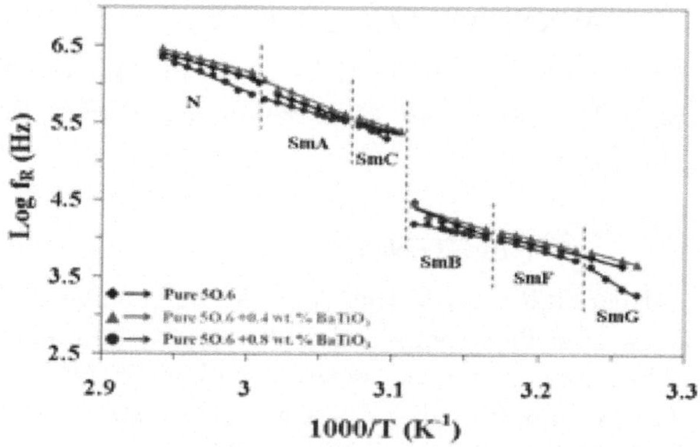

Figure 5.4: Variation of relaxation frequency (f_R) with inverse of absolute temperature (K^{-1}) for pure and dispersed 5O.6 samples, showing Arrhenius behavior of the observed relaxation mode.

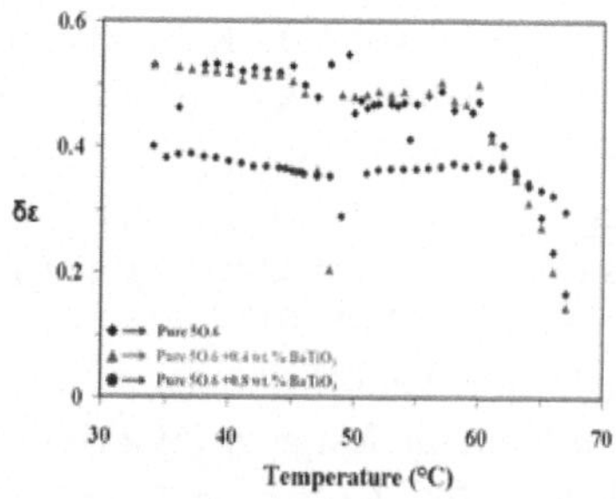

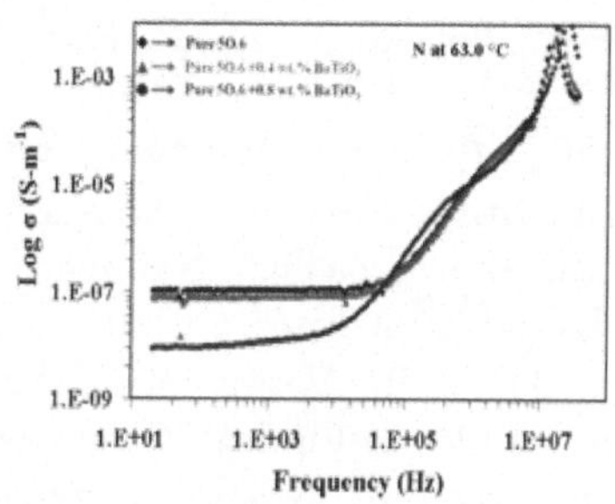

Figure 5.5: (a) Variation of dielectric strength ($\delta\varepsilon$) with temperature for pure and dispersed 5O.6 samples. **(b)** Variation of total conductivity along the director ($\sigma_{i\parallel}$) with frequency for pure and dispersed 5O.6 in nematic phase (at 63.0 °C).

III. Conclusion

Thus, with the dispersion of 0.4 wt. % of $BaTiO_3$ NPs in 5O.6, the longitudinal component of the dielectric permittivity increases, whereas the transverse component slightly decreases. Hence, dielectric anisotropy slightly decreases. However, in case of 0.8 wt. % dispersed sample, the longitudinal component decreases while transverse component slightly increases. Hence, dielectric anisotropy increases. The relaxation frequency corresponding to flip-flop motion of the molecules about the short axis has been found to be increased for 0.4 wt. % dispersed sample, whereas it decreases for 0.8 wt. % dispersed sample. The conductivity of lower concentration (0.4 wt. %) dispersed sample is almost identical to that of pure sample, but the conductivity decreases for higher concentration (0.8 wt. %) dispersed sample.

ACKNOWLEDGEMENT

One of the authors (RT) wishes to thank the Centre of Material Sciences, University of Allahabad, Allahabad and Department of Physics, NERIST, Nirjuli, for providing facilities.

REFERENCES

[1] Bhattacharjee A., Alapati P.R. and Verma A.L., A comparative study of the properties of the liquid crystalline compounds 5O.5 and 5O.6 in solid, solution and thin film forms by laser Raman spectroscopy, *Liq.Cryst.*, 29(5), 725 (2002).

[2] Rao, P.B., Rao, N.V.S. and Pisipati, V.G.K.M., The Smectic F. Phase in nO.m Compounds, *Mol. Cryst. Liq. Cryst.*, 206, 9 (1991).

[3] Benguigui, L., Dielectric Relaxation in Liquid Crystals, *Mol. Cryst. Liq. Cryst.*, 114, 51 (1984).

[4] Benguigui, L., Dielectric relaxation and nematic order in liquid crystals, *Phy. Rev. A*, 29, 2968 (1984).

[5] Alapati, P.R., Potukuchi, D.M., Rao, N.V.S., Pisipati, V.G.K.M., Paranjpe, A.S. and Rao, U.R.K., Phase Transition and Structural Investigations in 50.5, 50.6 and 50.7, *Liq. Cryst.,* 3, 1461(1988).

[6] Herrington, M.R., Buchnev, O., Kaczmarek, M. and Nandhkumar, I., The Effect of the Size of $BaTiO_3$ Nanoparticles on the Electro-Optic Properties of Nematic Liquid Crystals, *Mol. Cryst. Liq. Cryst.*, 527, 72 (2010).

[7] Paul, S.N., Dhar, R., Verma, R., Sharma, S. and Dabrowski, R., Change in Dielectric and Electro-Optical Properties of a Nematic Material (6CHBT) Due to the Dispersion of $BaTiO_3$ Nanoparticles, *Mol. Cryst. Liq. Cryst.*, 545, 105(2011).

[8] Vardanyan, K.K., Thiel, A., Fickas, B. and Daykin, A., Multicomponent nematic systems with doped gold nanoparticles, Liq. Cryst., 42, 445 (2015).

[8] Kumar, P., Kishore, A. and Sinha, A., Effect of Different Concentrations of Dopant Titanium Dioxide Nanoparticles on Electro-optic and dielectric Properties of Ferroelectric Liquid Crystal Mixture, *Adv. Mater. Lett.*, 7(2), 104 (2016).

[10] Ha, Y.S., Kim, H.J., Park, H.G. and Seo, D.S., Enhancement of electro-optic properties in liquid crystal devices via titanium nanoparticle doping, *Optics Express,* 20(6), 6448 (2012).

[11] Mishra, M., Dabrowski, R. and Dhar, R., Thermodynamical, optical, electrical and electro-optical studies of a room temperature nematic liquid crystal 4-pentyl-4′-cyanobiphenyl dispersed with barium titanate nanoparticles, *J. Mol. Liq.*, 213, 247 (2016).

[12] Singh, U. B., Dhar, R., Dabrowski, R. and Pandey, M. B., Enhanced electro-optical properties of a nematic liquid crystals in presence of $BaTiO_3$ nanoparticles, *Liq. Cryst.*, 41, 953 (2014).

[13] Basu, R. and Garvey, A., Effects of ferroelectric nanoparticles on ion transport in a liquid crystal, *Appl. Phys. Lett.,* 105, 151905 (2014).

[14] Dubey, R., Mishra, A., Singh, K.N., Dhar, R. and Alapati, P.R., Dielectric properties of a highly polar liquid crystalline material showing various types of layered structures, *Liq. Cryst.,* 43, 3(2016).

[15] Maier, W. and Maier, G., A simple theory of the dielectric properties of homogeneously oriented crystalline liquid phases of the nematic type, *Z. Naturforsch*, 16A, 262(1961).

[16] Lopatina, L.M. and Selinger, J.V., Theory of Ferroelectric Nanoparticles in Nematic Liquid Crystals, *Phys. Rev. Lett.*,102, 197802 (2009).

[17] Vardanyan, K.K., Walton, R.D., Bubb, D.M., Liquid crystal composites with a high percentage of gold nanoparticles, Liq. Cryst., 38, 1279(2011).

[18] Cole, K.S. and Cole, R.H., Dispersion and Absorption in Dielectrics I. Alternating Current Characteristics, *J. Chem. Phys.*, 9, 341(1941).

[19] Edwards, D.M.F. and Madden, P.A., A molecular theory of the dielectric permittivity of a nematic liquid crystal, *Mol. Phys.*, 48, 471(1983).

[20] Ouskova, E., Buchnev, O., Reshetnyak, V. and Reznikov, Yu., Dielectric relaxation spectroscopy of a nematic liquid crystal doped with ferroelectric Sn 2 P 2 S 6 nanoparticles, *Liq. Cryst.,* 30, 1(2003).

[21] Rosa, C.D. and Auriemma, F., Crystals and Crystallinity in Polymers: Diffraction Analysis of Ordered and Disordered Crystals, *Wiley & Sons* (2013).

[22] Shaydyuk, Y., Puchkovska, G., Goncharuk, A. and Lebovka, N., Aggregation of clay platelets in nematic liquid crystal, 5CB: microstructure, electrical conductivity and rheological investigations, *Liq. Cryst.*, 38, 155 (2011).

Enhanced Structural and Dielectric Features in Gadolinium Doped Lanthanum Ferrite

N. Saikia* and B.N. Parida

Department of Physics, Central Institute of Technology Kokrajhar (Deemed-to-be-University under MoE, Govt. of India), BTR, Assam, India

*E-mail: bichitra72@gmail.com

ABSTRACT: In the recent decades, $LaFeO_3$ *(LFO) as an excellent dielectric materials has received a lot of attention due to its huge applications in sophisticated technology and uses in optoelectronics, electronics, and a variety of other disciplines. In this current work tried to explain the structural and dielectric effect of* $(La_{0.50}Gd_{0.50})\ FeO_3$ *pervoskite. The traditional high temperature solid-state route method is used to synthesize the sample. The X-ray diffraction (XRD) characteristic reveals the formation of single phase new compound. With the variation of frequency (100 Hz–5 MHz) and temperature (25°C–300°C), dielectric behavior of the material is analyzed which may be beneficial for device application. While non-destructive impedance tool is implemented to evaluate electrical behaviors.*

Keywords: Pervoskite, XRD, Dielectric and Impedance Spectroscopy.

I. Introduction

An innovative way to generate new active materials for a diverse range of scientific purposes is the convergence of several physical features in one material. For example, one class of compounds that exhibits intriguing magnetic, optical, and electrical properties is the perovskite prospects with a configuration of $MFeO_3$ (M = La, Yb, Er, Sm, Y) rare-earth sesquioxides. Data storage devices, magnetostrictive alloys, magnetic refrigeration, cement additives, paints and coatings, and cathode ray tubes are just a few of the applications that rely heavily on the oxides of lanthanum, neodymium, dysprosium, samarium, and gadolinium [1-2].

Many substitutions have recently improved the multifunctional characteristics of lanthanum orthoferrites. Using a polymerization complex technique, Janbutrach *et al.* created lanthanum ferrite nanopowders with an Al substitution and investigated their ferromagnetic and optical properties [3]. Hunpratub *et al.* improved Ga-doped lanthanum ferrite nanoparticles' optical and magnetic behaviours using a complicated polymerization technique [4].

Further, Prasad *et al.* developed copper ion-doped lanthanum ferrite using the sol-gel method.

These qualities prompted us to create a "pervoskite" based on an LFO. We used a solid state technique to make the Gd-doped LFO compound $(La_{0.50}Gd0_{.50})FeO_3$. This study discusses structural, dielectric and electrical behaviour under various experimental conditions.

II. MATERIALS AND SYNTHESIS

The proportional quantity of the raw materials of Gadolinium (III) oxide (Gd_2O_3), iron oxide (Fe_2O_3), and lanthanum (III) oxide are determined by using the equation below:

$$0.5\ Fe_2O_3 + 0.375\ La_2O_3 + 0.125\ Fe_2O_3 \rightarrow Fe(La_{0.75}Gd_{0.25})O_3$$

After that, traditional solid state method is used to synthesize our studied compound.

III. RESULTS AND DISCUSSION

A. Structural Analysis

The spectrum differs from the original powder components by having numerous sharp, distinct peaks. This outlines how a unique compound is

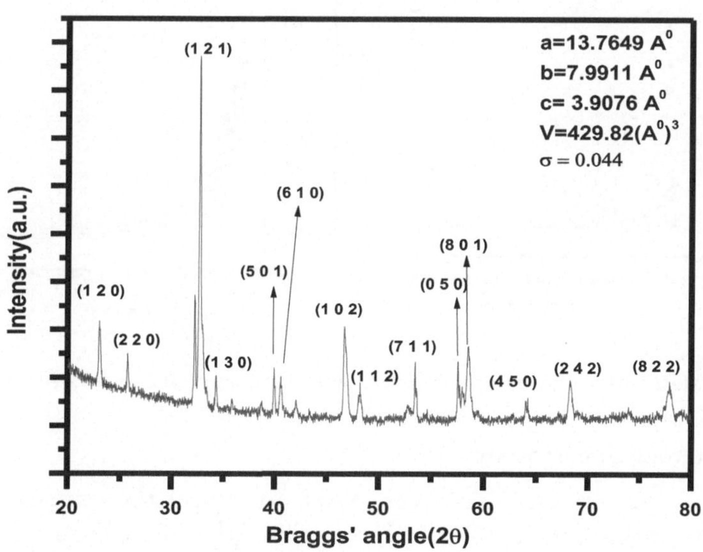

Figure 1: Crystalline XRD Structure of the $(La_{0.5}Gd_{0.5})FeO_3$

generated. The seven-crystal system's frame is utilized to arrange the collected XRD peaks, and the commercially available programme POWD is employed to identify the crystal structure. The synthesized sample crystallizes in an orthorhombic configuration, as evidenced by the measured least-square variation between the fitted and experimental values regarding the interplanar separation [1].

B. Dielectric Spectroscopy

At some selected temperatures, the change of dielectric constant and loss factor with frequency are shown in Figures 2(a) and 2(b). Dipolar and space charge polarizations among the several polarizations had a substantial role in the elevated ε_rValue at low frequency. The existence of several dipoles boosts dipolar polarization in the prepared sample. In contrast, the presence of the multiple valence ion Fe in the octahedral position improves space charge polarization through electron hopping across (Fe^{3+}- Fe^{2+}). Diploes and hopping electrons maintain their alignment towards the direction of the external field while the frequency is low, yielding higher ε_r Value, but when the frequency is high, they are unable to maintain their alignment and thus the ε_r value drops. Like dielectric constant, loss factor with frequency is also shown same pattern [1].

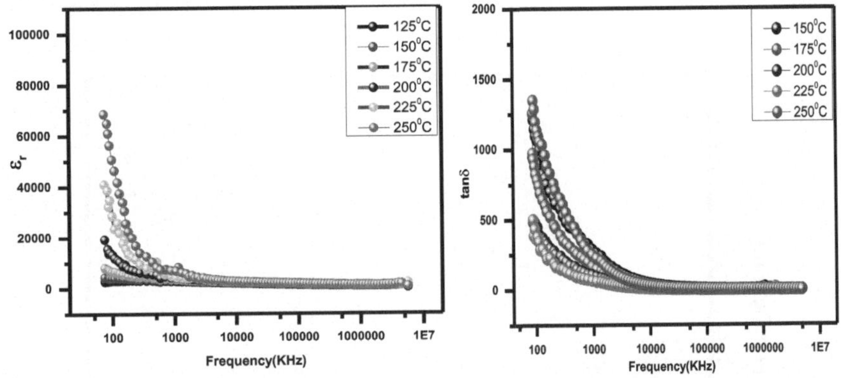

(a) (b)

Figure 2: (a)&(b) Variation of ε_r and tanδ with Frequency at Selected Temperatures

C. Impedance Spectroscopy

Impedance spectroscopy has been used to assess the impact of grain boundaries, intra- and inter-grain interactions plus the effects of electrodes on entire electrical and transport actions. The investigated compound shows semiconducting behaviour, followed by an NTCR response, as demonstrated by a decrease in Z′ value with the increasing temperature inside the higher frequency range (in figure 3(a)). At a particular frequency, the peak of Z″ was recorded, indicating a relaxation state for the LGFO. The Z''_{max} actually changes along with the increased frequency as well as temperature (in Figure 3(b)). That suggests that the process of relaxing is influenced by temperature. In the investigated sample, only single semicircle appears in the Nyquist plot (in Figure 3(c)), which indicates the presence of effect of grain. The next semicircle, which relates to the effect of grain boundary, could, nevertheless, exist at high temperatures [1].

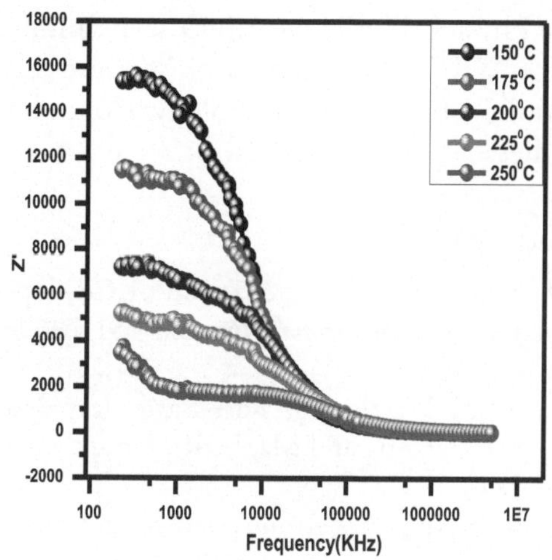

(a)

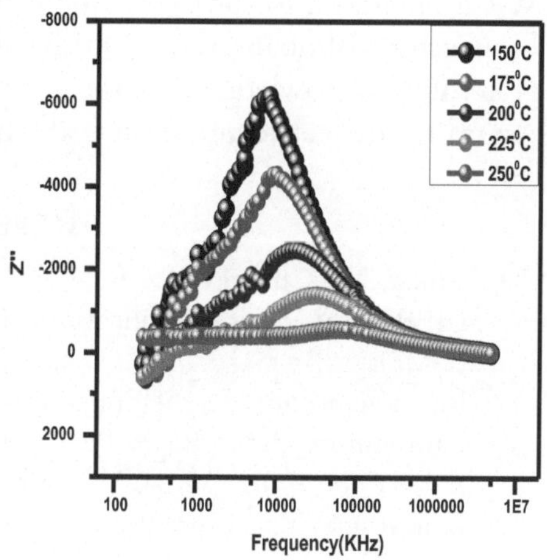

(b)

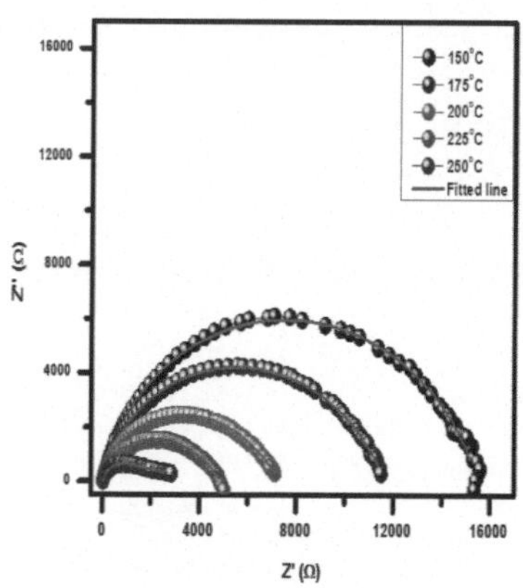

(c)

Figure 3: (a)&(b) Variation of Z' and Z" with Frequency at Selected Temperatures and (c) Nyquist Plot

IV. CONCLUSION

XRD analysis determined that the material has orthorhombic structure. It may be suitable for storage applications, according to the results of the dielectric investigation. We confirmed the participation of grain in the conduction mechanism via investigation of the impedance spectrum. In addition, it also demonstrated the non–Debye kind relaxation procedure. From impedance spectroscopy, NTCR behaviour of the material also predicted.

ACKNOWLEDGEMENT

We are thankful to the Department of Physics, ITER, Siksha 'O' Anusandhan University, Bhubaneswar, Odisha for XRD measurements. We are also gratefully acknowledged to the Central Institute of Technology Kokrajhar, Assam for dielectric and electrical studies.

REFERENCES

[1] Saikia, N., Chakravarty, R. *et al.*, Synthesis and characterization of Gd-doped $LaFeO_3$ for device application, Mater. *Sci. Semicond. Process.* 151,106969 (2022).

[2] Ramachandran, T., Natarajan, K. *et al.*, Fraction of Rare-Earth (Sm/Nd)-Lanthanum Ferrite-Based Perovskite Ferroelectric and Magnetic Nanopowders. *J. Electron. Mater.* 48, 1694–1703 (2019).

[3] Janbutrach, Y., Hunpratub, S. and Swatsitang, E. Ferromagnetism and optical properties of $La_{1-x}Al_xFeO_3$ nanopowders, *Nanoscale Res Lett.* 9, 498 (2014).

[4] Prasad, B.V., Rao, B.V. *et al.*, Preparation and characterization of perovskite Cu doped $LaFeO_3$ semiconductor ceramics. *IOP Conf. Ser. Mater. Sci. Eng.* 73, 012129 (2015).

SECTION-II

Atomic Molecular Physics and Spectroscopy

Spectroscopic and Molecular Characterization of Menthol

N. Bidyapani Chanu[1], Th. Joymati Devi[2] and Th. Gomti Devi[1,*]
[1]Department of Physics, Manipur University, Manipur–795003, India
[2]Department of Chemistry, Manipur University, Manipur–795003, India
*E-mail: tgd@manipuruniv.ac.in

ABSTRACT: The present work focuses on the spectroscopic study of Menthol utilizing FTIR, Raman and SERS spectroscopy and DFT method. DFT (density functional theory) is employed to probe the theoretical study of the compound under B3LYP/6-311+G(d, p) basis set. Natural bond orbital, Frontier molecular orbital and thermodynamic properties of the titled molecule are computed and analyzed. The vibrational assignments are carried out using VEDA software. Molecular docking study is performed against 6P9E receptor, showing effective bioactivity with binding energy -4.83kcal/mol.

Keywords: Menthol, DFT, FTIR, Docking, 6P9E Receptor.

I. Introduction

People are increasingly emphasising the importance of maintaining a healthy lifestyle with changing environment. Traditional medicinal plants have become the focus of the researchers seeking to discover new biomolecular complexes using various spectroscopic and theoretical methods [1-3]. Menthol is a primary component of the essential oil of peppermint and has been used since antiquity to alleviate several ailments [4].Its topical formulation on skin gives the cooling effect as well as its penetration enhancing the antibacterial, anti-inflammatory and analgesic outcome. The extensive literature survey found that a detailed quantum chemical and spectroscopic investigation of this low toxicity and popular topic for researchers and commercial industries has not yet been executed. In this regard, our work emphasizes the molecular characterization and spectral investigation of the titled compound both theoretical and experimental.

II. Materials and Methods

A. Experimental

Menthol is purchased from Sisco Research Laboratory. The FTIR spectra of the sample is analysed within the range of (400–4000) cm^{-1} in the solid state using PerkinElmer Spectrum two FT-IR spectrometer. The Raman and SERS spectra are obtained with a resolution of 1.1 cm^{-1} using the Horiba Xplora1 Micro-Raman system using a 785 nm diode laser.

B. Computational Methods

The molecule is optimized using DFT method at B3LYP/6-311+G(d, p) basis set with Gaussian 09 W software[5]. The theoretical vibrational modes along with their TED analysis are conducted with VEDA software and Gauss View 5.0 software. Gauss Sum 3.0 software is used for conversion of Raman intensity. A scaling factor of 0.9688 is used to account for the systematic errors arise in calculated vibrational data[6]. The Molecular docking analysis is carried out using Autodock 4.0 software and Biovia Discovery Studio 2021.

III. Results and Discussion

A. Molecular Geometry

The optimized molecular structure of menthol is shown in Figure 1and some of its geometrical parameters are presented in Table 1. It consists of nineteen C-H, ten C-C, one C-O and one O-H bonds with bond lengths (1.09–1.10) Å, (1.53-1.56) Å, 1.43Å and 0.96Å respectively. The computed bond angles have thirtytwo H-C-C, thirteen C-C-C, twelve H-C-H, two C-C-O, one H-C-O and one C-O-H angles.

Table 1: Selected Bond Lengths and Bond Angles of Optimized Menthol.

Bond	*Bond Length*	*Bond*	*Bond Angle*
C^6-C^{12}	1.55 Å	H^{13}-C^{12}-O^{29}	103.48
C^{12}-H^{13}	1.09 Å	C^{12}-O^{29}-H^{30}	108.18
C^{12}-C^{10}	1.53 Å	C^6-C^{12}-O^{29}	111.38
C^{12}-O^{29}	1.43 Å	C^6-C^{14}-C^{20}	111.45
O^{29}-H^{30}	0.96 Å	C^{12}-C^6-H^7	104.92

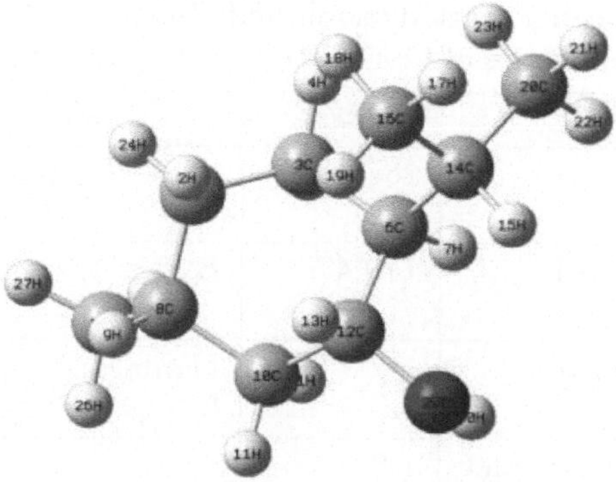

Figure 1: Optimized Structure

B. Natural Bond Orbital (NBO) Analysis

NBO analysis of menthol is performed and the lone pair orbital n_2 (O^{29}) is found to transfer the highest hyper conjugative energy of 22.80 and 21.09 kJ/mol to antibonding $\sigma^*(C^{10}-C^{12})$ and $\sigma^*(C^6-C^{12})$, showing the intramolecular charge transfer within the molecule [1] as listed in Table 3.

Table 3: Selected Second Order Perturbation Theory Analysis of Fock Matrix in NBO Basis for Menthol

Donor (i)	*Acceptor (j)*	*E (^{2})(kJ/mol)*	*E (j) – E (i)(a.u)*	*F_{ij}(a.u)*
$n_2(O^{29})$	$\sigma^*(C^6-C^{12})$	21.09	0.68	0.052
$n_2(O^{29})$	$\sigma^*(C^{10}-C^{12})$	22.80	0.69	0.055
$n_1(O^{29})$	$\sigma^*(C^{12}-H^{13})$	10.25	1	0.044
$\sigma(C^6-H^7)$	$\sigma^*(C^{12}-O^{29})$	4.60	0.76	0.026

C. Frontier Molecular Orbital (FMO) Analysis

The frontier molecular orbital analysis plays a vital role in determining the chemical reactivity, electronic behaviour and kinetic stability. The high HOMO-LUMO energy gap (-7.19 eV) of menthol in gas phase indicates its high stability and low reactivity[2]. The negativity of chemical potential implies that the molecule does not break down into its constituents instantaneously. The calculated values of thermodynamical and quantum parameters are presented in Table 4.

Table 4: Thermodynamic and Quantum Chemical Parameters of Optimized Structure of Menthol

SCF (hartree)	*-468.47*	*IE = - E_{HOMO}*	*7.28 (eV)*
Total Energy (thermal)(kcal mol^{-1})	187.35	EA = - E_{LUMO}	0.09 (eV)
Zero-point Vibrational Energy(kcal mol^{-1})	179.47	E_{HOMO} - E_{LUMO}	-7.19 (eV)
Chemical potential (μ) = ½(E_{HOMO} + E_{LUMO})	-3.69 (eV)	Hardness (η) = ½(E_{LUMO}–E_{HOMO})	3.60 (eV)
Global electro-philicity index (ω) = $\mu^2/2\eta$	1.89		

D. Analysis of Vibrational Assignment

The geometrical arrangement and presence of functional group appreciably influence the molecular vibrational spectra of a molecule. The experimental FTIR peak of O-H is measured at 3250 cm^{-1}, Raman peak at 3512 cm^{-1} and SERS at 3517 cm^{-1}. These values are found to be very close with the theoretical value of O-H stretching mode 3688 cm^{-1} and 3807 cm^{-1}. The asymmetric and symmetric vibrational peak of C-H stretching mode of menthol is calculated at 2888 cm^{-1}, 2936 cm^{-1}, 2984 cm^{-1} and 3080 cm^{-1}. The corresponding

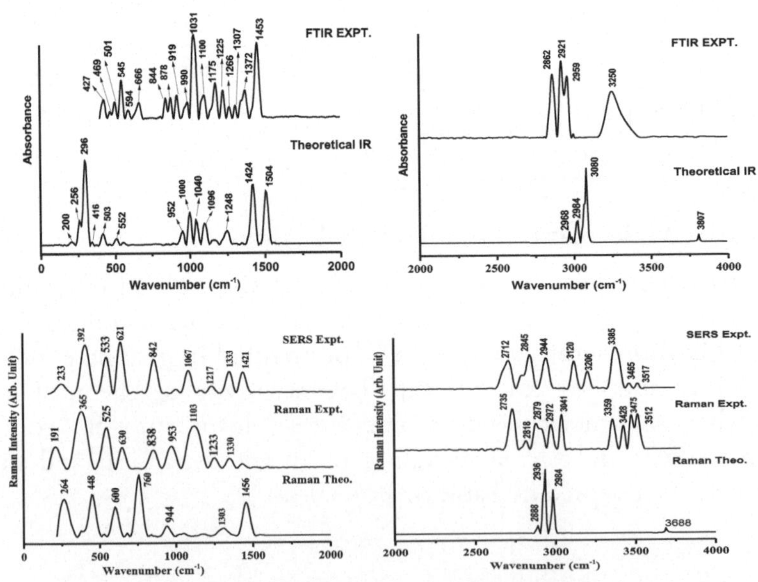

Figure 2: Theoretical and Experimental IR and Raman Spectra in The Range (0-2000) cm^{-1} and (2000-4000) cm^{-1}

experimentally predicted peaks are observed at 2862 cm^{-1}, 2921 cm^{-1} and 2959 cm^{-1} in the FTIR spectra. In Raman spectra, C-H vibrational peak are observed at 2735 cm^{-1}, 2819 cm^{-1}, 2879 cm^{-1}, 2972 cm^{-1}, 3041 cm^{-1} whereas it is detected at 2712 cm^{-1}, 2845 cm^{-1}, 2944 cm^{-1}, 3055 cm^{-1} and 3120 cm^{-1} in the SERS spectra. The peak of O-H and C-H values lie within the expected range (3400–3600) cm^{-1} and (3230-2890) cm^{-1} respectively [2]. It is apparent from the graphical representation of spectra in Figure 2 that the experimental and theoretical vibrational spectra are in strong coherence with each other.

E. Molecular Docking Analysis

The structural molecular biology and computer aided drug design heavily rely on molecular docking as an essential tool [7]. The current research aims to predict the potential binding configuration of menthol with single-chain cytokine present in human skin which has PDB ID 6P9E and has an X-ray resolution of 2.00 Å. The interaction exhibits binding energy of -4.83 kcal/mol which indicates that the ligand is reactive and interacts spontaneously with the selected receptor and is shown in Figure 3. The findings of our study elucidate the inhibitory activity of menthol against the receptor, indicating the potential of the ligand as an effective treatment for antimicrobial and analgesic.

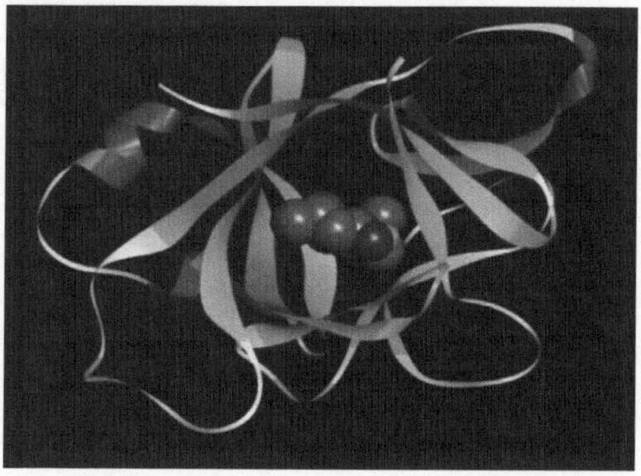

Figure 3: Interaction of Menthol with 6P9E Receptor

IV. Conclusion

The quantum chemical, thermodynamic and spectroscopic study of menthol is carried out. The calculated wave number of the molecule exhibits a strong correlation with the experimental Raman, SERS and FTIR spectra. NBO analysis confirms the occurrence of intramolecular charge transfer. High stability and low reactivity property is evident from the FMO analysis. The docking study of titled compound against the 6P9 Ereceptor affirms the ligand to be a viable applicant for biomedical and pharmaceutical applications with a high binding energy –4.83 kcal/mol.

ACKNOWLEDGEMENT

The authors are grateful to DST SERB, India for providing financial assistance during the research work.

REFERENCES

[1] Chetry, N. and Th. Gomti Devi, Intermolecular interaction study of L-Threonine in polar aprotic solvent: Experimental and theoretical study, *J. Molliq.*, 338, 116689 (2021).

[2] Saikia, J., Borah B. and Devi, T.G., Study of interacting mechanism of amino acid and alzheimer's drug using vibrational techniques and computational method, *J. Mol. Struc.*, 1227, 129664 (2020).

[3] Faik, G., The investigation of menthol's bioactivity as theoretical by using DFT (Density Functional Theory), *CBU J. Sci.*, 11, 363–364 (2015).

[4] Guy, P.P. Komatou, Ilze, V. and Alvaro M. Viljoen, Brian and Lawrence, M., Menthol: A simple monoterpene with remarkable with biological properties, *J. Phytochem.*, 96, 15–25 (2013).

[5] Frisch, M.J., Trucks, G.W., Schlegel, H.B. and Scuseria, G.E., *et al.*, Gaussian 09, Inc., Wallingford CT, (2009).

[6] Merrick, J.P., Damain, M. and Leo, R., An evaluation of harmonic vibrational frequency scale factors, *J. Phys. Chem. A.*, 111, 11683–11700 (2007).

[7] Saikia, J., Devi, T.G. and Karlo, T., Raman spectroscopy, AIM analysis, drug-likeness and molecular docking study of the hydrogen-bonded complex of carmustine with melatonin, *J. Matpr.*, 65, 2875–2882 (2022).

Electronic and Optical Properties of SnS_2/MoS_2 and SnS_2/WS_2 Heterostructure: A DFT and MD Study

Bipradip Chakraborty[1,*], Upasana Hazarika[2] and Nirab C. Adhikary[3]

[1]Department of Applied Sciences, Gauhati University, Assam-781014, India

[2]Department of Molecular Biology and Biotechnology, Tezpur University, Assam-784028, India

[3]Physical Sciences Division, Institute of Advanced Study in Science and Technology, Assam-781035, India

*E-mail: bipradipchax@gmail.com

ABSTRACT: *The electronic and optical properties of SnS_2/MoS_2 and SnS_2/WS_2 heterostructures are investigated in this study using Density Functional Theory (DFT). The calculated direct and indirect bandgap for SnS_2/MoS_2 heterostructure, using PBE functional, is 0.726 eV and 0.149 eV, respectively. Whereas, for SnS_2/WS_2 heterostructure, the calculated direct and indirect bandgap is 0.513 eV and 0.268 eV, respectively. The stability of the heterostructures is also checked using Ab-initio Molecular Dynamics (AIMD). The heterostructures are found to be stable at room temperature. Various optical parameters such as the absorption coefficient, extinction coefficient, and reflectivity of the heterostructures are investigated using the density functional perturbation theory (DFPT). A strong redshift has been observed in the absorption coefficient, extinction coefficient, and reflectivity. A strong absorption peak appears in the near UV and visible region. The excellent optical properties of the heterostructure make it a potential candidate for application in future optoelectronic devices.*

Keywords: Heterostructure, Density Functional Theory (DFT), Density Functional Perturbation Theory (DFPT), Ab-initio Molecular Dynamics (AIMD).

I. Introduction

Following the experimental synthesis of free-standing graphene in 2004, there was a surge of interest in other two-dimensional (2D) materials such as hexagonal boron nitride (h-BN) and transition metal dichalcogenides (TMDs) like MoS_2, $MoSe_2$, WS_2, WSe_2, etc. TMDs have garnered significant attention among these materials due to their potential practical applications in electronic and optoelectronic devices [1,2].

While relatively large band gaps characterize TMDs, controlling the band gap is essential for their use in nanotechnology applications. To accomplish this, various methods are employed to modify the electronic properties of TMDs, including dimension reduction, intercalation, alloying, and the creation of heterostructures. Among these methods, stacking layers along the vertical axis (vertical hetero-layers) has become the most promising approach. This method is theoretically and experimentally feasible, making it highly practical [3,4]. Therefore, SnS_2/MoS_2 and SnS_2/WS_2 heterostructures are designed in this work. Two different stacking patterns are considered. The pattern with the highest binding energy is considered for further calculations. Their electronic and optical properties are calculated. Our results suggest that heterostructure formation can effectively control the band gap and optical properties of TMDs to meet various requirements of electronic applications. Moreover, the calculated optical properties of the heterostructures show that they can be excellent candidates for optoelectronic devices.

II. Computational Details

All the simulations are performed by first-principles methods based on Density Functional Theory (DFT) as implemented in Atomistix Toolkit (ATK) software package [5]. To solve the Kohn-Sham equation, a Generalized Gradient Approximation (GGA) using Projected Augmented Wave (PAW) [6] is used for exchange and correlation effects. PBE functional (Perdew-Burke-Ernzerhof) is used to incorporate core correction of electrons. A tighter grid of $21 \times 21 \times 1$ is used for electronic and optical calculations for more accuracy. Geometry optimization is performed by keeping the atomic forces between each atom below 0.01 eV/Å and the energy difference within 10^{-5} eV. The energy cut-off is set to 400 eV.

III. Results and Discussions

The optimized lattice constants of hexagonal MoS_2 (3.20Å), WS_2 (3.18Å), and SnS_2 (3.68Å) unit cells (containing one metal atom (Mo, W, or Sn) and one S atom) are in close agreement with previously reported works [7]. There is a considerable difference of about 15% in the MoS_2 and WS_2 layer lattice constant values with the SnS_2 monolayer. However, previous experimental results show that stable SnS_2/MoS_2 and SnS_2/WS_2 hetero-structures can be fabricated [8]. The schematic model of MoS_2, WS_2, and SnS_2 monolayers is shown in Figures 1(a)&(b). Heterostructure is formed by combining MoS_2, WS_2, and SnS_2 monolayers and doing structural relaxation. Since it is known that the stability of heterostructure depends on the stacking pattern, two different stacking patterns are considered, as shown in Figure 1(c)&(d). It was found that the first pattern showed the highest binding energy. Therefore the first pattern is used for further calculations. To check for the thermal stability of the heterostructure, AIMD simulation is performed using the NVT Nose-Hoover canonical ensemble. The time step is kept at 1fs, and the simulation is performed for 5ps. The simulation snapshots, as shown in Figure 2, show the stability of the heterostructures at room temperature.

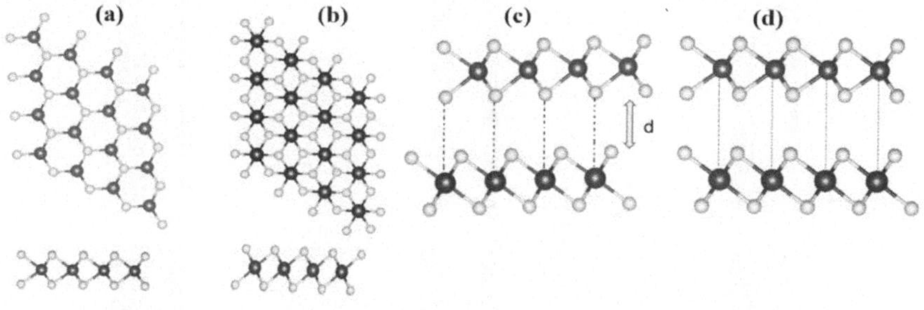

Figure 1: Schematic model of (a) MoS_2 and WS_2 monolayer top view and side view, (b) SnS_2 monolayer top view and side view. Green corresponds to Molybdenum and Tungsten atoms, Blue to Tin atoms, and Yellow to Sulphur atoms; (c) Stacking pattern 1 and (d) Stacking pattern 2.

The calculated direct and indirect bandgap for SnS_2/MoS_2 heterostructure, using PBE functional, is 0.726 eV and 0.149 eV, respectively. Whereas, for SnS_2/WS_2 heterostructure, the calculated direct and indirect bandgap is 0.513 eV and 0.268 eV, respectively.

The optical properties, such as the absorption coefficient, extinction coefficient, and reflectivity of the heterostructures, are calculated using DFPT and DFT-½ correction methods. The optical properties are plotted in Figure 3. Compared to the optical properties of pristine MoS_2, WS_2, and SnS_2 monolayer, the optical properties of SnS_2/MoS_2 and SnS_2/WS_2 heterostructure show a strong redshift, with the peaks predominantly in the UV-Visible range.

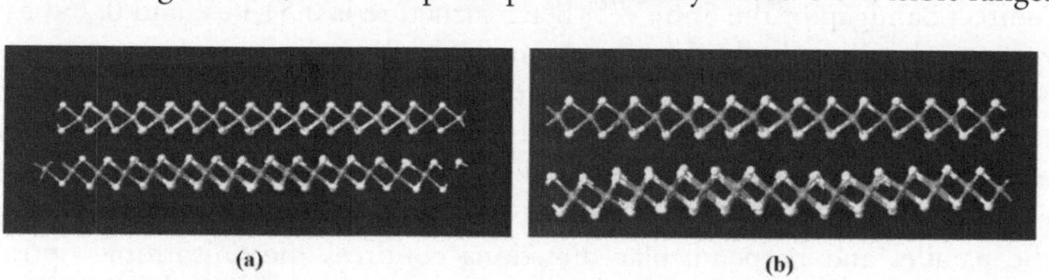

Figure 2: Snapshot of AIMD Simulation of (a) SnS_2/MoS_2 Heterostructure and (b) SnS_2/WS_2 Heterostructure after 5ps

It is well known that two-dimensional monolayers and heterostructures show anisotropic properties. Therefore, to confirm the anisotropic optical properties of SnS_2/MoS_2 heterostructure and SnS_2/WS_2 heterostructure, all the optical properties were plotted for both parallel (XX) and perpendicular (ZZ) directions. It was found that all of the optical properties showed a variation in values in the parallel (XX) and perpendicular (ZZ) directions, thus confirming the anisotropic optical behavior of the heterostructure.

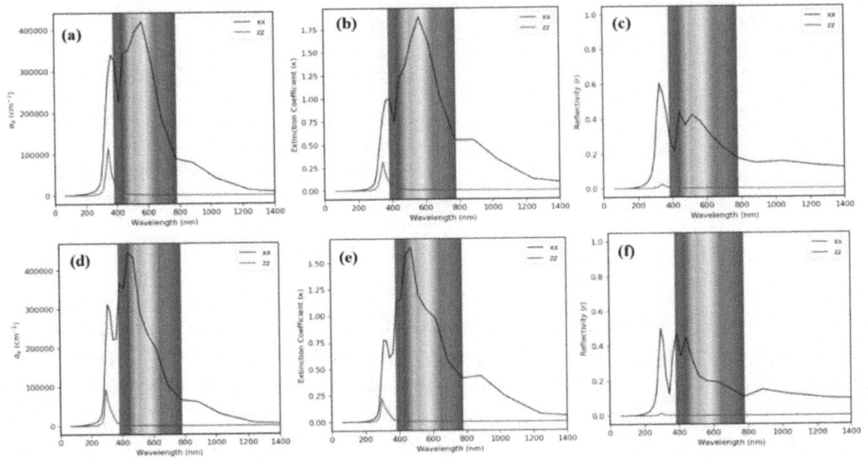

Figure 3: (a) Absorption Coefficient, (b) Extinction Coefficient, and (c) Reflectivity of SnS_2/MoS_2 heterostructure; (d) Absorption Coefficient, (e) Extinction Coefficient, and (f) Reflectivity of SnS_2/WS_2 heterostructure. The visible region of the spectrum is shown with a color bar.

IV. Conclusion

In conclusion, the electronic and optical properties of MoS_2, WS_2, and SnS_2 monolayer, and SnS_2/MoS_2 and SnS_2/WS_2 heterostructures were calculated and plotted. The direct and indirect bandgap of the SnS_2/MoS_2 heterostructure is 0.726 eV and 0.149 eV, respectively. Similarly, the direct and indirect bandgap of the SnS_2/WS_2 heterostructure is 0.513 eV and 0.268 eV, respectively. To check for the thermal stability of the heterostructures, AIMD is performed. It is found that the heterostructures are stable at room temperature. Optical properties such as the absorption coefficient, extinction coefficient, and reflectivity of the heterostructures are plotted for both parallel (XX) and perpendicular (ZZ) directions. Variation in the values in the parallel and perpendicular directions confirms the anisotropic optical behavior of the heterostructure. Moreover, a strong redshift is recorded in the optical properties values. A significant absorption peak appears in the

UV-visible region. All these findings make SnS_2/MoS_2 and SnS_2/WS_2 heterostructures ideal candidates for futuristic optical properties.

ACKNOWLEDGMENT

The authors would like to extend their sincere gratitude to the Department of Applied Sciences, Gauhati University, and E&ICT Academy, IIT Guwahati, for providing the computation facility for the simulation and support. Bipradip Chakraborty would like to acknowledge the MHRD, Govt. of India, for offering Research Assistantship under TEQIP-III Project.

REFERENCES

[1] Novoselov, K.S., Geim, A.K., Morozov, S.V., Jiang, D., Zhang, Y., Dubonos, S.V., Grigorieva, I.V. and Firsov, A.A. Electric field effect in atomically thin carbon films, Science, 306 (5696), 666–669 (2004).

[2] Novoselov, K., Jiang, D., Schedin, F., Booth, T., Khotkevich, V., Morozov, S. and Geim, A., Two-dimensional atomic crystals, Proc. Natl. Acad. Sci. U.S.A., 102 (30), 10451–10453 (2005).

[3] Gordon, R., Yang, D., Crozier, E., Jiang, D., and Frindt, R., Structures of exfoliated single layers of WS_2, MoS_2, and $MoSe_2$ in aqueous suspension, *Phys. Rev. B*, 65 (12), 125407 (2002).

[4] Joshi, P.D., Joag, D.S., Rout, C.S. and Late, D.J., Photosensitive field emission study of SnS_2 nanosheets, J. Vac. Sci. Technol. B Nanotechnol. Microelectron. *Mater. Process. Meas. Phenom.* 33 (3), 03C106 (2015).

[5] Smidstrup, S., Markussen, T., Vancraeyveld, P., Wellendorff, J., Schneider, J., Gunst, T., Verstichel, B., Stradi, D., Khomyakov and Vej-Hansen, P.A. Quantumatk: An integrated platform of electronic and atomic-scale modeling tools, *Journal of Physics: Condensed Matter*, 32(1):015901, (2019).

[6] Kresse, G. and Joubert, D., From ultrasoft pseudopotentials to the projector augmented wave method, Physical review b, 59(3):1758, (1999).

[7] Mabiala-Poaty, H.B., Douma, D.H., Malonda-Boungou, B.R., Mapasha, R.E., and M'Passi-Mabiala, B., First-principles studies of SnS2, MoS2, and WS2 stacked van der Waals hetero-multilayers, *Computational Condensed Matter* 16: e00303 (2018).

[8] Chang, R.J., Sheng, Y., Chen, T., Mkhize, N., Lu, Y., Bhaskaran, H. and Warner, J.H., Morphology Control of Two-Dimensional Tin Disulfide on Transition Metal Dichalcogenides Using Chemical Vapor Deposition for Nanoelectronic Applications, *ACS Appl. Nano Mater.*, 2, 4222–4231(2019).

Performance Optimization of $CH_3NH_3PbI_3$-based Perovskite Solar Cell Using SCAPS-1D Simulation

Parijat Borgohain[1,*] and Mahananda Baro[2]

[1]Institute of Advanced Study in Science and Technology, Guwahati-781035, Assam, India

[2]Department of Physics, Assam University, Silchar, Assam, India-788011

*E-mail: parijat.borgohain@gmail.com

ABSTRACT: Lead halide perovskite solar cells (PSC) are potential candidates among the existing solar cells, achieving maximum efficiency of 25.5%. In this work, the performance optimization of a planar heterojunction solar cell with the architecture FTO/SnO_2/$CH_3NH_3PbI_3$/Cu_2O/Au is studied. For the calculations, we have used one-dimensional Solar Cell Capacitance Simulator (SCAPS-1D). Some of the device key parameters effecting the photovoltaic performance are thickness, doping concentration and donor concentration of the absorber, electron transport layer (ETL), hole transport layer (HTL) and absorber defect density. We have observed an increase in the performance of PSC at absorber thickness of 500 nm, at doping concentration of 1×10^{16} cm^{-3} and absorber defect density of 1×10^{14} cm^{-3}. The PCE gradually decreases with increasing donor concentration of ETL, but increases with increasing acceptor concentration of HTL. After optimization, the efficiency of the solar cell has been enhanced by 4%approximately.

I. Introduction

Perovskite solar cells (PSCs) have gained significant attention in the scientific community as a promising alternative to traditional silicon-based solar cells due to their high efficiency, low cost, and ease of manufacturing. PSCs use perovskite materials as the light-absorbing layer, which absorbs photons from sunlight and converts them into electrical energy. Since the first demonstration of a PSC in 2009 by Kojima *et al.*, the efficiency of PSCs has improved rapidly, and they now have the potential to reach efficiencies similar to or even higher than those of silicon-based solar cells [1]. However, there are still challenges to be addressed, such as the stability and durability of perovskite materials in harsh environmental conditions, as well as the scalability of the manufacturing process. Inorganic hole transport layers (HTLs) in particular, copper oxide-based HTLs, such as cupric oxide (CuO) and cuprous oxide (Cu_2O), have been extensively studied due to their high absorption coefficient, environmental friendliness, low cost, abundance, thermal stability, and ease of preparation [2]. In perovskite solar cells (PSCs), the electron transport layer (ETL) plays a critical role in extracting and transporting electrons from the perovskite layer to the external circuit. Titanium dioxide (TiO_2) is a commonly used ETL due to its high electron mobility, chemical stability, and compatibility with perovskite materials. However, the high-temperature processing required to obtain the rutile phase of TiO_2 can limit its scalability and compatibility with low-temperature substrates. To address this issue, several low-temperature processed ETLs have been investigated, such as zinc oxide (ZnO) and tin oxide (SnO_2) [3]. Low-temperature processing of the ETL can also improve the stability of PSCs by reducing the thermal and mechanical stress on the device during manufacturing. Theoretical simulations of the PSCs provide a deeper understanding of the underlying physical processes and to explore a wider range of materials and device architectures than is possible experimentally.

- Top of Form
- Bottom of Form

In this work, the factors affecting the performance of a $CH_3NH_3PbI_3$ ($MAPbI_3$)-based perovskite solar cell have been analysed using SCAPS.

II. Methodology

We have simulated a perovskite solar cell using SCAPS-1D software with architecture FTO/SnO_2/$CH_3NH_3PbI_3$/Cu_2O/Auunder standard illumination conditions (AM1.5, 1000 W/m^2 and 300 K). The software simulates the performance of the device by solving the fundamental equations of semiconductors, namely, the Poisson's equation and the continuity equations for the electrons and holes [4]. The solar cell structure and energy band diagrams are shown in Figure 1. Table 1 represents the input parameters used for the simulation of each layer in the PSC structure [5].

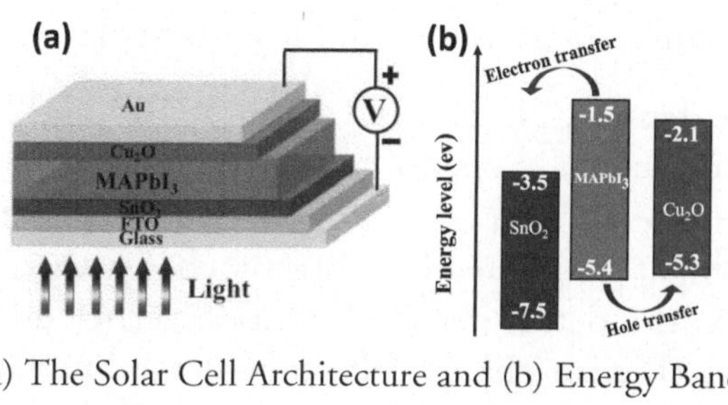

Figure 1: (a) The Solar Cell Architecture and (b) Energy Band Diagram

Table 1: Input Parameters of the Solar Cell

Parameters	FTO	SnO_2	$MAPbI_3$	Cu_2O
t (nm)	500	70	900	50
E_g (eV)	3.5	3.5	1.5	2.17
χ (eV)	4	4	3.9	3.2
ε_r	9	9	10	7.1
N_C (cm^{-3})	2.2×10^{18}	2.2×10^{17}	2.75×10^{18}	2.5×10^{20}
Nv (cm^{-3})	1.8×10^{19}	2.2×10^{17}	3.9×10^{18}	2.5×10^{20}
μ_n ($cm^2V^{-1}S^{-1}$)	20	20	10	200
μ_h ($cm^2V^{-1}S^{-1}$)	10	10	10	8600
N_D (cm^{-3})	1×10^{15}	1×10^{15}	1×10^{17}	-
N_A (cm^{-3})	-	-	-	1×10^{19}
N_t (cm^{-3})	1×10^{15}	1×10^{18}	1×10^{13}	1×10^{13}

III. Results and Discussion

Initial simulation of the device exhibited power conversion efficiency (PCE) of 24.32% with J_{SC} of 24.7 mA/cm^2, V_{OC} of 1.14 V and FF of 86.28%. In the preceding sections, various parameters that can affect the PV parameters of the PSC have been discussed and systematically optimized to achieve maximum PCE.

A. Influence of Thickness, Defect Density and Donor Concentration of the Absorber Layer

The thickness of the light-absorber perovskite layer plays a pivotal role in determining the performance of PSCs. The variation of the photovoltaic (PV) parameters, J_{SC}, V_{OC}, FF and PCE, with the absorber layer thickness, defect density (N_t) and doping concentration (N_D) is shown in Figure 2. J_{SC}, V_{OC}, and PCE have increases with increasing absorber thickness in the range 100–600 nm, and then decrease. FF gradually decreases.PCE increases for a thick absorber because of more material are exposed to light, causing generation of more electron-hole pair. The reduction in FF is probably because of increasing series resistance in the absorber layer.

When the defect density (N_t) is as low, at 1×10^{14} cm^{-3}, the PSC exhibits significant enhancement in performance (Figure 2b). However, as N_t increased beyond 1×10^{15} cm^{-3}, the PCE decreases. The decrease in PCE is attributed to a high recombination rate of charge carriers because of an increase in N_t. As the defect density increases, carrier diffusion length ($l_{n,p}$) decreases. But for efficient charge collection at the interfaces, it is necessary to have a diffusion length greater than or equal to the absorber thickness. At this value of N_t, 1×10^{14} cm^{-3}, the optimum absorber thickness is found to be 500 nm, which is smaller than the carrier diffusion length ($l_{n,p}$ = 1.1 μm), allowing most carriers to reach their respective electrodes before recombination occurs.

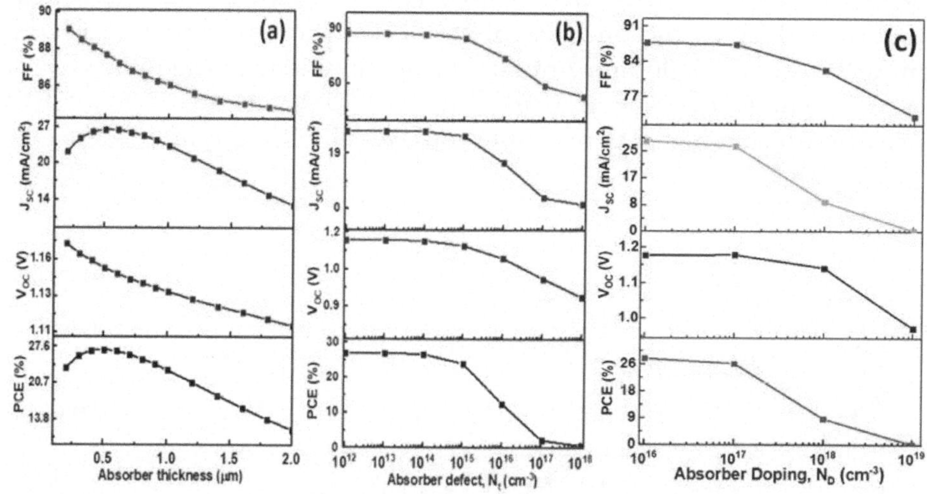

Figure 2: PV Parameters with the Variation of (a) Absorber Thickness and (b) Absorber Defect Density (N_t) and (c) Absorber Donor Concentration (N_D)

PV parameters, J_{SC}, V_{OC}, FF and PCE decrease with increasing N (Figure 2c). PCE becomes maximum at N_D of 10^{16} cm^{-3} and falls drastically at higher N_D of the absorber because of abrupt decrease in J_{SC}. PCE increases at lower N_D mainly due to decline in the dark current, keeping photocurrent (J_{ph}) nearly constant [6]. At higher N_D, the diffusion length of the minority carriers reduces, which result in decrease the photocurrent, J_{ph}. In addition to this, the recombination rate of charge carriers enhances at higher N_D values, causing to decrease J_{SC} and PCE [6]. The optimized donor concentration of absorber is 1×10^{16} cm^{-3}.

C. Influence of Doping Concentration of HTL and ETL

To enhance the performance of PSCs, ETL is doped with donor (N_D) type dopants and HTL can be doped with acceptor (N_A) type dopants. Figure 3 represents PV parameters obtained by changing the acceptor concentration (N_A) of HTL and donor density (N_D) of ETLs. J_{SC}, V_{OC}, FF and PCE increases with increasing N_A of HTL. As the donor concentration of ETL is increased beyond 1×10^{15} cm^{-3}, the carrier recombination rate increases, which resulted in a rapid decrease of J_{SC} and PCE. The optimized PCE of 28.67% is obtained at N_A of 1×10^{20} cm^{-3} and N_D of 1×10^{15} cm^{-3}.

Further, the thickness of the hole transport layer (HTL) and electron transport layer (ETL) have been optimized for the maximum PCE of the solar cell. It is observed that PCE almost unaffected by the variation of HTL thickness. A moderate thickness of 200 nm is considered in this work. There is a slight decrease in PCE as the thickness of the ETL increases. The optimal ETL thickness is set to be 30 nm. After device optimization, a maximum PCE of 28.67% with J_{SC} of 28.56 mA/cm^2, V_{OC} of 1.15 V and FF of 87.25% is obtained, which is higher than the previous work by Raoui *et al.* [7].

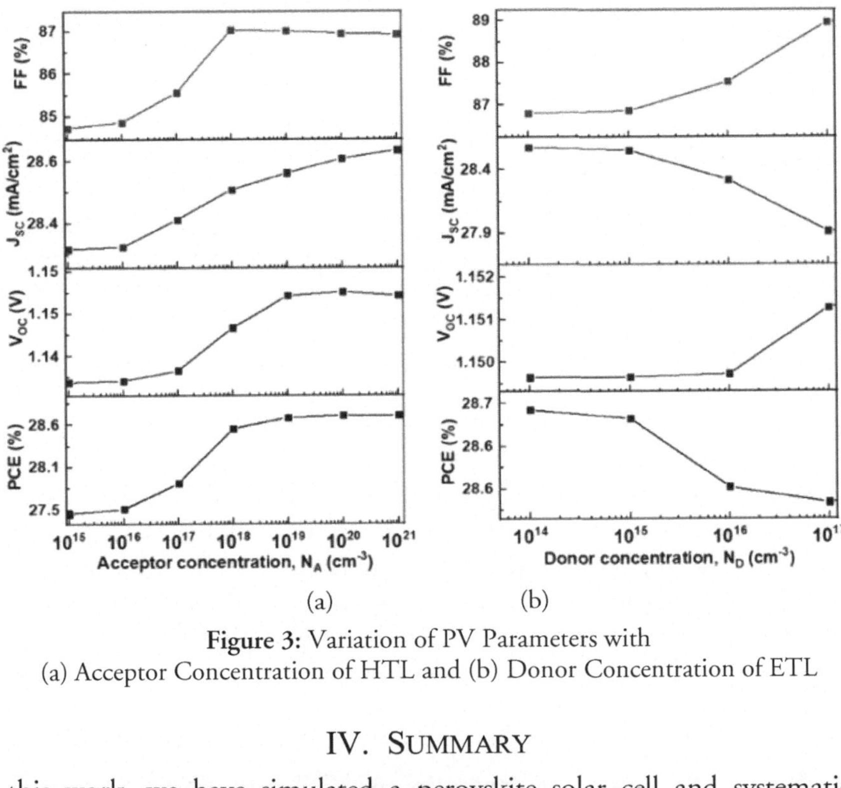

(a) (b)

Figure 3: Variation of PV Parameters with
(a) Acceptor Concentration of HTL and (b) Donor Concentration of ETL

IV. SUMMARY

In this work, we have simulated a perovskite solar cell and systematically optimized various parameters of the device to enhance its performance. Initial simulation of the device gave a PCE of 24.32%. The effect of various properties of the absorber layer, such as thickness, defect density and doping concentration, is studied on the performance of the solar cell and it is found that the best combination is an absorber thickness of 500 nm, a defect density of 10^{14} cm^{-3} and a doping concentration of 10^{16} cm^{-3}. The absorber thickness was optimized based on the diffusion length of the carriers, which was 1.1 μm corresponding to the defect density of 10^{14} cm^{-3}. The thickness of the hole transport layer (HTL) and electron transport layer (ETL) are also optimized for the maximum efficiency of the solar cell. Finally, the acceptor and donor concentrations of HTL and ETL are optimized, resulting in a maximum theoretical PCE of 28.67% with a J_{SC} of 28.56 mA/cm^2, a V_{OC} of 1.15 V and an FF of 87.25%. This study showed systematic optimization of device parameters of a $MAPbI_3$-based solar cell for improvement its performance theoretically.

REFERENCES

[1] Sahil, F., Werner, J., Kamino, B. A., Bräuninger, M., Monnard, R., Paviet-Salomon, B., Barraud, L., Ding, L., Diaz Leon, J.J., Sacchetto, D., Cattaneo, G., Despeisse, M., Boccard, M., Nicolay, S., Jeangros, Q., Niesen, B. and Ballif, C., Fully textured monolithic perovskite/silicon tandem solar cells with 25.2% power conversion efficiency, *Nat. Mater.*, 17, 820 (2018).

[2] Ye, T., Sun, X., Zhang, X. and Hao, Sue., Recent advances of Cu-based hole transport materials and their interface engineering concerning different processing methods in perovskite solar cells, *J. Energy Chem.* 62, 459 (2021).

[3] Wang, K., Olthof, S., Subhani, W. S., Jinag, X., Cao, Y., Duan, L., Wang, H., Du, M. and Frank Liu, S., Novel inorganic electron transport layers for planar perovskite solar cells: Progress and prospective, *Nano Energy,* 68, 104289 (2020).

[4] Burgelman, M., Nollet, P. and Degrave, S., Modelling polycrystalline semiconductor solar cells, *Thin Solid Films,* 361, 527 (2000).

[5] Jayan, K.D. and Sebastian, V., Comparative Study on the Performance of Different Lead-Based and Lead-Free Perovskite Solar Cells, *Adv. Theory Simul,* 4, 2100027(2021).

[6] Zhao, P., Liu, Z., Lin, Z., Chen, D., Su, J., Zhang, C., Zhang, J., Chang, J. and Hao, Y., Device simulation of inverted $CH_3NH_3PbI_{3-x}Cl_x$ perovskite solar cells based on PCBM electron transport layer and NiO hole transport layer, *Solar Energy,* 169, 11 (2018).

[7] Raoui, R., Ez-Zahraouy, H., Tahiri, N., El Bounagui, O., Ahmad, S. and Kazim, S., Performance analysis of $MAPbI_3$ based perovskite solar cells employing diverse charge selective contacts: Simulation study, *Sol. Energy,* 193, 948 (2019).

Quantum Phases of Two-Component Dipolar Bosonic Mixtures in Optical Lattices

Bhumika Thoudam*, Kohenjit Pebam, Laxmi Angom and D. Angom

Manipur University, Canchipur, Imphal, Manipur-795003, India

*E-mail: bhumika.th@gmail.com

ABSTRACT: We study the quantum phases of two-component ultracold quantum gases in two-dimensional optical lattices. Several experimental and theoretical investigations have examined the properties of such systems of bosons using single atomic species. To simulate and understand more complexed condensed matter systems, two-component quantum gases is considered. Two-component quantum gases can be described by the Bose-Hubbard model with appropriate modifications. They show additional quantum phases due to the interspecies interactions. Such quantum gases may also display miscible phase where the two species form a homogeneous solution or immiscible phase where the species separate spatially. To relate with the condensed matter systems the nearest neighbour interaction is introduced in the system. In experiments this correspond to dipolar bosonic mixtures. Theoretically, such a system is described by the extended Bose-Hubbard model. This generates two more phases: the density wave phase and the supersolid phase. We consider both intraspecies and interspecies nearest neighbour interaction in this study to correctly depict the physics of two-component dipolar bosonic mixtures.

Keywords: Quantum Phases, Optical Lattices, Bose-Hubbard Model, Density Wave, Supersolid.

I. Introduction

In 1995, the Bose-Einstein condensates (BECs) of a dilute gas of alkali atoms was achieved using sophisticated laser cooling and trapping techniques. These techniques also led to the realization of trapping ultracold atoms in optical lattices [1, 2]. These are versatile quantum many-body systems. It is possible to tune the inter-particle interaction, geometry, spatial dimension, number of particles, depth of the potential, number of components, disorder strength and introduce strong synthetic magnetic fields. Thus, optical lattices are ideal platform to simulate complex systems in condensed matter physics [3, 4]. The Bose-Hubbard model (BHM) describes the physics of such a system of ultracold atoms in optical lattices. Understanding the ground state quantum phases of these systems and their dynamics are essential to gain better insights on their properties. There have been several experimental and theoretical investigations on the different aspects of optical lattices with single atomic species. However, the simplest analogue of electrons in optical lattices require two atomic species to mimic the two spin states of electrons. The two components could be two different hyperfine spin states of ^{87}Rb, different atomic species and of different isotopes of the same atomic species. And, these have been experimentally realized. This provide rich physics which are not accessible with the single species. Binary mixture could also exhibit additional quantum phases which emerge due to competition between the intra and inter-species interactions. They can be used to study novel phenomena such as phase separation, pattern formation, collective excitations, etc. Phase separation is a phenomenon unique to two components. This phenomenon occurs when the interspecies interaction exceeds a critical value [5]. The two components are in the miscible phase when the interspecies interaction is weak, but it is in the immiscible phase if repulsion between the two components is sufficiently strong. To study more complex systems such as dipolar ultracold quantum gases in optical lattice we consider the extended Bose-Hubbard model (eBHM). This model is an extension of the BHM that includes additional interactions beyond the contact interaction, such as long-range interactions. For simplicity, we consider the nearest neighbour (NN) interactions to describe the effect of long-range interactions. This system generates two more phases: the density wave (DW) phase and supersolid (SS) phase. The DW phase is an insulating phase with crystalline order and the SS is a compressible phase with both diagonal and off-diagonal long-range orders [6].

II. THEORETICAL METHOD

The Bose-Hubbard Hamiltonian for two-component ultracold gas confined in a square optical latticeat zero temperature is given as,

$$H_{\mathrm{TBHM}} = -\sum_{p,q,k}\Big[\big(J_x^k \hat{b}_{p+1,q}^{k\dagger}\hat{b}_{p,q}^k + \mathrm{H.c.}\big) + \big(J_y^k \hat{b}_{p,q+1}^{k\dagger}\hat{b}_{p,q}^k + \mathrm{H.c.}\big) - \frac{U_{kk}}{2}\hat{n}_{p,q}^k\big(\hat{n}_{p,q}^k - 1\big) + \mu_{p,q}^k\hat{n}_{p,q}^k\Big] + \sum_{p,q} U_{12}\hat{n}_{p,q}^1\,\hat{n}_{p,q}^2 \qquad \ldots (1)$$

where k = 1, 2 represents the 1st and 2[nd]components, $b_{p,q}^k(b_{p,q}^{k\dagger})$ are the bosonic annihilation (creation) operators respectively, J_x^k (J_y^k) is the hopping strength along $x(y)$ direction, U_{kk} is the intra-species interaction, and $U_{12}$is the interspecies interaction between the two components [5]. For the mean field computations, we shall consider a two-dimensional square optical lattice system comprising of 10 × 10 latticesites.

The extended Bose-Hubbard Hamiltonian for two-component ultracold gas confined in a square optical latticeat zero temperature is given as,

$$H_{\text{eTBHM}} = H_{\text{TBHM}} + \sum_{k=1,2} \sum_{p,q} V_k\, \hat{n}^k_{p,q}\left(\hat{n}^k_{p-1,q} + \hat{n}^k_{p+1,q} + \hat{n}^k_{p,q-1} + \hat{n}^k_{p,q+1}\right) \quad \ldots (2)$$

where, V_k is the nearest neighbour intraspecies interactions [6]. To solve the two-component BHM Hamiltonian in Eq.(1), we use the Gutzwiller mean field approximation. In this approach the bosonic operators are decomposed as,

$$\begin{aligned} \hat{b}^k_{p,q} &= \langle \hat{b}^k_{p,q} \rangle + \delta \hat{b}^k_{p,q} \\ &= \phi^k_{p,q} + \delta \hat{b}^k_{p,q} \end{aligned} \quad \ldots (3)$$

$$\begin{aligned} \hat{b}^{k\dagger}_{p,q} &= \langle \hat{b}^{k\dagger}_{p,q} \rangle + \delta \hat{b}^{k\dagger}_{p,q} \\ &= \phi^{k*}_{p,q} + \delta \hat{b}^{k\dagger}_{p,q} \end{aligned} \quad \ldots (4)$$

where, $\phi^k_{p,q}$ $(\phi^{k*}_{p,q})$is the mean field and $\delta \hat{b}^k_{p,q} (\delta \hat{b}^{k\dagger}_{p,q})$ is the corresponding fluctuation operator. Using these definitions and neglecting the second order term in fluctuations, we obtain the mean-field Hamiltonian. The Hamiltonian is diagonalized for each site separately to obtain the ground state of the system. The ground state Gutzwiller wave function at $(p, q)^{\text{th}}$ site is,

$$|\psi\rangle_{p,q} = \sum_{n_1,n_2} c^{(p,q)}_{n_1,n_2} |n_1, n_2\rangle_{p,q} \quad \ldots (3)$$

where, n_1 and n_1 are the Fock states for the 1st and 2nd components respectively, $|n_1, n_2\rangle_{p,q}$ are the occupation number basis states at site (p, q), $c^{(p,q)}_{n_1,n_2}$ are complex coefficient of the groundstate $|\psi\rangle_{p,q}$ with the normalization condition $\sum_{n_1,n_2} \left|c^{(p,q)}_{n_1,n_2}\right|^2 = 1$. We can calculate the superfluid order parameter and lattice occupancy for the two components as,

$$\phi^k_{p,q} = \langle \psi | \hat{b}^k_{p,q} | \psi \rangle = \sum_{n_1,n_2} \sqrt{n_k}\, c^{(p,q)*}_{n_1-1,n_2} c^{(p,q)}_{n_1,n_2},$$

$$¢ﬂ¥\ \phi^k = \frac{1}{L} \sum_{p,q} \phi^k_{p,q} \quad \ldots (4)$$

$$\rho^k_{p,q} = \langle \psi | \hat{n}^k_{p,q} | \psi \rangle = \sum_{n_1,n_2} n_k \left| c^{(p,q)}_{n_1,n_2} \right|^2,$$

$$¢ﬂ¥?\rho^k = \frac{1}{L} \sum_{p,q} \rho^k_{p,q}, \quad \ldots (5)$$

where L is the total number of lattice sites.

III. Phase Diagram

The MI-SF phase boundary is identified based on the superfluid order parameter. The superfluid order parameter, ϕ^k is zero and average lattice occupancy, $\rho = \rho^1 + \rho^2$ is integer commensurate for the MI phase at each lattice site. For the SF phase, ϕ^k is nonzero and ρ is real commensurate. We consider the hopping strength $J_x^k = J_y^k = J$, the chemical potential $\mu_{p,q}^1 = \mu_{p,q}^2 = \mu$ and the intraspecies interactions $U_{kk} = U$. All energies are scaled with U and the phase diagram is defined in the $J/U - \mu/U$ plane.

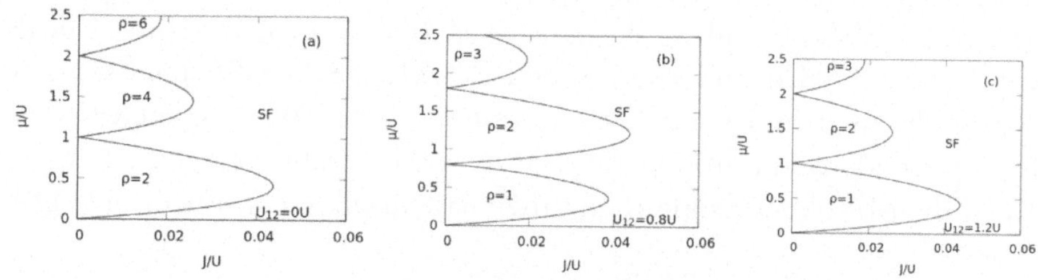

Figure 1: Phase Diagram of TBHM by Varying the Interspecies Interaction Strength U_{12}

A. Phase Diagram of Two-Componentbose-Hubbard Model

The phase diagram for different values of U_{12} are shown in Figure 1. In Figure 1(a), when the interspecies interaction $U_{12} = 0$, the phase diagram is identical to the case of single species, but the occupancy $\rho = 2n$ with $\rho^1 = \rho^2 = n$ and $n \in \mathbb{N}$. Then we consider two cases where $U_{12}^2 < U_{11}U_{22}$, the miscible phase and $U_{12}^2 > U_{11}U_{22}$, the immiscible phase. In the miscible domain, with the introduction of interspecies interaction ($U_{12} \neq 0$) the half-filled lobe like $\rho = 1$ with $0 < \rho^1 < 1$ and $\rho^2 = 1 - \rho^1$ emerges in the phase diagram. As we increase U_{12}, the Mott lobes with odd-integer occupancies grow in size and the lobes with even-integer

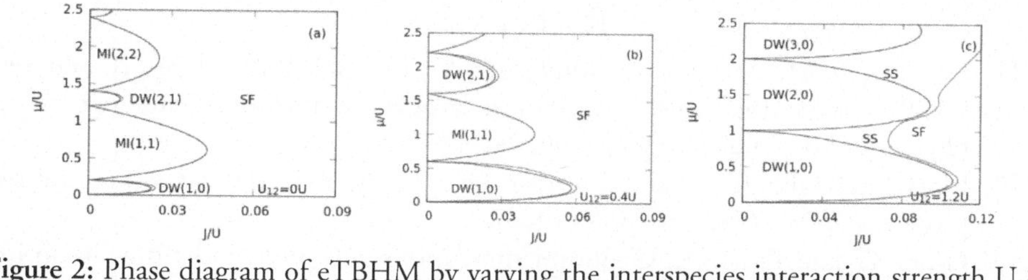

Figure 2: Phase diagram of eTBHM by varying the interspecies interaction strength U_{12} and for NN interactions $V_k = 0.05U$. Around the density wave phase, a supersolid phase exists and the boundary between SS phase and SF phase is represented by green lines.

occupancies shifts to higher μ/U value. In the immiscible domain, the Mott lobes have the same sizes as that of $U_{12} = 0$ but the occupancies $\rho = n$ with $n \in \mathbb{N}$ and $\rho^k = n/2$. Phase separation is observed in this domain.

B. Phase Diagram of Two-Component Bose-Hubbard Model

The nearest-neighbour (NN) interactions in the eTBHM introduce two more phases, density wave and supersolid phase. We set the intraspecies NN interaction strength $V_k = 0.05U$ and the interspecies NN interaction strength $V_{12} = 0$ then vary the interspecies onsite interaction strength U_{12}. Figure 2(a) shows the phase diagram when $U_{12} = 0$. It is identical to the single-species eBHM and consists of DW, MI, SS and SF phases. As we increase the interspecies onsite interaction U_{12}, the DW lobe increases and the supersolid region around the density wave region enlarges. At $U_{12} = 1.2U$, the MI lobes disappear, and the phase diagram consists of only DW, SS and SF phases.

IV. Conclusion

The Bose-Hubbard Hamiltonian is described for two-component ultracold quantum gases and the ground state solution is obtained using Gutzwiller ansatz. Then the extended Bose-Hubbard Hamiltonian is discussed for two-component ultracold quantum gases considering the nearest neighbour intraspecies interactions only. Then the phase diagrams of TBHM and eTBHM are provided with different interspecies onsite interactions.

Acknowledgement

Bhumika Thoudam acknowledge DST Inspire for providing financial support during this study under grant number DST/INSPIRE/Fellowship/2020/IF200260.

References

[1] Lewenstein, M., Sanpera, A., Ahufinger, V., Damski, B., Sen(De), A. and Sen, U., Ultracold atomic gases in optical lattices: mimicking condensed matter physics and beyond, *Adv. Phys.*, 56, 243 (2007).

[2] Bloch, I., Dalibard, J. and Zwerger, W., Many-body physics with ultracold gases, *Rev. Mod. Phys.*, 80, 885 (2008).

[3] Gross, C. and Bloch, I., Quantum simulations with ultracold atoms in optical lattices, Science, 357, 995 (2017).

[4] Greiner, M. and Fölling, S., Optical lattices. *Nature,* 453, 736 (2008).

[5] Bai, R., Gaur, D., Sable, H., Bandyopadhyay, S., Suthar, K. and D. Angom, Segregated quantum phases of dipolar bosonic mixtures in two-dimensional optical lattices, *Phys. Rev.,* A 102, 043309 (2020).

[6] Suthar, K., Sable, H., Bai, R., Bandyopadhyay, S., Pal, S. and D. Angom, Supersolid phase of the extended Bose-Hubbard model with an artificial gauge field, *Phys. Rev.,* A 102, 013320 (2020).

Impact of Detergents on Water Bodies

Manisha Phukan
Department of Physics, Lalit Chandra Bharali College, Guwahati-781014, India
E-mail: mphukan09@gmail.com

ABSTRACT: The detergents are a complex group of substance and the principal ingredients are surfactants, building agents and additives along with stabilizers. The surfactants are the surface active agents which have both hydrophilic and hydrophobic parts. The building agents are inorganic compounds such as sodium phosphate or sodium polyphosphate which are to bind the surfactants together. The surfactants can have serious impact on water quality. Some of the commercial detergents contains a high amount of inorganic phosphates. These phosphates enter into water streams and raise the nutrient level of such water which leads to eutrophication. In this work a few commercial detergents available in the market were studied. Dissolved oxygen (DO), Biochemical Oxygen Demand (BOD), and P^H *were studied which are a good indicator to show if biodegradable organic matters present in water bodies.*

I. INTRODUCTION

All types of wastes affect the quality of receiving waters. The wastes may be either inorganic or organic chemicals, biological matter or radioactive materials. The role of organic matter is particularly important because its presence exhausts the oxygen resources of rivers and creates unpleasant tastes, odours and general septic conditions. The oxygen shortage is considered to be the most critical impact of organic matter in water. The Biochemical Oxygen Demand (BOD) is usually exerted by dissolved and colloidal organic matter. If a particular water sample despite having a large concentration of organic matter if exhibits lower BOD value then the type of inference to be drawn is that most of the organic matter present is non-biodegradable i.e., environmentally persistent. Phosphates are commonly used detergent builder. Phosphates containing detergents mixed with Nitrogen from the decaying organic matter act as nutrients which stimulate the growth of aquatic vegetation can finally lead to eutrophication. The presence of turbidity and cyanobacterial blooms is an important environment issue in many European countries, U.S.A. and Japan.

II. Experiment

A. DO and its Determination

In this method DO is determined by the modified Winkler method. This is an iodometric method based on the oxidizing property of DO. The DO value is obtained from the relation.

$$1 \text{ ml } 0.25 \text{ N } Na_2S_2O_3 = 1 \text{ mg DO/L}$$

B. BOD and its Determination

The principle of the method is to keep a 300 ml BOD bottle filled with the sample to overflow and keeping it incubated at 20°C for 5 days. DO is measured initially and after incubation and the difference between initial and final Do gives the BOD. A BOD incubator (model NSW 75.2) at 20°C was used for this work. A sample of domestic water was used for seeding the detergent solutions for BOD determination and necessary corrections were made in BOD calculations. The BOD values were calculated using the following formula,

$$\text{BOD} \frac{200C}{5d} \text{ Mg/L} = \frac{DOi - DOf}{P}$$

Where,

DO_i = DO of the detergent solution immediately after preparation in mg/L.

DO_f = DO of the detergent solution after incubation.

P = Decimal volumetric fraction of sample used.

C. P^H and its Determination

In this work the p^H of the detergent solutions were determined immediately after preparation and after 5 days using a digital P^H meter (model systronics 335). The P^H was calibrated with standard buffer solution before use every time.

III. Results and Discussion

A. Detergents

The packets of detergent powders were collected from the local market and they were named from A to E. The price range of the detergents selected for the work is as follows.

Detergent	*Price (Rs per Kg)*
A	140/-
B	80/-
C	65/-
D	45/-
E	30/-

B. DO Content

The results of determination of dissolved oxygen of the different detergent solutions at different concentrations (100, 80, 60, 40, 20, ppm) immediately after preparation and after 5 days are given in Tables 1 to 5. The following observations can be made based on these tables. All the detergents have a reducing influence. In all cases the DO values of 20 to 100 ppm solutions of the detergents had lower DO content than the usual value in case of ordinary water at room temperature. All the detergents also have a decreasing influence on the DO content of water if the detergents containing water is allowed to remain at a particular temperature. This shows that the detergents have some biodegradable components which are responsible for the decrease in DO level. However from the data it is seen that all the detergents do not reduce DO content to the same extent. The detergents A, D and C seem to have more biodegradable content than the others with respect to DO depletion.

C. BOD Values

The BOD values corresponding to different concentrations of the detergent solutions are shown in the Tables 1 to 5. The following conclusions can be made from BOD determination of the detergent solutions.

- All the detergents contribute to the BOD load.
- All the detergents have biodegradable surfactants as principal components.
- A few detergents are less bio degradable than the others.

Table 1: DO and BOD Values for Detergent a at Different Dilution

Detergent Concentration	*Initial DO (mg/L)*	*Final DO (mg/L)*	*BOD (mg/L)*
100	2.76	1.93	3.32
80	2.94	1.84	5.50
60	3.12	1.84	8.53
40	3.22	1.74	14.80
20	3.31	1.56	35.00

- The BOD values of all the detergents showed increasing trend with dilution and in the range of concentrations under study no detergent solutions showed the true BOD load.

Table 2: Do and BOD Values for Detergent B at Different Dilution

Detergent Concentration	*Initial DO (mg/L)*	*Final DO (mg/L)*	*BOD (mg/L)*
100	2.97	2.4	2.28
80	3.07	2.4	3.35
60	3.07	2.3	5.13
40	3.16	2.2	9.6
20	3.26	1.82	28.8

Table 3: DO and BOD Values for Detergent C at Different Dilution

Detergent Concentration	*Initial DO (mg/L)*	*Final DO (mg/L)*	*BOD (mg/L)*
100	2.66	2.11	2.20
80	2.76	1.93	4.15
60	2.85	1.84	6.73
40	2.94	1.65	12.90
20	2.94	1.47	29.40

Table 4: DO and BOD Values for Detergent D at Different Dilution

Detergent Concentration	*Initial DO (mg/L)*	*Final DO (mg/L)*	*BOD (mg/L)*
100	2.88	1.92	3.84
80	3.07	1.72	6.75
60	3.16	1.54	10.80
40	3.16	1.35	18.10
20	3.26	1.35	38.20

Table 5: DO and BOD Values for Detergent E at Different Dilution

Detergent Concentration	*Initial DO (mg/L)*	*Final DO (mg/L)*	*BOD (mg/L)*
100	2.69	2.40	1.16
80	2.88	2.30	2.90
60	2.98	2.11	5.80
40	3.07	1.92	11.50
20	3.26	1.82	28.80

P^H Values

The PH values obtained for different detergents at 5 different concentrations immediately after preparation and after 5 days are shown in Table 6. The results reveal that all the detergents yield slightly alkaline solutions excepting the detergent A which gave near neutral solution at all the concentrations immediately after preparation. For this detergent also the PH turned alkaline after 5 days. One common feature of the PH values is that the PH exhibited a slight decrease with dilution of the detergent solution in all cases. The detergent solutions also showed a perceptible increase in PH after 5 days.

Table 6: P^H Values of the Detergent Solutions at Different Dilutions at Room Temperature

Detergent Types	*P^H*									
	After Preparation of Solution (ppm)					*After Five Days (ppm)*				
	100	*80*	*60*	*40*	*20*	*100*	*80*	*60*	*40*	*20*
A	7.62	7.53	7.46	7.38	7.29	7.69	7.60	7.54	7.43	7.34
B	7.45	7.43	7.40	7.38	7.37	7.66	7.64	7.62	7.61	7.60
C	7.45	7.40	7.36	7.26	7.20	7.46	7.41	7.39	7.32	7.27
D	7.32	7.29	7.28	7.24	7.23	7.41	7.38	7.33	7.29	7.27
E	7.57	7.53	7.45	7.43	7.40	7.63	7.56	7.49	7.46	7.43

IV. CONCLUSION

The present work demonstrated in a qualitative way the differences existing among the detergents with respect to their impact on pH, DO and BOD. The properties of the dilute solutions of the detergents indicate that may have considerable influence on water quality exerting sufficient BOD load and hence depleting the DO levels of receiving water bodies.

REFERENCES

[1] Rebhun, Mand L., M. (1993). *Wat, Sci, Tech*, 27, 1–20.

[2] Pattusamy, V., Nandini, N. and Bheemappa, K., *International Journal of Advanced Reasearch*, 2013, Volume 1, Issue 3, 129–133. ISSN No 2320-5407.

[3] Johar Sarabjeet Singh, Johar G.S., Tiwari, A. and Dixit A.K., *Universal Journal of Environmental Research and Technology,* 2014, eISSN 2249 0256, Volume 4, Issue 4: 199–207.

[4] Farmer, A. (2004). Phosphate Pollution, a Global Overview of the problem in: E, Valsani-Jones (Ed.). Phosphorus in environmental Technologies: Principles amd Applications. London: IWA.

[5] Holt, M.S. Mitchell, G.C. and Watleinson, R.J. (1992). In the Handbook of Environmental Chemistry, Vol. 3, Detergents, pp. 89-144, Springer, Berlin.
[6] European Commission (14 December 2011). 'EP supports ban of Phosphates in consumer detergents' Ńeuorpeeu. Retrieved 20 March 2017.

Quantum Phases of Bosons in Optical Lattices with Disorder and Longrange Interactions

Kohenjit Pebam*, Laxmi Angom, Bhumika Thoudam and D. Angom
Department of Physics, Manipur University, Manipur-795003, India
*E-mail: kohenjitpebam123@gmail.com

ABSTRACT: Optical lattices are crystals of light formed by counter-propagating laser beams which can trap ultracold atoms in a clean and controllable manner. Such systems are described by the Bose-Hubbard model. Long range inter-atomic interactions can be introduced using dipolar atoms, molecules, or Rydberg atoms and disorder can be introduced using speckle lasers. The former is studied using the extended Bose-Hubbard modeland the later using the disordered Bose-Hubbard model. We consider square optical lattice with both the long-range interactions and disorder simultaneously. The theoretical method to study our system of interest is the single-site Gutzwiller mean field theory and the ground state phase diagrams generated using it are presented.

Keywords: Bose-Hubbard Model, Extended Bose-Hubbard Model, Disorder Bose-Hubbard Model, Single-Site Gutzwiller Mean Field Theory.

I. Introduction

Ultracold atoms in optical lattices are ideal to study the physics of quantum many-body systems. Due to recent experimental developments in the techniques to manipulate ultracold bosonic atoms in optical lattices, new quantum phases can be observed in the strongly correlated regime. Such a system of bosonic atoms can be studied using Bose-Hubbard model (BHM) and the model in it's simplest form exibits two quantum phases: superfluid (SF) and Mott insulator (MI) phases. Long-range interaction in optical lattices is introduced when dipolar atoms, molecules and Rydberg atoms are used [1]. The introduction of long-range interaction leads to two new quantum phases. These are the supersolid (SS) and density wave (DW) phases. Research on the SS phase, which is characterized by simultaneous appearance of superfluidity and crystalline order, has increased in the recent years [7]. The SS phase breaks two continuous symmetries, the global U(1) symmetry to create superfluid (SF) and translational invariance to form crystal. The DW phase is an insulating state characterized by the existence of long-range crystalline order. The signatures of the SS phase has been observed in recent experiments. Theoretically, long-range interaction in the optical lattices is described by the extended Bose-Hubbard model (eBHM) [1], which is an extension of the BHM. The relevant theory and method used in the present study is the single-site mean field theory [2, 3]. Further, optical lattice can be used to study impurities in real solids through the disordered Bose-Hubbard model (DBHM) [4]. Introduction of disorder in optical lattices gives rise to a quantum glassy phase called as Bose glass (BG) phase [4]. Simultaneous introduction of long-range interaction and disorder in optical lattices give rise to a new quantum phase called as disordered solid (DS) and this is studied through disordered extended Bose-Hubbard model (DEBHM) [5,6]. This paper focusses on the phase diagrams of the various quantum phases obtained due to the simultaneous introduction of long-range interaction and disorder in optical lattices.

II. EXTENDED BOSE-HUBBARD MODEL

The model considers a system of ultracold bosonic atoms with long-range interactions in square optical lattice. Since the temperature of the system is low, the atoms occupy the lowest Bloch band. The physics of such a system is well described by eBHM and the corresponding Hamiltonina is,[1]

$$\hat{H}_{\text{eBHM}} = -J\sum_{<i,j>}\left(\hat{b}_i^{\dagger}\hat{b}_j + \hat{b}_j^{\dagger}\hat{b}_i\right) + \sum_i \frac{U}{2}\,\hat{n}_i(\hat{n}_i - 1) - \sum_i \mu\,\hat{n}_{i,} + V\sum_{<i,j>}\hat{n}_i\,\hat{n}_{j,} \quad \ldots (1)$$

where, $< i,j >$indicatessummation over the nearest neighbours and the term represents nearest-neighbourhopping with strength J,$\hat{b}_i(\hat{b}_i^{\dagger})$ is bosonic annihilation(creation) operator, $\hat{n}_i = \hat{b}_i^{\dagger}\hat{b}_i$is the number operator of the bosons, μ is the chemical potential, $U > 0$ is the on-site interaction energy and V is the long range potential energy. This model exhibits four quantum phases-MI, SF, DW, and SS. In MI, there is commensurate integer fillings of the lattice sites and have zero compressibility. The SF phase is compressible, gapless and atoms in the system are itenerant. The SS phase is characterized by the coexistence of superfluidity and long-range crystalline order [7] whereas DW phase is an insulating phase with crystalline order.

III. DISORDERED BOSE-HUBBARD MODEL

Optical lattices are clean and controllable. However, condensed matter systems are never free of impurities and introduction of disorder in optical lattice provide a way to study impurities in real solids. The presence of disorder in optical lattice give rise to quantum glassy phases like the BG phase. It is characterized by finite compressibility and zero superfluid stiffness. The Hamiltonian of the DBHM is given by [4],

$$\hat{H}_{\mathrm{DBHM}} = -J\sum_{<i,j>}\left(\hat{b}_i^{\dagger}\hat{b}_j + \hat{b}_j^{\dagger}\hat{b}_i\right) + \sum_i \frac{U}{2}\,\hat{n}_i(\hat{n}_i - 1) - \sum_i \tilde{\mu}\,\hat{n}_i,$$

where the quantities are as defined in Eq. (1), and $\tilde{\mu} = \mu - \epsilon_{p,q}$ is the local chemical potential. The disorder is introduced through the random energy off set $\epsilon_{p,q}$, which are uniformly distributed random numbers $\epsilon_{p,q} \in [-D, D]$ and$D$ is the strength or bound of the random numbers. The DBHM exhibits three quantum phases-MI, BG and SF phases. The BG phase acts as an intermediate between the MI and SF phases, and it has finite compressibility, zero superfluid stiffness and insulating [4].

IV. Disordered Extended Bose-Hubbard Model (DEBHM)

Simultaneous introduction long-range interaction and disorder in optical lattice is studied through DEBHM. The Hamiltonian of this model is [5,6],

$$\hat{H}_{\mathrm{DEBHM}} = -J\sum_{<i,j>}\left(\hat{b}_i^{\dagger}\hat{b}_j + \hat{b}_j^{\dagger}\hat{b}_i\right) + \sum_i \frac{U}{2}\,\hat{n}_i(\hat{n}_i - 1) - \sum_i \tilde{\mu}\,\hat{n}_i + V\sum_{<i,j>}\hat{n}_i\,\hat{n}_j.$$

This model exhibits six quantum phases-MI, SF, SS, DW, BG, and DS. The identifying properties of the DS phase are zero superfluid stiffness, non-zero compressibility, and non-zero structure factor [5,6].

V. Single-Site Gutzwiller Mean Field Theory

The single-site Mean Field Theory (SGMF) is a mean-field theory suitable to obtain the ground states of lattice systems like the DEBHM in the present work. In the SGMF theory, the many-body wave function of the ground state is given by [2,3],

$$|\Psi\rangle = \prod_i^{M}\sum_n^{N_b} c_n|n\rangle,$$

where, M is the total number of lattice sites, N_b is the Fock space dimension and c_n is the coefficients of linear combination of the Fock states. The bosonic operators can be written as the sum of their expectation values and respective quantum fluctuations [2,3],

$$\hat{b}_i = \phi + \delta\hat{b}_i,$$
$$\hat{b}_i^\dagger = \phi^* + \delta b_i^\dagger.$$

With these approximations, the total Hamiltonian of the system can be written as the sum of single -site mean field Hamiltonian,

$$\hat{H}_{\text{MF}} = \sum \hat{h}^{\text{MF}}.$$

We diagonalize this single-site Hamiltonian $\hat{h}^{l\ e}$, and retain the lowest eigenstate and eigenvalue which corresponds to the ground state of the system. The various order parameters used to distinguish the quantum phases from each other are superfluid order parameter, relative average occupancy, superfluid stiffness and structure factor.

VI. Results and Phase Diagram

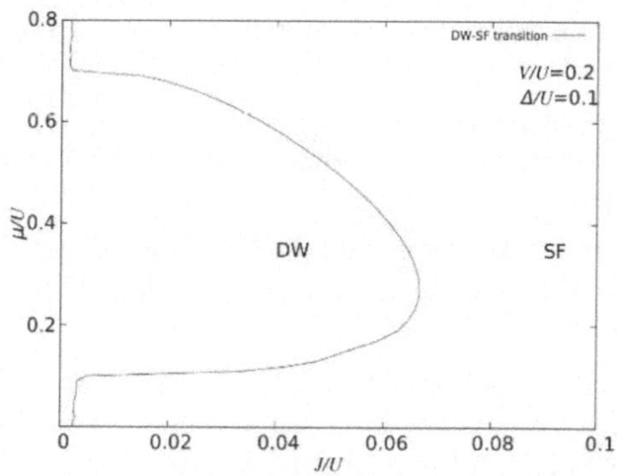

Figure 1: Coarse-grained phase diagram for DW to SF quantum phase transition in disordered extended Bose-Hubbard model at long-range interaction potential $V = 0.2$ and disorder strength $\Delta = 0.1$.

The coarse-grained phase diagram presented here is for the case of disordered extended Bose Hubbard model. Using the disorder averaged superfluid order parameter, we can distinguish the DW-SF phase boundary. The phase diagram is obtained as a disorder average of 50 realizations. We can distinguish the BG-SS and SS-SF phase boundary using superfluid stiffness and structure factor respectively.

VII. Conclusion

We have simultaneously introduced long-range interaction and disorder in the optical lattice and the quantum phases that arises are observed. We found six quantum phases MI, DW, BG, SS, SF, DS. In future, we will work on different values of long-range interaction and disorder strengths with different system sizes and the phase diagrams obtained will be analysed.

ACKNOWLEDGEMENT

Kohenjit Pebam acknowledge Manipur University non-NET fellowshipfor providing financial support during this study.

REFERENCES

[1] Suthar, K., Hrusikesh, S., Rukmani, B., Soumik, B., Sukla, P. and Angom, D., Supersolid phase of the extended Bose-Hubbard model with an artificial gauge field, *Phys. Rev.* A 102, 013320 (2020).

[2] Sheshadri, K., Krishnamurthy, H.R., Pandit, R. and Ramakrishnan, T.V., Superfluid and insulating phases in an interacting-boson model: Mean-field theory and the RPA, EPL 22, 257 (1993).

[3] Rokhsar, D.S. and Kotliar, B.G., Gutzwiller projection for bosons, *Phys. Rev.* B 44, 10328 (1991).

[4] Pal, S., Bai, R., Bandyopadhyay, S., Suthar, K. andAngom, D., Enhancement of the bose glass phase in the presence of an artificial gauge field, *Phys. Rev.* A 99, 053610 (2019).

[5] Ji-Ming Gao, Rong—An Tang and Ju-Kui Xue, Phase diagram of disordered two-dimensional extended Bose-Hubbard model, *Europhy. Lett.* 117, 60007 (2017).

[6] Zhang, C. and Rieger, H., Phase diagrams of the disordered Bose- Hubbard model with cavity-mediated long-range and nearest-neighbor interactions, arXiv:1908.11165v1 [cond-mat.dis-nn] 29 Aug. 2019.

[7] Boninsegni, M. and Nikolay, V. Prokof'ev, Colloquium: Supersolids: What and where are they?, *Rev. Mod. Phys,* 84, 759 (2012).

Ultracold Quantum Gases in Disordered Optical Lattices

Laxmi Angom*, Bhumika Thoudam, Kohenjit Pebam and D. Angom
Department of Physics, Manipur University, Manipur-795003, India
*E-mail: angomlaxmi308@gmail.com

ABSTRACT: Optical lattices are periodic intensity patterns formed by counter-propagating laser beams. Their dimension is ~1000 times larger than the crystals in condensed matter systems and provides a platform to simulate or emulate condensed matter systems using a simplified model. The optical lattices, however, are devoid of imperfections and lattice vibrations within the crystal. In contrast, disorders like impurities and defects are unavoidable in the condensed matter systems. It is possible to introduce disorder to the optical lattices using additional lasers. Thus, ultracold atoms in disordered optical lattice mimics condensed matter systems with impurities. Such a system can be studied using disordered Bose-Hubbard model. A new quantum phase, referred to as the Bose glass emerges with the introduction of disorder. There are several methods to study disordered Bose-Hubbard model. The Gutzwiller mean field method is one of the simplest and numerically inexpensive methods. It is also applicable to different optical lattice systems ranging from homogeneous to spatially varying ones.

Keywords: Disordered Bose-Hubbard Model, Gutzwiller Mean Field Theory, Bose Glass.

I. Introduction

Optical lattices are clean and highly tunable [1,2]. On the other hand, impurities are unavoidable in the condensed matter systems. So, understanding the role of disorder is important to materials research and condensed matter physics. It is, however, difficult to study these systems since the lattice constant is ~10^{-10} m and not amenable to modifications. Ultracold atoms in disordered optical lattice provide a means to understand impurities in real solids through simplified models. Optical lattices are created by counter-propagating laser beams travelling in opposite directions which results in standing interference pattern of bright and dark intensities. Thus, the lattice constant is of the order of half the optical wavelength, which is ~10^{-10} m. The properties and dynamics of atoms in optical lattices with disorder can be studied using disordered Bose-Hubbard Model (DBHM). Without disorder there are two distinct ground states near zero temperature depending on the ratio J/U, where J is the hopping strength and U is the strength of the onsite interaction. These are the Mott insulator (MI) and superfluid (SF) states. The introduction of disorder leads to the emergence of new quantum phases in the system.

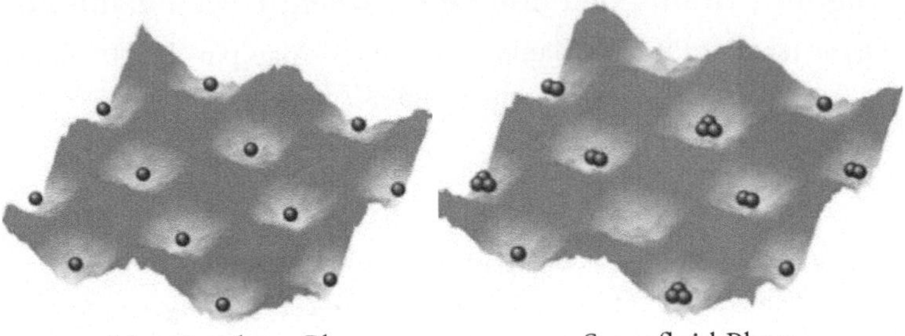

Figure 1: Ultracold atoms, shaded spheres, in optical lattice with disorder. The distributions represent (left panel) commensurate distribution in Mott insulator quantum phase, and (right panel) superfluid quantum phase.

In the optical lattices J/U is varied by changing the laser intensities. At higher laser intensities the depth of optical potential is increased and J/U is lowered. As a result, the atomic wave functions are localized and there is a strong onsite interaction. The system is then in the MI phase. It is characterized by zero compressibility and a gap in the particle-hole excitation. For higher J/U, the system is in the SF phase. The SF phase is compressible and gapless. The SF-MI transition is a quantum phase transition well described by the Bose-Hubbard Model (BHM) [3] and it has been observed in experiments [4]. However, in the presence of disorder direct transition between SF and MI is not possible. A new quantum phase, referred to as the Bose glass (BG), acts as an intervening phase between the two phases. This phase is gapless and compressible phase but insulating. The compressibility of the BG phase is due to the existence of rare but arbitrarily large SF islands. The DBHM Hamiltonian, which describes optical lattices with disorder is essentially the BHM Hamiltonian with the addition of disorder. To study ultracold quantum gases in disordered optical lattices we shall use the single-site Gutzwiller mean field theory.

II. Theoretical Method

The DBHM Hamiltonian for bosons confined in a 2D square optical lattice with disorder, in the grand canonical ensemble, at zero temperature is given by [5],

$$\hat{H} = -\sum_{p,q}\left[J_x\left(\hat{b}^{\dagger}_{p+1,q}\hat{b}_{p,q} + \text{H.c}\right) + J_y\left(\hat{b}^{\dagger}_{p,q+1}\hat{b}_{p,q} + \text{H.c}\right)\right] + \sum_{p,q}\hat{n}_{p,q}\left[\frac{U}{2}\left(\hat{n}_{p,q} - 1\right) - \tilde{\mu}_{p,q}\right] \quad \dots (1)$$

where $p(q)$is the lattice index along the $x(y)$ axis, $\hat{b}_{p,q}(\hat{b}^{\dagger}_{p,q})$ is the creation (annihilation) operator for a boson at the (p,q) lattice site and $\hat{n}$ is the number operator. The other important parameters are the nearest neighbour hopping strength $J_x(J_y)$ along the $x(y)$ axis and the onsite interaction strength $U > 0$. The disorder is subsumed in the local chemical potential $\tilde{\mu}_{p,q} = \mu - \epsilon_{p,q}$. Here, $\epsilon_{p,q} \in [-D, D]$ is a random number with univariate distribution. It represents quenched diagonal disorder and D is the disorder strength. The above Hamiltonian can describe three phases depending on the ratio J/U and D/U. With moderate disorder $D/U{\sim}0.2$, the system is in the MI and SF phases in the domains $J/U \to 0$ and $J/U \gtrsim 0.05$, respectively. The BG phase is present in a small region of the intermediate domain [5] and the actual values depends on μ.

A. Single-site Gutzwiller Mean-Field Theory

The eigenstates of DBHM can be obtained using the single site Gutzwiller mean-field theory. In this theory, $\hat{b}_{p,q}(\hat{b}^{\dagger}_{p,q})$ at a particular lattice site (p,q) is decomposed as,

$$\hat{b}_{p,q} = \langle\hat{b}_{p,q}\rangle + \delta\hat{b}_{p,q} = \phi_{p,q} + \delta\hat{b}_{p,q} \quad \dots (2)$$

$$\hat{b}^{\dagger}_{p,q} = \langle\hat{b}_{p,q}\rangle + \delta\hat{b}^{\dagger}_{p,q} = \phi^{*}_{p,q} + \delta\hat{b}^{\dagger}_{p,q} \quad \dots (3)$$

where $\phi_{p,q}(\phi^{*}_{p,q})$ are mean fields and $\delta\hat{b}_{p,q}(\delta\hat{b}^{\dagger}_{p,q})$ are the fluctuation operators. Using these definitions in Eq. (1) and considering terms only up to linear in fluctuation operators, we obtain the mean-field Hamiltonian as,

$$\hat{H}_{\text{MF}} = -\sum_{p,q}\Big\{J_x\left[\left(\hat{b}^{\dagger}_{p+1,q}\phi_{p,q} + \phi^{*}_{p+1,q}\hat{b}_{p,q} - \phi^{*}_{p+1,q}\phi_{p,q}\right) + \text{H.c.}\right] - J_x\left[\left(\hat{b}^{\dagger}_{p+1,q}\phi_{p,q} + \phi^{*}_{p+1,q}\hat{b}_{p,q} - \phi^{*}_{p+1,q}\phi_{p,q}\right) + \text{H.c.}\right] + \hat{n}_{p,q}\left[\frac{U}{2}\left(\hat{n}_{p,q} - 1\right) - \tilde{\mu}_{p,q}\right]\Big\} = \sum_{p,q}\hat{h}_{p,q} \quad \dots (4)$$

where $\hat{h}_{p,q}$ is the single-site Hamiltonian. We can therefore diagonalize the Hamiltonian for each lattice site separately to obtain the ground state $|\psi\rangle_{p,q}$. Then the ground state of the system is the direct product of the single site wavefunctions,

$$|\Psi_{\text{GW}}\rangle = \prod_{p,q}|\psi\rangle_{p,q} = \prod_{p,q}\sum_{n=0}^{N_b} c_n^{(p,q)}|n\rangle_{p,q}, \quad \dots (5)$$

with the normalization condition,

$$\sum_n \left|c_n^{(p,q)}\right|^2 = 1$$

and N_b is the highest occupation number single-site Fock state. The mean field or SF order parameter is,

$$\phi_{p,q} = \langle \Psi_{GW} | \hat{b}_{p,q} | \Psi_{GW} \rangle \qquad \ldots (6)$$

Considering the above expression, it is evident that $\phi_{p,q}$ is 0 for MI phase as only one occupation number state $|n\rangle$ contributes to $|\psi\rangle_{p,q}$. It is finite for SF and BG phase. The occupancy and number fluctuation are given by,

$$n_{p,q} = \langle \hat{n}_{p,q} \rangle = \Sigma_n \left|c_n^{(p,q)}\right|^2 n \qquad \ldots (7)$$

$$\delta n_{p,q} = \sqrt{\langle \hat{n}_{p,q}^2 \rangle - \langle \hat{n}_{p,q} \rangle^2} \qquad \ldots (8)$$

In the MI phase $\delta n_{p,q} = 0$ and thus, it is a number coherent state. Whereas, the SF phase has $\delta n_{p,q} \neq 0$ and it is phase coherent. Thus, it requires finite energy to alter the phase of the state and this leads to a non-zero superfluid stiffness ρ_s. To compute ρ_s, the state of the system is subjected to a twisted boundary condition (TBC). If the TBC is applied along the *x* direction, the hopping term in the DBHM is transformed as,

$$J_x\left(\hat{b}_{p+1,q}^{\dagger} \hat{b}_{p,q} + \text{H.c}\right) \rightarrow J_x\left(\hat{b}_{p+1,q}^{\dagger} \hat{b}_{p,q} e^{i2\pi\varphi/L} + \text{H.c}\right) \qquad \ldots (9)$$

where, φ is the phase shift or twist applied to the periodic boundary, L is the size of the lattice along the *x* direction, and $2\pi\varphi/L$ is the phase shift of an atom when ithops between nearest neighbours. Accordingly, ρ_s is computed as,

$$\rho_s = \frac{L}{8\pi} \frac{\partial E_0}{\partial \varphi^2} \Big|_{\varphi = 0}, \qquad \ldots (10)$$

where E_0 is the ground state energy with TBC. The BG phase, on the other hand, lacks phase coherence across the system and hence $\rho_s = 0$.

Table 1: Classification of Ground State Quantum Phases with Order Parameters Supported by DBHM

Quantum Phases	*Order Parameter*
Superfluid	$\rho_s > 0, \phi \neq 0$
Bose glass	$\rho_s = 0, \phi \neq 0$
Mott insulator	$\rho_s = 0, \phi = 0$

III. CONCLUSION

We scale the parameters of the DBHM Hamiltonian with respect to the interaction strength U to compute the ground state of the system and determine the phase diagram. The relevant parameters of the model are J/U, μ/U and D/U. We then determine the phase diagram of the DBHM in the $J/U - \mu/U$ plane for $D/U = 0.02$, and the BG phase emerges. This phase is incompressible and gapless in nature. To construct the phase diagram, we consider a 50×50 square lattice superimposed with a homogeneous disorder distribution.

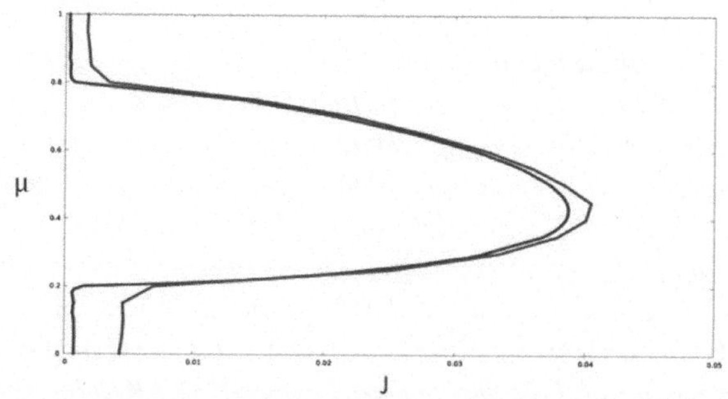

Figure 2: Phase Diagram for Disordered Bose-Hubbard Model for Disorder Strength D/U = 0.2 for 50 Different Realizations

ACKNOWLEDGEMENT

Laxmi Angomg ratefully acknowledges Department of Science and Technology (DST), India under Grant No. DST/INSPIRE Fellowship/ [IF200263] for providing financial support.

REFERENCES

[1] Morsch, O. and Oberthaler, M., Dynamics of Bose-Einstein condensates in optical lattices, *Rev. Mod. Phys.*, 78, 179 (2006).

[2] Bloch, I., Dalibard, J. and Zwerger, W., Many-body physics with ultracold gases, *Rev. Mod. Phys.*, 80, 885 (2008).

[3] Jaksch, D., Bruder, C., Cirac, J.I., Gardiner, C. W. and Zoller, P., Cold bosonic atoms in optical lattices, *Phys. Rev. Lett.* 81, 3108 (1998).

[4] Greiner, M., Bloch, I., Mandel, O., H¨ansch, T.W. and Esslinger, T., Exploring phase coherence in a 2D lattice of Bose-Einstein condensates, *Phys. Rev. Lett.*, 87, 160405 (2001).

[5] Pal, S., Bai, R., Bandyopadhyay, S., Suthar, K. and Angom, D., Enhancement of the bose glass phase in the presence of an artificial guage field, *Phys. Rev.*, A 99, 053610 (2019).

An Overview of the Electronic Properties of Titanium Nitride

Purbajyoti Bhagowati* and Munima B. Sahariah
Computational Materials Science Lab, Physical Science Division,
Institute of Advanced Study in Science and Technology, Assam, India
*E-mail: purbajyoti.bhagowati40@gmail.com

ABSTRACT: Transition metal nitrides (TMNs) are known for their high melting point, chemical stability, hardness, and often known as refractory materials. They have been used as protective coatings for wear and corrosion resistance, decorative coatings, cutting tools and many machinery applications in hostile environments. Nowadays these metal nitrides are also used in plasmonic applications due to the easy tuning of their optical properties and superiority over the conventional metals like gold and silver. One such member of TMN family is the titanium nitride (TiN), which has rock-salt structure in their bulk phase. Compared to the bulk and films, the nanocluster systems of TiN had not been explored yet. Keeping this in mind, we have studied the electronic properties of these nanoclusters using first-principles method. In this chapter we briefly discuss the electronic properties of bulk TiN as well as the nanoclusters, obtained by using density functional theory (DFT) methods.

Keywords: Electronic Properties, Nanocluster, Metal Nitrides, Titanium Nitride.

I. Introduction

Transition metal nitrides (TMNs) are known for their excellent electrical conductivity, high melting point, electrolytic activity, fascinating chemical stability in a hostile environment, and excellent mechanical robustness [1-2]. They can be found forming cubic rocksalt-type structures. The TMNs can be found in an ocean of applications such as protective and anti-corrosive coatings, decorative coatings, cutting tools, and machine equipment [3-9]. Some other applications of the TMNs include super-conducting devices, optoelectronic devices, diffusion barriers, and many more. These TMNs are also used as plasmonic materials due to their superior properties over conventional ones like gold and silver. Among the TMNs, titanium nitride (TiN) is the most studied member. Due to its excellent properties and wide usage, it is often considered "technologically the most important nitride."

In this chapter, we will briefly highlight the various properties and applications of titanium nitride. We will also shed light on the electronic properties of TiN, studied using density functional theory. Along with this, we will concentrate on the TiN nanocluster systems, a relatively unexplored region.

II. TITANIUM NITRIDE: PROPERTIES AND APPLICATIONS

TiN combines the transition metal of group IVb of the periodic table and non-metal nitrogen. Titanium has four valence electrons with the electronic configuration $[Ar]3d^2 4s^2$. Several phases of nitrides are found in titanium, such as Ti2N and Ti3N4. However, in bulk format, cubic rock-salt δ-TiN or B1-TiN structure is the most stable [10]. The colour of TiN is bright yellow and looks almost like gold.

Over the years, TiN has been fabricated using different techniques. Some of the major processes of preparing TiN films involve magnetron sputtering, chemical vapour deposition, pulsed laser deposition, cathodic vacuum arc, atomic layer deposition, ion beam assisted deposition, high power impulse magnetron sputtering, etc. Using these standard methods, the underlying mechanism of TiN film growth, grain size, orientation, and many other inherent properties become understandable.

Other than experimental techniques, there are a few hundred reports exploring various properties of TiN using theoretical and simulation models. These include numerical methods like discrete dipole approximation, classical Mie scattering theory, finite difference time domain method, and modern-day *ab-initio* methods like density functional theory (DFT). Discussions of all the properties and applications of TiN will be out of the scope of this book chapter. So, we have restricted ourselves to discussing only the electronic properties of TiN using DFT calculations.

The advancement in computational facilities helped the researchers to explore the materials by going beyond the conventional length scale available for experimental techniques. During the end of the 20th century, Marlo *et al.* studied the surface properties of bulk TiN using various exchange-correlation functionals [11]. They took PW91, PBE, and RPBW functionals and found that all of these give good surface properties of TiN. At the start of the 21st century, Dridi *et al.* studied the effect of the vacancy on the structural and electronic properties of TiN_x [12]. Their results clearly showed that the introduction of vacancies varies the electronic properties like band structure and density of states (DOS), which ultimately results in a decrease in the hardness of the material. Didziulis *et al.* reported the electronic and bonding properties of some small clusters of TiN using the DFT technique [13]. Their results showed the covalent mixing present between the titanium and nitrogen. By taking the framework of the linear muffin-tin-orbital (LMTO) method, Allmaier *et al.* studied the electron correlation effects in TiN [14]. They analyzed the DOS plot and found that the peculiar pseudogap found near the Fermi level indicates a metal-insulator transition. Yu *et al.* reported some stable structures of TiN at different external pressures and interpreted the correlation between chemical bonding and mechanical properties [15]. Their results showed that the increase in nitrogen content in the structure results in more covalent Ti-N bonding and less Ti-Ti metallic bonding. This leads to an increase in the hardness of the system and hence can be useful in technological applications. A similar kind of study was done by Yang *et al.*, where they reported some high-pressure structures of TiN and investigated their electronic properties [16]. They found metallic characters in TiN and Ti_2N, while Ti_3N_2 shows a semiconducting character. Their results showed that the N-Ti bond is made up of covalent and ionic characters; with increasing nitrogen content, the covalent nature becomes stronger due to the shortening of Ti-N bonds. Farhadizadeh *et al.* studied the effect of metal transition dopants on the electronic and mechanical properties of TiN [17]. They took the dopants from the transition metals like V, Cr, Mn, Zr, Nb, Mo, Hf, Ta, W, Re, and Os. In all the cases, the metallic behaviour is observed due to the contribution of Ti-*3d* and TM-*d* electrons. Apart from the various reports on bulk TiN, its nanocluster systems are hardly explored. In one of our recent reports, we tried to discuss the electronic properties of some small nanoclusters of TiN using DFT methods. In the following section, we will focus on these results.

A. Titanium Nitride Nanoclusters

The nanocluster geometries of a system are more reactive than their bulk counterpart due to the large surface-to-volume ratio. In the case of nanoclusters, the stability and other physical properties also vary from the bulk system. The low-dimensional quantum mechanical properties become prominent in the case of nanoclusters.

In a recent study, we generated some very small structures of TiN nanocluster using DFT and studied their electronic properties [18]. Our study included the binary nanoclusters Ti_nN_n, where n = 4-10. We performed Bader charge analysis on the nanoclusters and found that the charge lost by Ti and gained by N is not equal. There is a possibility of sharing of charge in the nanoclusters. Also, from the electron localization function (ELF) analysis,

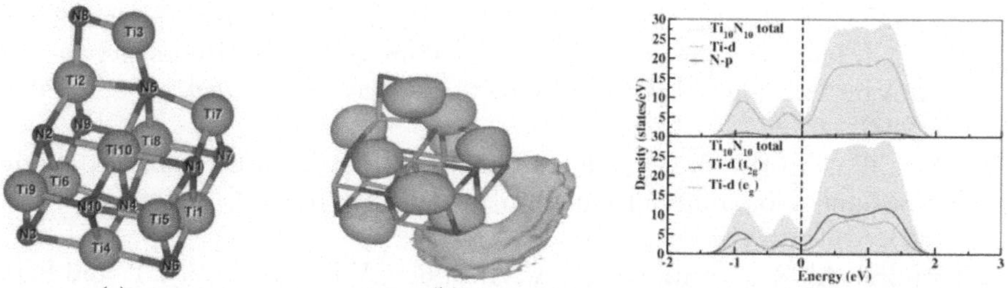

(a) (b) (c)

Figure 1: (a) Global Minima Structure, (b) ELF Plot, and (c) DOS Plot of $Ti_{10}N_{10}$

it was found that these nanoclusters have a mixed type of Ti-N bonding, with characters of both covalent and metallic bonds. The nature of the Ti-N bond tends towards ionic behaviour when we move from the lower to the higher geometries. From the DOS plots, it was found that the Ti-*d* electrons are responsible for the conductivity of these nanoclusters, and the reason for Ti-N bonding is the hybridization between the Ti-*d* and N-*p* electrons. These results from DOS also match with the bulk TiN. In Figure 1, we have shown the structure, ELF plot, and DOS plot of the $Ti_{10}N_{10}$ nanocluster for reference.

III. Conclusion

In conclusion, we have reviewed some of the reports available on the electronic properties of TiN, studied by using DFT methods. We also briefly discussed a rather unexplored field comprising very small nanoclusters of TiN. The nanoclusters of our study can give an insight into the low-dimensional properties and their difference from the bulk counterpart. This will help in further analysis and synthesis of small nanoclusters for application purposes.

Acknowledgement

The first author acknowledges DST for providing financial support through INSPIRE Fellowship during this study under grant number IF170146.

References

[1] Gao, B., Li, X., Ding, K., Huang, C., Li, Q., Chu, P.K. and Huo, K., Recent progress in nanostructured transition metal nitrides for advanced electrochemical energy storage, *Journal of Materials Chemistry A*, 7(1), 14–37 (2019).

[2] Pierson, H.O., Handbook of Refractory Carbides and Nitrides, Properties, Characteristics, Processing and Applications. *Elsevier Science* (1996).

[3] Hiltunen, L., Leskelä, M., Mäkelä, M., Niinistö, L., Nykänen, E. and Soininen, P., Nitrides of titanium, niobium, tantalum and molybdenum grown as thin films by the atomic layer epitaxy method, *Thin Solid Films,* 166(C), 149–154 (1988).

[4] Fix, R., Gordon, R.G. and Hoffman, D.M., Chemical vapor deposition of vanadium, niobium, and tantalum nitride thin films, *Chemistry of Materials,* 5(5), 614–619 (1993).

[5] Andrievski, R.A., Anisimova, I.A., Anisimov, V.P., Makarov, V.P. and Popova, V. P., Grain size and recrystallization of TiN, ZrN, NbN, and CrN alloyed and multilayer films, *Thin Solid Films,* 261(1–2), 83–86 (1995).

[6] Soriano, L., Abbate, M., Pen, H., Prieto, P. and Sanz, J.M., The electronic structure of TiN and VN: X-ray and electron spectra compared to band structure calculations, *Solid State Communications,* 102(4), 291–296 (1997).

[7] Mitterer, C., Mayrhofer, P.H., Waldhauser, W., Kelesoglu, E., and Losbichler, P., The influence of the ion bombardment on the optical properties of TiN_x and ZrN_x coatings, *Surface and Coatings Technology,* 108–109, 230–235 (1998).

[8] Budke, E., Krempel-Hesse, J., Maidhof, H., and Schüssler, H., Decorative hard coatings with improved corrosion resistance, *Surface and Coatings Technology,* 112(1–3), 108–113 (1999).

[9] Lévy, F., Hones, P., Schmid, P.E., Sanjinés, R., Diserens, M. and Wiemer, C., Electronic states and mechanical properties in transition metal nitrides, *Surface and Coatings Technology,* 120–121, 284–290 (1999).

[10] Patsalas, P., Kalfagiannis, N., Kassavetis, S., Abadias, G., Bellas, D.V., Lekka, C. and Lidorikis, E., Conductive nitrides: Growth principles, optical and electronic properties, and their perspectives in photonics and plasmonics, *Materials Science and Engineering: R: Reports,* 123, 1–55 (2018).

[11] Marlo, M. and Milman, V., Density-Functional Study of Bulk and Surface Properties of Titanium Nitride Using Different Exchange-Correlation Functionals, *Phys. Rev. B* 62, 2899 (2000).

[12] Dridi, Z., Bouhafs, B., Ruterana, P. and Aourag, H., First-Principles Calculations of Vacancy Effects on Structural and Electronic Properties of TiC_x and TiN_x, *J. Phys. Condens. Matter,* 14, 10237 (2002).

[13] Didziulis, S.V., Butcher, K.D. and Perry, S.S., Small Cluster Models of the Surface Electronic Structure and Bonding Properties of Titanium Carbide, *Vanadium Carbide, and Titanium Nitride, Inorg. Chem.,* 42, 7766 (2003).

[14] Yu, S., Zeng, Q., Oganov, A.R., Frapper, G. and Zhang, L., Phase Stability, Chemical Bonding and Mechanical Properties of Titanium Nitrides: A First-Principles Study, *Phys. Chem. Chem. Phys.* 17, 11763 (2015).

[15] Yang, R., Zhu, C., Wei. Q. and Du, Z., Investigations on Structural, Elastic, Thermodynamic and Electronic Properties of TiN, Ti_2N and Ti_3N_2 under High Pressure by First-Principles, *J. Phys. Chem. Solids,* 98, 10 (2016).

[16] Farhadizadeh, A.R., Amadeh, A.A. and Ghomi, H., The Effect of Metal Transition Dopant on Electronic and Mechanical Properties of Titanium Nitride: First Principle Method, *Comput. Mater. Sci.* 141, 82 (2018).

[17] Allmaier, H., Chioncel, L. and Arrigoni, E., Titanium Nitride: A Correlated Metal at the Threshold of a Mott Transition, *Phys. Rev. B* 79, 235126 (2009).

[18] Bhagowati, P., Saikia, U. and Sahariah, M.B., Low-energy structures and electronic properties of small titanium nitride nanoclusters, *Nanotechnology*, 32(50), 505718 (2021).

Study of the Molecular Interaction between Favipiravir and Uracil Using DFT Method

Suraj Singh Pukhrambam, Willingson Lamthaka and Th. Gomti Devi*

Department of Physics, Manipur University, Canchipur, India

*E-mail: tgd@manipuruniv.ac.in

ABSTRACT: The present study focuses on the interacting mechanisms of drug and RNA nucleobase and we choose drug, Favipiravir and nucleobase, Uracil for the study. We use Gauss view 5.0 software and Gauss09W to obtain the optimized structure and frequency of the molecules performing at B3LYP/6311-++g(d, p) basis set. Vibrational assignments are carried out using VEDA. We performed Natural Bond Orbital (NBO) analysis, Autodocking of the biomolecular complex.

Keywords: Favipiravir, Uracil, DFT, NBO, AIM, Autodocking.

I. Introduction

Drug protein binding in blood stream determines the fate of drugs once it enters the blood-circulation by controlling the distribution, excretion and toxicity in the blood [1]. Due to amphiphilic properties of drug molecules, living organism and drug molecules exhibit electrostatic interaction. So, it becomes quintessential to understand the interplay of drugs and proteins for improving the efficiency of the drugs. For this, repurposed drugs are being pressed into action. Favipiravir is one such drug.

Favipiravir is an antiviral drug used for the treatment of mild to moderate influenza including A(H1N1)pdm09, A(H7N9) avian virus etc. [2]. It reduces the virus load in the body by stopping the virus from multiplying. It inhibits viral RNA-dependent RNA polymerase (RdRp) activity. Error prone viral RdRp assimilate Favipiravir-RTP into the nascent viral RNA leading to chain termination and viral mutagenesis. Favipiravir has been short listed for trials against Influenza, Ebola and other pathogenic RNA viruses.

Uracil is one of the four nitrogenous bases in RNA, the other three being Adenine (A), Guanine (G) and Cytosine (C). It belongs to pyrimidine family and is one of the RNA nucleobase of the RNA virus. Hence the study of the interacting mechanism of Favipiravir-Uracil biomolecular complex could help in screening of the potential Active pharmaceutical ingredients which have the potential in treating the cold related diseases.

II. Materials and Methods

A. Computational Method

The optimized structure of Favipiravir, Uracil and Favipiravir + Uracil is performed in the gas phase by the Gaussian 09 W software at the level of B3LYP/6-311++G (d, p) basis set [3]. The complete theoretical vibrational assignments were carried out using VEDA software [4] and Gauss View 5.0 molecular visualization software [5]. A scaling factor of 0.9668 [3] is multiplied to the calculated theoretical vibrational frequencies to account the systematic error and thereby improving the correlation between the calculated and the experimental result. The NBO analysis is carried out using the NBO 5.0 software in Gaussian to understand the intermolecular and intramolecular charge transfer behaviour in the complex [6]. The AIM parameters are evaluated using Multiwfnsoftware [3].

III. Results and Discussions

A. Geometrical Parameters

The optimized molecular structures of Favipiravir, Uracil and their interaction are shown in Figure 1(a) (b) and (c) respectively. Table 1 shows some of the computed bond lengths of Favipiravir, Uracil and Favipiravir + Uracil. The N-H bond lengths of the interaction are found to be in agreement with the N-H bond lengths of Favipiravir and Uracil (1.01Å) except for N_{13}–H_{15}(1.02Å) and N_{23}–H_{24}(1.03Å) which are elongated by 0.02Å each as compared to that of Favipiravir and Uracil respectively. The C–C bond length of the interaction is in between 1.39 Å–1.51Å. The C=C bond length of the interaction is reduced by 0.01Å in comparison to Uracil. The C=O bond length is increased by 0.01Å as compared to that of C=O bond length in Uracil. The bond length of C–N lies between 1.33Å– 1.41Å in the complex biomolecules.

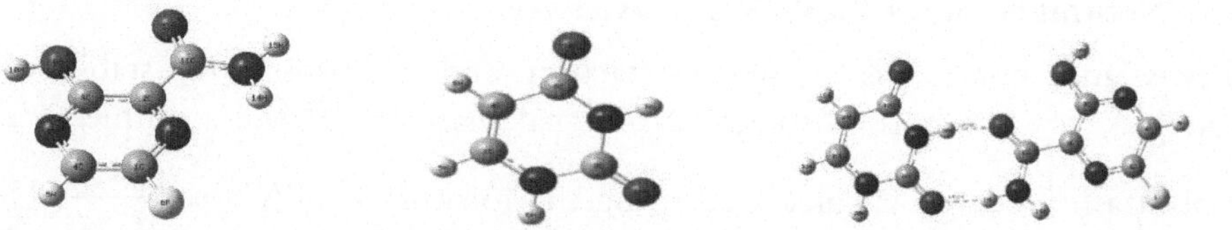

(a) (b) (c)

Figure 1: Optimized Molecular Structure of: a) Favipiravir b) Uracil c) Favipiravir + Uracil

Table 1: Some of the Selected Bond Lengths of Uracil, Favipiravir and Uracil + Favipiravir

Uracil			*Favipiravir*			*Uracil + Favipiravir*		
Sl. No.	*Bond*	*Bond Length (Å)*	*Sl. No.*	*Bond*	*Bond Length (Å)*	*Sl. No.*	*Bond*	*Bond Length (Å)*
1.	C_2–N_{12}	1.39	1.	C_{11}=O_{12}	1.21	1.	C_3=O_{10}	1.22
2.	C_2=O_{10}	1.21	2.	C_{11}–N_{13}	1.36	2.	C_3–N_{12}	1.39
3.	N_{12}–H_6	1.01	3.	N_{13}–H_{15}	1.01	3.	C_2=C_1	1.34
4.	C_1=C_4	1.35	4.	C_3=C_2	1.36	4.	N_{20}–H_{21}	1.02

B. Vibrational Spectral Analysis

It is being studied to understand the interacting mechanism of the biomolecular complex. The vibrational spectra are influenced by the presence of functional groups and geometrical structure. Uracil and Favipiravir contains 12 and 15 atoms so they have 3N-6 i.e. 30 and 39 vibrational modes respectively. The interacting complex have 75 vibrational modes (27 atoms). Some of the selected computational vibrational spectra is present in Table 2. The N–H, C=O, C–H stretching modes of the interaction all lies within their expected region of 3000–3500 cm^{-1}, 1550–1850 cm^{-1}, 2800–3000 cm^{-1} respectively.

Table 2: Some Selected Vibrational Assignments of Uracil + Favipiravir

Mode	*DFT Wave Number*	*Scaled Wave Number*	*Vibrational Assignments*
1	3551	3433	$\upsilon(N_{26}$–$H_{20})100$
2	3355	3244	$\upsilon(N_{13}$–$H_5)90$
3	3187	3081	$\upsilon(C_4$–$H_5)20$
4	1775	1716	$\upsilon(C_{17}$=$O_{18})71$

C. Natural Bond Orbital (NBO) Analysis

Lewis and non-Lewis orbital interaction and corresponding stabilization energies is achieved through NBO calculation of the complex. The stabilization energy is calculated using the formula $E^2 = \Delta E_{ij} = q_i\frac{(F_{ij})^2}{\varepsilon_j-\varepsilon_i}$. Here E^2, ε_j, ε_i and q_i denotes the stabilization energy, diagonal elements and the donor orbital occupancies respectively. The Fock matrix diagonal elements are represented by F_{ij}. The high value of E^2 denotes higher stability and a strong interaction between the donor and the acceptor. Overlapping of bonding and anti-bonding orbitals within the molecule lead to the intramolecular interaction which causes intramolecular charge transfer (ICT) and leads to stabilization. The hyper conjugation energy values of 27.49 and 41.04 kJ/mol are observed due to charge transfer between $n_1(O_{19})$ and $\sigma^*(N_{23}–H_{24})$, $n_2(O_{12})$ and $\sigma^*(N_{23}–H_{24})$ orbital. Table 3 shows the Second order perturbation theory analysis of Fock matrix in NBO basis for Favipiravir + Uracil.

Table 3: Second Order Perturbation Theory Analysis of Fock Matrix in NBO Basis for Favipiravir + Uracil

Donor (i)	*Acceptor (j)*	*E^2 (KJ/mol)*	*E(j) –E(i) (kcal/mol)*	*F_{ij} (a.u)*
$n_1(O_{12})$	$\sigma^*(N_{23}–H_{24})$	27.49	1.10	0.076
$n_2(O_{12})$	$\sigma^*(N_{23}–H_{24})$	41.04	0.68	0.075
$n_1(O_{25})$	$\sigma^*(N_{13}–H_{15})$	19.71	1.16	0.066
$n_2(O_{25})$	$\sigma^*(N_{13}–H_{15})$	26.32	0.73	0.062

D. Atom in Molecule (AIM) Analysis

The AIM parameters which are evaluated is shown in table 4. The table reveals that all the Laplacian values of electron density are positive indicating electrostatic interaction. The positive values of H(r) indicate weak hydrogen bonds. Thus, it can be summarised that all the interaction in Favipiravir + Uracil is electrostatic with weak Hydrogen bond. From the table the electron density (ρ_{BCP}) of the molecules (0.03 au) lies within the range of hydrogen bond (0.002–0.04 au).

Table 4: AIM Parameters of Favipiravir + Uracil

Hydrogen Bond	ρ_{BCP}	∇^2_{BCP}	*H(r)*	*G(r)*	*V(r)*
$O_{12}–H_{24}–N_{23}$	0.0304	0.1046	0.0012	0.0250	–0.0238
$H_{15}–O_{25} = C_{16}$	0.0255	0.0946	0.0023	0.0213	–0.0190

IV. Molecular Docking

The ligand-protein receptor interaction can be studied by molecular docking using Autodock tools [3] and visualized using Biovia Discovery Studio 2021. It predicts the probable binding site of the interaction. Here the molecular docking of Favipiravir + Uracil is carried out against the 6LU7 receptor. The receptor used is downloaded from the RCSB PDB website and the docking is performed by following the its protocols.

6LU7 is MPro in Covid-19 shares 96% similarity with the MPro of SARS-CoV. The binding energy of 6LU7 is –5.06 kcal/mol (Table 5) indicating that the ligand has high binding affinity for 6LU7. Figure 2 shows the 2D representation of the ligand-receptor interaction. This reflects the potentiality of Favipiravir + Uracil as potent anti cold diseases.

Table 5: Binding Energies and Bond Distance of Favipiravir + Uracil with 6LU7 Receptor

Ligand	*Receptor*	*Binding Energy (kcal/mol)*	*Bond Distance and Residue*
Favipiravir + Uracil	6LU7	–5.06	2.94Å (HIS 163), 2.95Å (PHE 140) 2.91Å, 2.87Å (LEU 141), 2.98Å, 2.06Å (SER 144), 5.47Å, 3.20Å (CYS 145), 4.35Å (MET 49), 3.38Å (MET 165), 3.06Å (GLU 166)

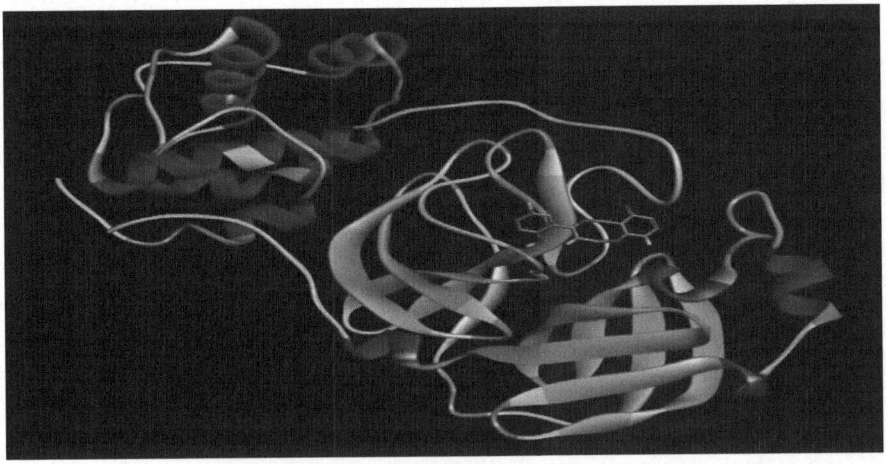

Figure 2: 2D Representation of the Ligand-Receptor Interaction

V. Conclusion

The charge transfer within the atoms is observed while studying NBO. The AIM analysis shows that the interaction is electrostatic with weak Hydrogen bond in nature. From autodocking, the binding energy of the complex is reported to be –5.06 kcal/mol. The above analysis confirms biologically active nature and medicinal fulfillment of the complex.

References

[1] Hage, D.S., Jackson, A., Sobansky, M.R., Schiel, J.E., Yoo, M.J. and Joseph, K.S., Characterization of drug-protein interactions in blood using high-performance affinity chromatography. *Journal of Separation Science*, 32(5–6), 835–853 (2009).

[2] Agrawal, U., Raju, R. and Udwadia, Z.F., Favipiravir: A new and emerging antiviral option in COVID-19. *Medical Journal Armed Forces India*, 76, 370–376 (2020).

[3] Chetry, N. and Thongam, G.D., Intermolecular interaction study of L-Threonine polar aprotic Solvent: Experimental and theoretical study. *Journal of Molecular Liquids*, 116689, 338 (2021). https://doi.org/10.1016/j.molliq.2021.116689.

[4] Sakia, J. and Thongam, G.D., Study of the molecular interaction between hormone and anti-cancer drug using DFT and vibrational spectroscopic method. *Journal of Molecular Structure*, 131889, 1250 (2022). https://doi.org/10.1016/j.molstruc.2021.131889

[5] Frish, A., Nielsen, A.B. and Holder A.J., Gauss View User Manual, Gaussian Inc., Pittsburg, PA, 2001.

[6] Borah, B. and Thongam, G.D., Molecular property analysis of the interacting state of L-Threonine and Metformin: An experimental and computational approach. *Journal of Molecular Structure*, 128819, 1221 (2020). https://doi.org/10.1016/j.molstruc.2020.128819.

SECTION-III

High Energy Physics

Study on Minimal Modification of Tri-bimaximal Neutrino Mixing Matrix and its Phenomenological Implications

Animesh Barman*, Ng. K. Francis and Hrishi Bora
Department of Physics, Tezpur University, Tezpur-784028, India
*E-mail: animesh@tezu.ernet.in

ABSTRACT: We study and analyze the potential variations of the most recent global fit to neutrino data with the simplest modification of the tri-bimaximal (TBM) neutrino. With minimal modification of the TBM mixing matrix, it makes possible to predict all the neutrino oscillation parameters in terms of our model parameters. The modified TBM can reproduce non-zero reactor mixing angle θ_{13} *which is consistent with the recent experimental results. In our study we also explore phenomenological implications such as Jarlskog rephasing invariant* J_{CP} *and Dirac Phase* δ_{CP}.

Keywords: Tribimaximal Mixing, Reactor Mixing Angle, Neutrino Oscillation, Jarlskog Rephasing Invariant J_{CP}, Dirac Phase δ_{CP}.

I. Introduction

Although the Standard Model (SM) of the elementary particle physics has been remarkably successful, there are still unanswered questions regarding the origin of flavor structure, as well as the masses and mixings among different generations of matter particles. The neutrino mass and mixing is one of the most important evidence of beyond Standard Model physics. Many experiments show that neutrinos have tiny masses and their mixing is sill mysterious. The discovery of the neutrino masses and the lepton mixing has stimulated the work of the flavor symmetries.

Recent experiments in neutrino oscillation have entered a new phase of precise determination of mixing angles and mass squared differences. Based on these findings, the tri-bimaximal mixing [1-3] paradigm has been proposed as a leading form of the leptonic mixing in the lepton sector. Many of the recent flavor models have aimed at the tri-bimaximal mixing as a leading form of the leptonic mixing. Experiments on neutrino oscillations confirmed that neutrinos are massive. Several neutrino oscillation experiments like SK, SNO, MINOS, T2K, DOUBLE CHOOZ, DAYABAY etc. have established the fact that neutrinos produced in a well-defined flavor eigenstate can be detected as a different flavor eigenstate while they propagate.

The mixing of fermions remains one of the most enigmatic phenomena in particle physics. The tribimaximal (TBM) mixing pattern [4–5] is one of the most extensively used lepton mixing patterns obtained utilizing discrete non-Abelian symmetries,

$$U_{TBM} = \begin{pmatrix} -\sqrt{\frac{2}{3}} & \sqrt{\frac{1}{3}} & 0 \\ \frac{1}{\sqrt{6}} & \frac{1}{\sqrt{3}} & -\frac{1}{\sqrt{2}} \\ \frac{1}{\sqrt{6}} & \frac{1}{\sqrt{3}} & \frac{1}{\sqrt{2}} \end{pmatrix}$$

But as TBM has been excluded due to a non-zero reactor mixing angle, one of the mosteffective ways to achieve realistic mixing is through either its extensions or through modifi-cations by introducing a simple perturbation matrix into TBM matrix [6] to make it compatible with the present experimental data. An analysis of the relationship in between CP violation and μ-τ symmetry breaking can be read in [7–8].

The broadly acknowledged Pontecorvo-Maki-Nakawaga-Sakata (PMNS) matrix encodes the mixing in between the neutrino flavour eigenstates and their mass eigenstates. In a three flavoured paradigm, three mixing angles and three CP phases are used to parameterize this PMNS matrix.

$$U_{PMNS} = \begin{pmatrix} c_{12}c_{13} & s_{12}c_{13} & s_{13}e^{-i\delta} \\ -s_{12}c_{23} - c_{12}s_{23}s_{13}e^{i\delta} & c_{12}c_{23} - s_{12}s_{23}s_{13}e^{i\delta} & s_{23}c_{13} \\ s_{12}s_{23} - c_{12}c_{23}s_{13}e^{i\delta} & -c_{12}s_{23} - s_{12}c_{23}s_{13}e^{i\delta} & c_{23}c_{13} \end{pmatrix} . U_{Maj}$$

Where, $c_{ij} = \cos\theta_{ij}$, $s_{ij} = \sin\theta_{ij}$. The diagonal matrix $U_{Maj} = \mathrm{diag}(1, e^{i\alpha}, e^{i(\alpha+\beta)})$ contains the Majorana CP phases, α, β which become observable in case the neutrinos behave as Majorana particles. To show that neutrinos are Majorana particles, it will most likely require neutrinoless double beta decay to be discovered.

II. Framework of the Model

For non-zero θ_{13}, we modify TBM neutrino mixing matrix, by multiplying TBM with a perturbation matrix U_x which is as follows:

$$U_x = \begin{pmatrix} 1 & 0 & 0 \\ 0 & c_x & s_x e^{i\delta} \\ 0 & -s_x e^{-i\delta} & c_x \end{pmatrix}$$

The modified TBM neutrino mixing matrix is given by $U'_{TBM} = U_{TBM}.U_x$, which takes the form as follows:

$$U'_{TBM} = \begin{pmatrix} \frac{\sqrt{6}}{3} & \frac{\sqrt{3}}{3}c_x & \frac{\sqrt{3}}{3}s_x e^{i\delta} \\ \frac{-\sqrt{6}}{6} & -\frac{\sqrt{2}}{2}s_x e^{-i\delta} + \frac{\sqrt{3}}{3}c_x & \frac{\sqrt{3}}{3}s_x e^{i\delta} + \frac{\sqrt{2}}{2}c_x \\ \frac{-\sqrt{6}}{6} & \frac{\sqrt{2}}{2}s_x e^{-i\delta} + \frac{\sqrt{3}}{3}c_x & \frac{\sqrt{3}}{3}s_x e^{i\delta} + \frac{\sqrt{2}}{2}s_x \end{pmatrix}$$

We carefully analyze for the case of the modification of tribimaximal mixing matrix for non-zero θ_{13} and all other neutrino oscillation parameters and an explicitillustrations on the results is presented by numerical calculations and figures. The strength of CP violation is fully determined by the Jarlskog invariant $J_{cp} = \frac{1}{8}\cos\theta_{13}\sin 2\theta_{12}\sin 2\theta_{13}\sin\delta$, which is expressed in the standard parametrization, though it does not depend on any concrete parametrization. The neutrino mass matrix M_ν can be diagonalized by the PMNS matrix U as,

$U^\dagger M_\nu U^* = \text{diag}\ (m_1, m_2, m_3)$. The neutrino oscillation parameters θ_{12}, θ_{13}, θ_{23} and δ_{CP} can be obtained from U as$s_{12}^2 = \frac{|u_{12}|^2}{1-|u_{13}|^2}$, $s_{23}^2 = \frac{|u_{23}|^2}{1-|u_{13}|^2}$, $s_{13}^2 = |u_{13}|^2$and δ may be given by $\delta = sin^{-1}\frac{8Im(h_{12}h_{23}h_{31})}{P}$, where $p = (m_2^2 - m_1^2)(m_3^2 - m_2^2)(m_3^2 - m_1^2)\sin 2\theta_{12}\sin 2\theta_{23}\sin 2\theta_{13}\cos\theta_{13}$

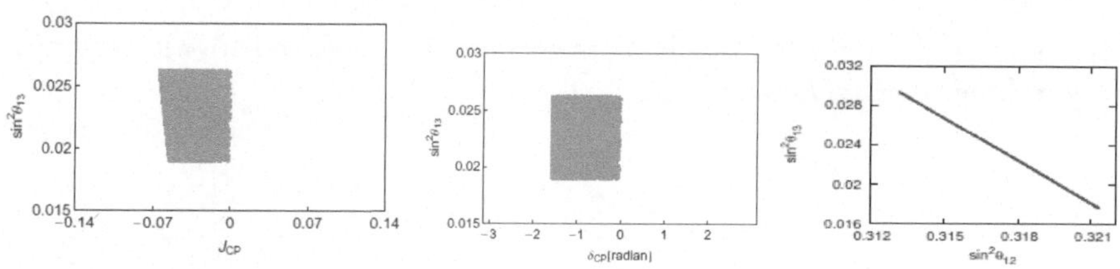

III. CONCLUSION

We modified TBM by introducing a simple perturbation matrix into TBM matrix and using the advantage of the experimental results of the mixing angles, we then can obtain non-zero θ_{13} along with other neutrino oscillation parameters which is in agreement with the present experimental results. The Dirac phase δ_{13} and the Jarlskog rephasing invariant J_{cp} are also discussed.

ACKNOWLEDGEMENT

Animesh Barman is grateful to CSIR, New Delhi, India for CSIR Senior Research Fellowship (file no 09/796(0072)/2017-EMR-1). Hrishi Bora is grateful and acknowledges the financial support provided by the Tezpur University, India for Institutional(Non-NET) Research Fellowship (F.14-5/3/2003(Acad)/1755-A. The research of Ng. K. Francis is funded by DST-SERB, India under Grant no. EMR/2015/001683.

REFERENCES

[1] Lin, Y., Tri-bimaximal Neutrino Mixing from A4 and $\theta_{13}\sim\theta_{C}$., *Nuclear physics* B, 824.1–2, 95–110 (2010).

[2] Kang, S.K. and Tanimoto, M., Prediction of leptonic CP phase in A_4 symmetric model., *Physical Review D*, 91.7, 073010 (2015).

[3] Chao, W. and Zheng, Y., Relatively large theta13 from modification to the tri-bimaximal, bimaximal and democratic neutrino mixing matrices., *Journal of High Energy Physics,* 2013.2, 1–21 (2013).

[4] He, X. and Zee, A., Minimal modification to tribimaximal mixing, *Physical Review* D, 84.5, 053004 (2011).

[5] Low, C.I. and Volkas, R.R., Tribimaximal mixing, discrete family symmetries, and a conjecture connecting the quark and lepton mixing matrices, *Physical Review D*, 68.3 033007 (2003).

[6] Damanik, A., Nonzero θ13 and neutrino masses from modified neutrino mixing matrix, *International Journal of Modern Physics A,* 27.17, 1250091 (2012).

[7] Xing, Z.Z., Zhang, H. and Zhou, S., Nearly tri-bimaximal neutrino mixing and CP violation from μ–τ symmetry breaking, *Physics Letters B* 641.2, 189–197 (2006).

[8] Luo, S. and Xing, Z.Z., Resolving the octant of θ 23 via radiative μ– τ symmetry breaking, *Physical Review D*, 90.7, 073005 (2014).

Realization of Minimal Left-Right Symmetric Model with A_4 Modular Symmetry

A. Kakoti*, B.B. Boruah and M.K. Das
Department of Physics, Tezpur University, Tezpur-784028, India
*E-mail: ankitak@tezu.ernet.in

ABSTRACT: The present work focusses on the construction of Left-Right Symmetric Model (LRSM) with modular symmetry. The modular group used here is $\Gamma(3)$ *which is isomorphic to* A_4 *discrete symmetry group. The Yukawa couplings here are represented in terms of modular forms* (Y_1, Y_2, Y_3) *and their values have been calculated and are plotted against the sum of the neutrino masses and the results are found to be well within the experimental bounds. The advantage of using modular symmetry over other symmetries is the non-requirement for the use of flavons. Many new physics contributions to neutrinolessdouble beta decay* ($0\nu\beta\beta$) *within LRSM can be analysed within the model and phenomenology like lepton flavor violation (LFV) can also be studied within the model. However, the present work only portrays the realization of Left-Right Symmetric Model within the framework of* A_4 *modular symmetry and hence study its implications on the model.*

Keywords: Left-Right Symmetric Model, Modular Symmetry, Modular Forms, Neutrinolessdouble Beta Decay.

I. Introduction

Left-Right Symmetric Model (LRSM) is a simple extension of the Standard Model (SM) of particle physics, which is considered to be one of the well-established theory to date. But, we need to consider its extensions because of the discrepancies within the model like its inability to explain neutrino masses, neutrino mass hierarchy, existence of baryon asymmetry of the universe, existence of dark matter etc. Within these extensions of SM it becomes possible to explain neutrino masses, phenomenology like $0\nu\beta\beta$, baryogenesis via leptogenesis, dark matter etc.

The gauge group of LRSM is $SU(3)_C \otimes SU(2)_L \otimes SU(2)_R \otimes U(1)_{B-L}$[1-3][5], where it is evident that unlike SM, the gauge group consists of a right-handed component for the group *SU(2)* and also the hypercharge corresponding to *U(1)* gauge group has been replaced by baryon number minus lepton number. In the model, both the right handed and left handed particles are doublets under the *SU(2)* gauge group. The particle content and their corresponding charge assignments under the gauge group of the model has been dictated in table (2).

The scalar sector of the model consists of a Higgs bidoublet $\phi(1,2,2,0)$ and two scalar triplets given by, $\Delta_R(1,1,3,2)$ and $\Delta_L(1,3,1,2)$[5] under the above mentioned gauge group, which provides masses to the gauge bosons and fermions within the model. In LRSM, the type-I and type-II seesaw masses appear naturally. So, the light neutrino mass is given as the sum of the type-I and type-II seesaw masses. The Yukawa Lagrangian in the lepton sector is given as,

$$\mathcal{L} = h_{ij}\overline{l_L}\phi l_R + \tilde{h}_{ij}\overline{l_L}\tilde{\phi} l_R + f_{L,i}\psi_L^T C i\sigma_2 \Delta_L \psi_{L,j} + f_{R,i}\psi_R^T C i\sigma_2 \Delta_R \psi_{R,j} \quad \ldots (1)$$

The light neutrino mass is then given as,

$$M_\nu = M_\nu^1 + M_\nu^2 \quad \ldots (2)$$

The first term in the above equation refers to type-I seesaw mass given by,

$$M_\nu^1 = M_D M_R^{-1} M_D^T \quad \ldots (3)$$

The second term in the equation gives the type-II seesaw term given by,

$$M_\nu^2 = \sqrt{2} v_L f_L \quad \ldots (4)$$

A. Modular Symmetry

Modular symmetry has gained much importance in model building because of the fact that it does not require the use of any extra particles (flavons) for the realization. In using modular symmetry, one has to express the Yukawa couplings in terms of modular forms, and the number of modular forms associated in the model is dependent on the level of the modular group taken ($\Gamma(N)$). Using modular symmetry, each of the particles incorporated within the model must be assigned a particular modular weight. An element q of the modular group acts on a complex variable τ which belongs to the upper-half of the complex plane given as [4],

$$q\tau = \frac{a\tau + b}{c\tau + d}$$

where, a, b, c, d are integers and ad-bc = 1.

Table 1: No. of Modular Forms Corresponding to Modular Weight 2 k

N	*No. of Modular Forms*	$\Gamma(N)$
2	k+1	S_3
3	2k+1	A_4
4	4k+1	S_4
5	10k+1	A_5
6	12k	
7	28k – 2	

B. LRSM with Modular Symmetry

The Lagrangian of the model with the incorporation of modular symmetry is given as,

$$\mathcal{L} = \bar{l}_L \phi l_R Y + \bar{l}_L \tilde{\phi} l_R Y + l_{L,i}^T C i\sigma_2 \Delta_L l_{L,j} Y + l_{R,i}^T C i\sigma_2 \Delta_R l_{R,j} Y \quad \dots (5)$$

The model contains the actual particle content of LRSM as shown below.

Table 2: Particle Content of the Model

Gauge group	Q_L	Q_R	l_L	l_R	ϕ	Δ_L	Δ_R
$SU(3)_C$	3	3	1	1	1	1	1
$SU(2)_R$	1	2	1	2	2	1	3
$SU(2)_L$	2	1	2	1	2	3	1
$U(1)_{B-L}$	–1/3	–1/3	–1	–1	0	2	2
A_4	3	3	3	3	1	1	1
k_I	0	–2	0	–2	0	–2	2

The modular form Y used in the Lagrangian is assigned a modular weight of 2 under A_4 discrete symmetry group.

The Yukawa couplings expressed as modular forms (Y_1, Y_2, Y_3) are given as,

$$Y_1 = 1 + 12q + 36q^2 + 12q^3 + 84q^4 + 72q^5 + 36q^6 + 96q^7 + 180q^8 + 12q^9 + 216q^{10}$$

$$Y_2 = -6q^{1/3}(1 + 7q + 8q^2 + 18q^3 + 14q^4 + 31q^5 + 20q^6 + 36q^7 + 31q^8 + 56q^9)$$

$$Y_3 = -18q^{2/3}(1 + 2q + 5q^2 + 4q^3 + 8q^4 + 6q^5 + 14q^6 + 8q^7 + 14q^8 + 10q^9)$$

The Dirac and Majorana mass matrices are determined from the Lagrangian by using the multiplication rules for A_4 group, given as,

$$M_D = v\begin{pmatrix} 2Y_1 & -Y_3 & -Y_2 \\ -Y_2 & -Y_1 & 2Y_3 \\ -Y_3 & 2Y_2 & -Y_1 \end{pmatrix}$$

where, v is the VEV of the Higgs bidoublet.

$$M_R = v_R\begin{pmatrix} 2Y_1 & -Y_3 & -Y_2 \\ -Y_3 & 2Y_2 & -Y_1 \\ -Y_2 & -Y_1 & 2Y_3 \end{pmatrix}$$

where, v_R is the VEV of the right-handed scalar triplet Δ_R.

The neutrino masses are then calculated using the mass terms for type-I and type-II seesaw masses.

II. Numerical Analysis and Results

We have determined the values of the modular forms (Y_1, Y_2, Y_3) and then plotted it against the sum of the neutrino masses for both normal and inverted hierarchies and the results obtained were found to be well within the experimental bounds.

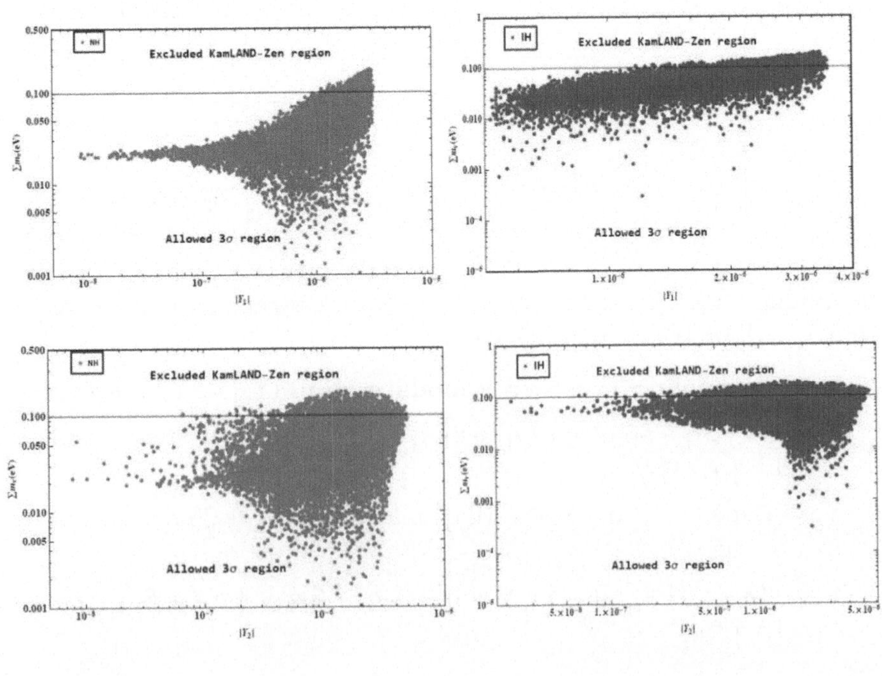

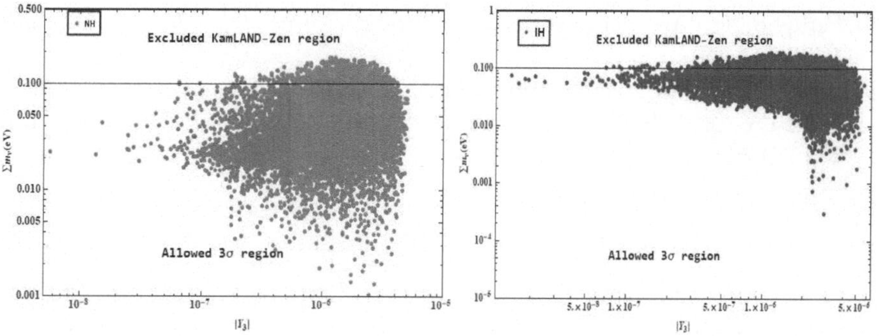

Using modular symmetry, the neutrino mass matrices and hence the sum of the neutrino masses can be expressed in terms of the modular forms. So, the above plots are for Yukawa couplings against the sum of the neutrino masses and the results are found to be respecting the experimental bounds.

III. Conclusion

The discovery of neutrino oscillations paved the gateway for physics beyond the Standard Model. In our work, we have realized LRSM with the help of modular A_4 symmetry. Using modular symmetry provides the advantage of using no extra particles called 'flavons'. The Yukawa couplings are represented as modular forms expressed as expansions of q. The values of the Yukawa couplings (Y_1, Y_2, Y_3) are calculated using 'Mathematica'. Some plots concerning the work have been shown above. Some of our results have been summarized as under:

(a) The Dirac and Majorana mass matrices are expressed in terms of the modular forms (Y_1, Y_2, Y_3), and the Majorana mass matrix is also found to be symmetric under the realization.

(b) The absolute value of the modulus was found to be within the range 1.073 to 1.197, which is greater than unity, that is the desired result.

(c) The value of the Yukawa coupling ranges from 10^{-9} to 10^{-6}.

(d) The sum of the neutrino masses are found to be within the range of 10^{-4} to 10^{-1}.

References

[1] Pati, J.C. and Salam, A., Lepton Number as fourth color, *Phys. Rev. D*, 10275–289 (1974).

[2] Mohapatra, R.N. and Pati, J.C., A Natural Left-Right Symmetry, *Phys. Rev. D*, 11 2558 (1975).

[3] Senjanovic, G., Spontaneous Breakdown of Parity in a class of gauge theories, other thesis, 1(1979).

[4] Feruglio, F., Are neutrino masses modular forms? 227–266 (2019) [arXiv: 1706.08749]

[5] Bhupal Dev, P.S., Mohapatra, R.N., Rodejohann, W. and Xu, X.J., Vacuum Structure of Left-Right Symmetric Model, *JHEP*, 02154 (2019) [arXiV: 1811.06869]

Quantifying Systematic Uncertainties on Neutrino Oscillation Parameters in Resonance Process

Paramita Deka*, **Nilavjyoti Hazarika**[1] **and Kalpana Bora**[1]
Department of Physics, Gauhati University, Assam-781014, India
*E-mail: deka.paramita@gmail.com

ABSTRACT: The long baseline (LBL) neutrino experiments use heavy nuclear targets for neutrino scattering in which nuclear effects give rise to complications in measuring the neutrino oscillation parameters up to high precision. These nuclear effects are not yet fully understood and therefore need to be quantified as they contribute to the systematic uncertainties. In this work, we investigate the effects of the resonance (RES) interaction process using two models for carbon target for $\nu_\mu \rightarrow \nu_\mu$ *disappearance channel of the NOνA experiment on neutrino oscillation parameter sensitivity. To quantify the systematic uncertainties due to RES interactions, we compare the sensitivity analysis for two models of RES processes - Rein-Sehgal and Berger-Sehgal. We observe that RES processes contribute significantly, and should be included carefully in models while extracting neutrino oscillation parameters.*

Keywords: Neutrino Oscillation, Resonance Interaction, Nuclear Effects.

I. INTRODUCTION

The use of heavy nuclear targets in long baseline (LBL) neutrino experiments for neutrino scattering introduces complications in accurately measuring the neutrino oscillation parameters due to nuclear effects, which are not yet fully comprehended. These effects contribute to systematic uncertainties and alter the true/reconstructed neutrino energy relationship and final-state particle kinematics. This study examines these effects in the resonance (RES) interaction process by utilizing two models for carbon targets in the NOνA experiment's $\nu_\mu \rightarrow \nu_\mu$ disappearance channel [1]. The NOνA (NuMI Off-Axis ν_e Appearance) [2] experiment is a long baseline neutrino oscillation experiment designed to measure the $\nu_e(\overline{\nu}_e)$ and $\nu_\mu(\overline{\nu}_\mu)$ probabilities, utilizing the NuMI beam at the Fermi laboratory. It consists of two Near Detector (ND) and Far Detector (FD) units, functionally identical to each other. The ND unit, located 105 meters underground and 1 km from the NuMI beam, has a mass of roughly 0.3 kilotons, while the FD unit is placed near Ash River, Minnesota, 810 km from the NuMI target. Both detectors are situated 14.6 mrad off-axis from the center of the NuMI beam to provide a narrow beam neutrino flux peaked at approximately ~2 GeV and to enhance sensitivity to the disappearance and appearance processes.

II. Physics and Simulation Details

Neutrino energy E_ν^{cal} in calorimetric method [3] can be reconstructed as,

$$E_\nu^{cal} = E_l + \sum_i T_i^n + \epsilon_n + \sum_m E_m$$

where E_l is the outgoing lepton energy, T_i^n is the kinetic energy of outgoing nucleons, ϵ_n denotes the single-nucleon separation energy and E_m is the total energy of any other produced particles. In this work, only disappearance $\nu_\mu \rightarrow \nu_\mu$ oscillation channel is accounted for analysis. This channel is believed to be mainly suited for θ_{23} and Δm_{32}^2 measurements. The relevant oscillation probability for NOνA for disappearance channel can be expressed as,

$$P(\nu_\mu \rightarrow \nu_\mu) = 1 - sin^2 2\theta_{23} sin^2 \Delta_{32} + 4 sin^2\theta_{23} sin^2\theta_{13} cos^2 2\theta_{23} sin^2 \Delta_{32}$$

where $\Delta_{ij} = \frac{\Delta m_{ij}^2 L}{4E_\nu}$, Δm_{ij}^2 are the mass differences and θ_{ij} are the mixing angles and E_ν is the incoming neutrino energy.

We use GENIE-3_00_06 [4] to generate the neutrino (antineutrino)-nucleus interaction cross sections for RES interaction processes with Carbon as target. The pion production models are based on the phenomenological approach of Rein and Sehgal (RS) [5] and Berger and Sehgal (BS) [6] model which is an extension of the RS model that includes the effects of lepton mass. We have generated 1 million events for neutrino antineutrino at ND and FD in the energy range 0-5 GeV using NOνA $\nu_\mu(\overline{\nu}_\mu)$ flux. For demonstration and comparison, we analysed 100% detector efficiency. 3.5% muon energy resolution and 25% hadron energy resolution respectively is used. We used Feldman and Cousin's [7] technique for the χ^2 analysis.

III. Results and Discussions

In Figure 1, we have shown the comparison of the contribution of resonance pion production to the cross-section of carbon target for BS and RS model. We have also shown the contribution of the QE process to the cross-section, to show the relative importance of the two interaction processes. In Figures 2-3, we represent the comparison of 1σ, 2σ and 3σ sensitivity contours in the θ_{23}-Δm_{32}^2 plane for different cases: no detector-BS (solid blue line), no detector RS (dashed blue line), efficiency = 100%-BS (solid red line), and

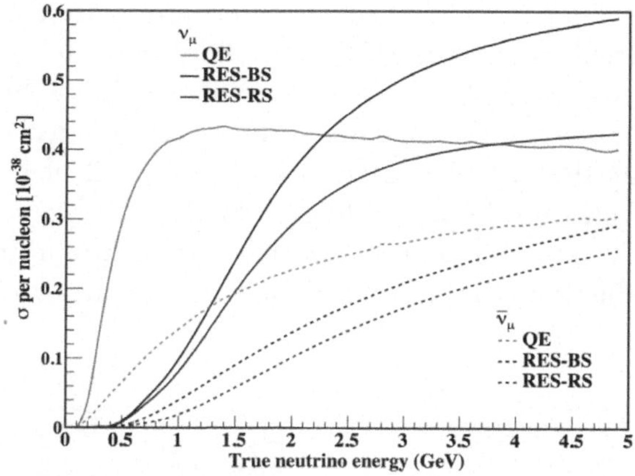

Figure 1: Total per nucleon cross-sections as a function of true neutrino energy for QE and RES processes for both neutrino (solid line) and antineutrino (dashed line) for two models of RES processes: Berger-Sehgal and Rein-Sehgal.

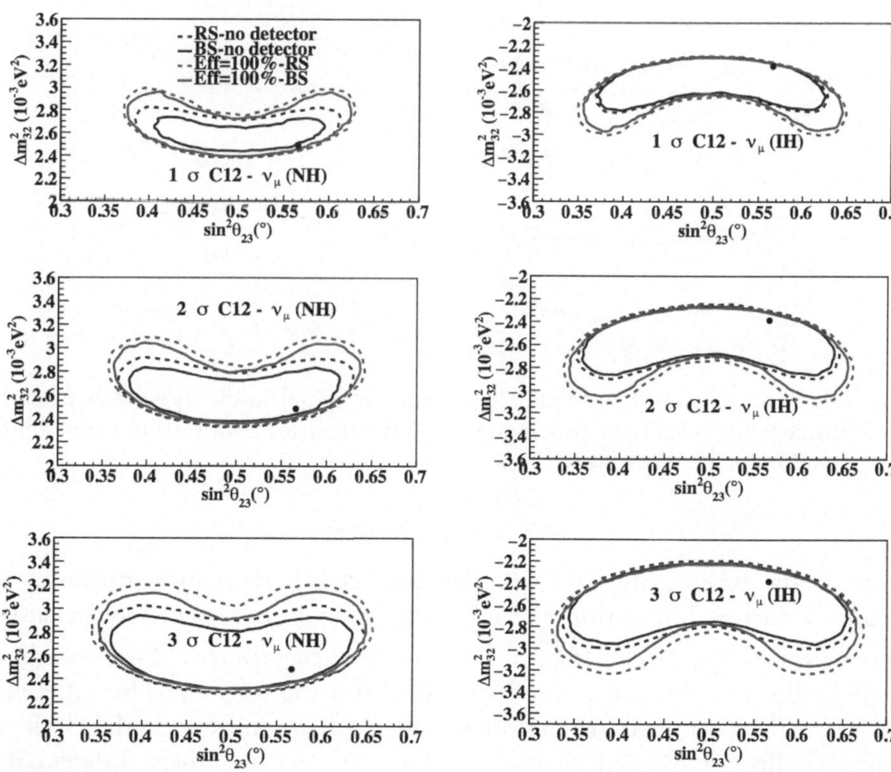

Figure 2: Comparison of 1σ, 2σ and 3σ confidence regions in the θ_{23}-Δm^2_{32} plane for RES processes for NH (right panel) and IH (left panel) for antineutrino with/without detector effect for 100% efficiency.

RS (dashed red line) for both neutrino and antineutrino for both normal (NH) and inverted hierarchies (IH). We show how the detector response changes when we overestimate the detector efficiency by 100%. These contour areas represent the analysis where nuclear effects are taken into account by using the carbon migration matrices to calculate the fitted rates. Comparing the area inside the confidence levels, we observe that out of the RS and BS models, the BS model shows better precision in the measurement of neutrino oscillation parameters as it has a lesser area inside the contour than the RS.

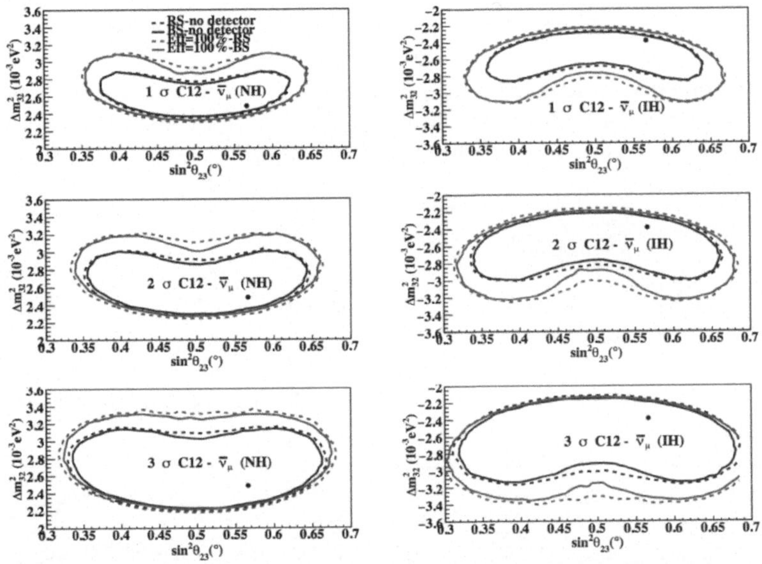

Figure 3: Comparison of 1σ, 2σ and 3σ confidence regions in the θ_{23}-Δm^2_{32} plane for RES processes for NH (right panel) and IH (left panel) for antineutrino with/without detector effect for 100% efficiency.

IV. Conclusion

This study extensively analyses the impact of resonance processes in neutrino-nucleus interaction on the sensitivity analysis of neutrino oscillation parameters - θ_{23} and Δm^2_{32} in the NOνA experiment. Two resonance models, BS and RS, are utilised to investigate this impact. The difference between these models is non-negligible, which indicates the lack of understanding of theoretical models. However, ignoring these differences is not possible as they affect the determination of neutrino oscillation parameters. The results show that RES processes have a significant contribution to cross-section and sensitivity analysis. The BS model provides higher precision than the RS model in extracting neutrino oscillation parameters. Additionally, sensitivity contours with 100% detector efficiency are found to be closer to no detector analysis.

REFERENCES

[1] Deka, P., Hazarika, N. and Bora, K., Uncertainties in neutrino oscillation parameter sensitivity due to resonance processes at NOvA, *Int. J. Mod. Phys.*, A 37, no.31n32, 2250200 (2022).

[2] Adamson, P. *et al.* [NOvA], First measurement of muon-neutrino disappearance in NOvA, *Phys. Rev.,* D 93, No. 5, 051104 (2016).

[3] Ankowski, A.M., Coloma, P., Huber, P., Mariani C., and Vagnoni, E., Missing energy and the measurement of the CP-violating phase in neutrino oscillations, *Phys. Rev., D* 92, No. 9, 091301 (2015).

[4] Andreopoulos, C., *et al.*, The GENIE Neutrino Monte Carlo Generator, *Nucl. Instrum. Meth.,* A 614, 87–104 (2010).

[5] Rein, D. and Sehgal, L.M., Neutrino-excitation of baryon resonances and single pion production, *Annals Phys.,* 133, 79–153 (1981).

[6] Berger., C. and Sehgal., Lepton mass effects in single pion production by neutrinos, *L.M., Phys. Rev.,* D 76, 113004 (2007).

[7] Feldman, G.J. and Cousins, R.D., A Unified Approach to the Classical Statistical Analysis of Small Signals, *Phys. Rev.,* D 57, 3873–3889 (1998).

Extraction of Gluon Saturation Momentum Using an Analytical Solution of the Balitsky-Kovchegov Equation

R. Saikia[1, *], P. Phukan[2] and J.K. Sarma[1]

[1]Department of Physics, Tezpur University, Tezpur-784028, India

[2]Department of Physics, Moran College, Moranhat-785670, India

*E-mail: ranjans@tezu.ernet.in

ABSTRACT: Gluon saturation is a phenomenon where at very small-x (Bjorken x), high-density gluons inside hadrons start overlapping and get saturated. This phenomenon is required to understand the implicit physics of high-density quantum chromodynamics (QCD). It is imperative to know the scale at which gluons get saturated. In this work, we extract the saturation momentum $Q_s^2(Y)$ at which gluons get saturated using an analytical solution of the Balitsky-Kovchegov (BK) equation. Saturation momentum is a function of rapidity, Y. Extracted saturation momentum plotted against various rapidities. We found that the saturation momentum is an increasing function of rapidity, as predicted by the numerical solution of the BK equation.

Keywords: Gluon Saturation, Small-x, BK Equation.

I. Introduction

Since the discovery of QCD, one of the greatest hurdles has been to comprehend the hadronic wave functions at high energy, or equivalently, at small-x.The hadronic wave function at high energy comprises a large number of gluons with a small fraction of longitudinal momentum x. AtHERA (Hadron Electron Ring Accelerator) accelerator facilities and via various parton distribution functions (PDFs) parametrization groups, the rapid expansion of gluons within hadrons at small-x has been proven. In addition, the rapid increase of small-x gluons within hadrons cannot continue indefinitely since it violates the unitarity of the theory. In QCD, the total hadronic cross sections are constrained or capped so that they cannot surpass the upper limit provided by Froissart and Martin [1]. How, therefore, shall we overcome this infraction? Then, it was observed that gluons within hadrons not only emit gluons but also recombine [2-3]. At small-x, gluon recombination becomes significant because it decreases the number of gluons in the hadronic wave function. At some times, gluons within hadrons begin to overlap and become saturated. This phenomenon is known as gluon saturation, and it is necessary to comprehend the underlying physics of high density QCD. The momentum scale at which the nonlinear effect of gluons becomes important or at which gluons start to saturate is known as saturation momentum.

To understand gluon saturation at small-x, the BK equation [4-6] is a widely studied nonlinear QCD evolution equation. We have recently proposed an analytical solution to the BK equation using a method called as the homotopy perturbation method (HPM) [7]. We get the saturation momentum from the analytical solution of the BK equation in this work. In the following section, we will explore the BK equation and the dipole description of DIS (Deep Inelastic Scattering). Finally, the saturation momentum is extracted via the analytical solution of the BK equation in section 3. We plot the extracted saturation momentum at different rapidities and compare it to the saturation momentum obtained in the numerical analysis of the BK equation. Lastly, we summarise our results in the conclusion section.

II. Dipole Description of DIS and BK Equation

The high gluon density picture of QCD can be explained with the color dipole description of DIS (Deep Inelastic Scattering). In the dipole description of DIS [8], a DIS process is considered in the rest frame of a target proton or a nucleus. In this frame, any DIS process can be factorized into several steps where a virtual photon first changes into a quark-antiquark dipole, which then collides with the target proton or nuclear target. After colliding with the target, this dipole scatters off the target and recombines to form the final-state particles. We can argue that the quark-antiquark dipole rescatters many times on the target, so that the quantum corrections to this picture of rescattering are included. In this picture, the initial quark-antiquark dipole develops a cascade of gluons before colliding with the target proton. In the large-N_c (Number of color) limit, this dipole cascade resummation leads to the famous Balitsky-Kovchegov (BK) equation, which is a non-linearization of the BFKL (Balitsky-Fadin-Kuraev-Lipatov) equation. The BK equation is an evolution equation for the imaginary part of the forward dipole-target scattering amplitude given by,

$$\frac{\partial}{\partial Y}N(x_{10},Y) = \frac{\alpha_s N_c}{2\pi^2}\int d^2x_2 \frac{x_{10}^2}{x_{20}^2 x_{21}^2} \times [N(x_{12},Y) + N(x_{20},Y) - N(x_{10},Y) - N(x_{12},Y)\,N(x_{20},Y)] \quad \dots (1)$$

The simple meaning of the above BK equation can be drawn as follows. At fixed rapidity Y, the initial colorless dipole with transverse size x_{10} transforms into two dipoles with transverse sizes x_{12} and x_{20}. Among these dipoles, either one can take part in the evolution and interact with the target while the other dipole stays a spectator. The nonlinear term of the equation would become important and lead to gluon saturation when rapidity increases. The minus sign in front of the nonlinear term denotes that the dipole amplitude N would begin to decrease rapidly when rapidity Y increases, taming the rapid growth of gluons. The original BK equation, which was proposed in coordinate space as an integro-differential equation, can be transformed into momentum space to yield a partial differential equation. To obtain the BK equation in momentum space, the following Fourier transformation is performed,

$$N(x,Y) = x^2 \int \frac{d^2k}{2\pi} e^{i\vec{k}.\vec{x}} N(k,Y) \qquad \dots (2)$$

so that we can write [5],

$$\partial_Y N(k,Y) = \overline{\alpha}\chi(-\partial_L)N - \overline{\alpha}N^2 \qquad \dots (3)$$

where $\overline{\alpha} = {}^{\alpha_s N_c}/_{\pi}$ and χ is the BFKL kernel. The equation (3) is useful for obtaining approximate solutions to the BK equation. Because of the complex nature of the BK equation, we do not see any exact solution of the equation; rather the equation is studied numerically many times. With some change of variables [9], equation (3) can be transformed to the FKPP (Fisher-Kolmogorov-Petrovsky-Piscounov) equation, a partial differential equation.

III. Extraction of Saturation Momentum from BK Solution

We recently proposed an approximate analytical solution to the BK equation (3) in momentum space using a method called the homotopy perturbation methodgiven by [7],

$$N(k,Y) = \frac{e^{Y-k^2/Q_{s0}^2}}{1-e^{-\frac{k^2}{Q_{s0}^2}} + e^{Y-k^2/Q_{s0}^2}} \qquad \dots (4)$$

Q_{s0}^2 is the initial saturation scale at which gluons start to saturate, which comes from the initial condition that we have used to solve the BK equation and evaluated by fitting the HERA data. Now, we shall use the solution (4) in order to extract the saturation momentum $Q_s^2(Y)$. The saturation momentum $Q_s^2(Y)$ can be defined by requiring the condition [10].

$$N(k = Q_s(Y), Y) = \text{constant [of order 1]} \qquad \dots (5)$$

we use $N(k, Y) = 1/2$ for $k = Q_s$ and extract the saturation momentum using the above condition. The extracted saturation momentum $Q_s^2(Y)$ using the BK solution is expressed in the following table:

Table 1: The Saturation Momentum Q_s^2 Extracted from Eq. (5) at Various Rapidities *Y*

Y	*0*	*0.5*	*1*	*1.5*	*2*
$Q_s^2(GeV^2)$	0.1664	0.3658	0.4679	0.6644	0.7959

In the numerical analysis of the BK equation [11], the saturation momentum Q_s^2 is estimated as a function of rapidity *Y* as: $Q_s^2(Y) = \Lambda^2 \exp(\Delta' \sqrt{Y + X})$, where $X = (\Delta')^{-2} \ln(Q_0^2/\Lambda^2)$ with $\Delta' = 3.2$. We plot the saturation momentum obtained in this work and compare it with the one estimated in the numerical analysis in Figure 1.

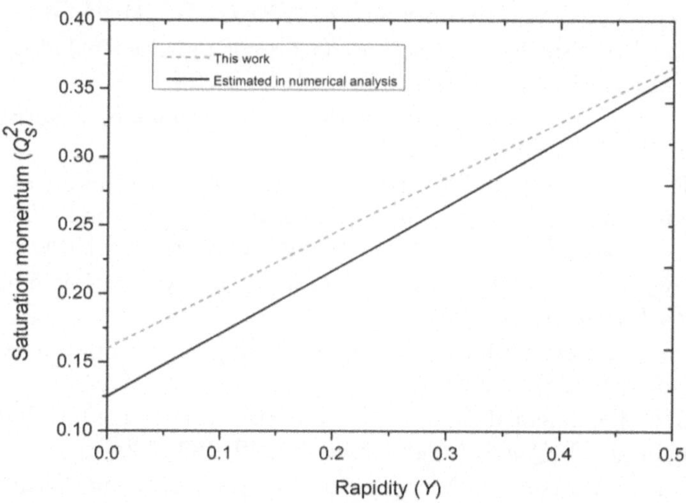

Figure 1: The Estimation of Saturation Momentum Q_s^2 at Various Rapidities *Y*

IV. Conclusion

In this work, we have extracted the saturation gluon momentum Q_s^2 at various rapidities *Y* using an analytical solution of the BK equation. The extracted saturation momentum is plotted and compared with the saturation momentum estimated by the numerical solution of the BK equation. We found a good agreement between our result and the numerical analysis of the

BK equation. The saturation momentum is an increasing function of rapidity, which is seen in the result plotted in Figure 1. Our solution of the BK equation can be helpful for further studies at high-density QCD. The results obtained in this work can be tested at future accelerator facilities such as the EIC (Electron Ion Collider). We must wait till future accelerator facilities are operational in order to accurately observe and confirm phenomena at small-x. The EIC will address phenomena such as gluon saturation at small-x with unprecedented precision in the near future.

REFERENCES

[1] Froissart, M., Asymptotic behavior and subtractions in the Mandelstam representation, *Physical Review*, 123(3), 1053 (1961).

[2] Mueller, A.H. and Qiu, J., Gluon recombination and shadowing at small values of x, *Nuclear Physics B*, 268(2), 427–452 (1986).

[3] Mueller, A.H., Small-x behavior and parton saturation: A QCD model, *Nuclear Physics B,* 335(1), 115–137 (1990).

[4] Balitsky, I., Operator expansion for diffractive high-energy scattering, In AIP Conference Proceedings, *American Institute of Physics*, 407(1), 953–957 (1997).

[5] Kovchegov, Y.V., Small-x F_2 structure function of a nucleus including multiple Pomeron exchanges, *Physical Review D*, 60(3), 034008 (1999).

[6] Balitsky, I., Effective field theory for the small-x evolution, *Physics Letters B*, 518(3-4), 235-242 (2001).

[7] Saikia, R., Phukan, P. and Sarma, J.K., An analytical solution of Balitsky-Kovchegov equation using homotopy perturbation method, *International Journal of Modern Physics A* 37, 2250190 (2022) [arXiv:2204.10111].

[8] Kovchegov, Y.V. and Levin, E., Quantum chromodynamics at high energy, *Cambridge University Press* (2012).

[9] Munier, S. and Peschanski, R., Geometric scaling as traveling waves, *Physical Review Letters,* 91(23) 232001 (2003).

[10] Iancu, E. and Venugopalan, R., The color glass condensate and high energy scattering in QCD, *Quark–Gluon Plasma,* 3, 249–363 (2004).

[11] Albacete, J.L., Armesto, N., Milhano, J.G., Salgado, C.A. and Wiedemann, U. A., Numerical analysis of the Balitsky-Kovchegov equation with running coupling: Dependence of the saturation scale on nuclear size and rapidity, *Physical Review D*, 71(1), 014003 (2005).

Δ54 Model for Dirac Neutrinos: Inverse Seesaw

Hrishi Bora*, Ng. K. Francis and Animesh Barman

Department of Physics, Tezpur University, Napaam-784028, India

*E-mail: hrishibora1997@gmail.com

ABSTRACT: Our analysis involves enhancing the Δ54 discrete flavor symmetry model with inverse seesaw mechanism through the incorporation of distinct flavons with two Higgs singlet. Additionally, we introduce supplementary $Z_2 \otimes Z_3 \otimes Z_4$ *symmetries to eliminate any undesirable components within our investigation. The inclusion of additional flavons leads to the deviation of exact tri-bimaximal neutrino mixing pattern by producing a non-zero* θ_{13} *which is consistent with the recent experimental results. It was found that for our model the data occupies the lower octant for normal hierarchy case.*

Keywords: Neutrinos, Flavor Symmetry, Tri-bimaximal, Hierarchy, Inverse Seesaw.

I Introduction

While particle physics experiments and observations have validated the standard model, numerous questions still remain unanswered. Additionally, it remains unclear whether the neutrino masses are of the normal hierarchy (NH) or the inverted hierarchy (IH) and whether the neutrinos are of the Dirac or Majorana type. Neutrino oscillation experiments are limited in their ability to detect only mass squared differences (Δm^2) and leptonic mixing angles (θ_{ij}). This study introduces a new approach, the inverse seesaw, using the Δ54 flavour symmetry framework. A previous proposal by the authors of [1] suggested Type I and inverse seesaw for Dirac neutrinos with A4 flavour symmetry. Lately, endeavors have been made in this particular area, particularly concerning the type I see-saw, type II seesaw and type I+II for Dirac neutrinos within the Δ54 symmetry [2]. The current work presents a more concise version of the inverse seesaw, utilizing only two Higgs singlets. The research also includes additional discrete symmetries, which help to clarify the hierarchy of the various terms in the neutrino mass matrix. Several recent reviews on neutrino physics can be found in references [3-5].

The Δ54 discrete flavor symmetry model serves as the foundation for our model. To deviate from the exact TBM neutrino mixing pattern, we introduced extra flavons $\chi, \eta,, \xi, \zeta, \phi$ and Φ_s under Δ54 and expanded the flavon sector of the model. In addition, we incorporated a supplementary $Z_2 \otimes Z_3 \otimes Z_4$ symmetry into our model to prevent unwanted terms and facilitate the construction of specific coupling matrices. The standard inverse seesaw model features a 9 × 9 matrix comprising the entire neutral fermion mass. This matrix is structured in the (ν_L, N_R, S_R) basis. We modified the neutrino mass matrix M_ν and conducted symmetry-based studies. Unlike other similar works that added perturbative terms to M_ν arbitrarily without calculating them from the Lagrangian, we employed the Inverse seesaw mechanism and depart from the tri-bimaximal mixing pattern. This approach allows us to thoroughly examine M_ν. We also predicted the mixing angles and mass squared differences with our model. The approach we take sets our work apart from that of others.

II. Framework of the Model

The multiplication rules are as follows:

$$3_{1(1)} \otimes 3_{1(1)} = 3_{1(2)} \oplus 3_{1(2)} \otimes 3_{2(2)}$$
$$3_{1(2)} \otimes 3_{1(2)} = 3_{1(1)} \oplus 3_{1(1)} \otimes 3_{2(1)}$$
$$3_{2(1)} \otimes 3_{2(1)} = 3_{1(2)} \oplus 3_{1(2)} \otimes 3_{2(2)}$$
$$3_{2(2)} \otimes 3_{2(2)} = 3_{1(1)} \oplus 3_{1(1)} \otimes 3_{2(1)}$$
$$3_{1(1)} \otimes 3_{1(2)} = 1_+ \oplus 2_1 \oplus 2_2 \oplus 2_3 \oplus 2_4$$
$$3_{1(2)} \otimes 3_{2(1)} = 1_- \oplus 2_1 \oplus 2_2 \oplus 2_3 \oplus 2_4$$
$$3_{2(1)} \otimes 3_{2(2)} = 1_+ \oplus 2_1 \oplus 2_2 \oplus 2_3 \oplus 2_4$$
$$3_{1(1)} \otimes 3_{2(2)} = 1_- \oplus 2_1 \oplus 2_2 \oplus 2_3 \oplus 2_4$$

Field	L	l	H_1	H_2	ν_R	N_R	N_L	S_R	S_L	χ	η	ζ	ξ	Φ_s	ϕ
SU(2)	2	1	1	1	1	2	2	1	1	1	1	1	1	1	1
Δ54	$3_{1(1}$	$3_{1(1)}$	1_-	1_+	$3_{2(2)}$	$3_{1(1)}$	$3_{2(2)}$	$3_{1(1)}$	$3_{2(2)}$	$3_{2(1)}$	$3_{2(2)}$	1_-	$3_{2(1)}$	$3_{2(1)}$	1_-
Z_2	1	1	1	1	-1	-1	1	-1	1	1	1	-1	-1	-1	-1
Z_3	ω^2	ω	1	1	1	1	1	1	1	1	1	1	ω	ω	ω
Z_4	1	1	1	-1	i	1	-i	i	1	-1	1	1	1	-1	i

The invariant Yukawa Lagrangian is as follows:

$$\mathcal{L} = \frac{\mathrm{y}}{\Lambda}(Ll)\chi H_1 + \frac{\mathrm{LH_1N_R}}{\Lambda}\left(y_\xi + y_s\Phi_s\right) + \frac{\mathrm{LH_2N_R}}{\Lambda} y_a\Phi_s$$
$$+\frac{\mathrm{y_{RN}}}{\Lambda}\nu_\mathrm{R}\mathrm{N_L}\eta\zeta + \mathrm{y_{NS}}\mathrm{S_R}\mathrm{N_L}\zeta + \mathrm{y_{NS}}'\mathrm{S_L}\mathrm{N_R}\zeta + \frac{\mathrm{y_s}}{\Lambda}\mathrm{S_L}\mathrm{S_R}\phi$$

The <VEV> is assumed as;

$$< \Phi_s \geq (v_s, v_s, v_s) < \chi \geq (v_\chi, v_\chi, v_\chi)$$
$$< \xi >= (v_\xi, v_\xi, v_\xi,)$$
$$< \eta >= (v_\eta) < \zeta >= (v_\zeta)$$
$$< \phi >= (v_\phi)$$

The charged lepton mass matrix is given as,

$$M_l = \frac{v_\chi H}{\Lambda}\begin{pmatrix} y_e & 0 & 0 \\ 0 & y_\mu & 0 \\ 0 & 0 & y_\tau \end{pmatrix}$$

A. Effective Neutrino Mass Matrix

$$M_{RN} = \frac{y_{RN}}{\Lambda}\begin{pmatrix} v_\eta v_\zeta & 0 & 0 \\ 0 & v_\eta v_\zeta & 0 \\ 0 & 0 & v_\eta v_\zeta \end{pmatrix}$$

$$M_{NS} = \frac{y_{NS}}{\Lambda}\begin{pmatrix} v_\zeta & 0 & 0 \\ 0 & v_\zeta & 0 \\ 0 & 0 & v_\zeta \end{pmatrix}$$

$$M'_{NS} = \frac{y'_{NS}}{\Lambda}\begin{pmatrix} v_\zeta & 0 & 0 \\ 0 & v_\zeta & 0 \\ 0 & 0 & v_\zeta \end{pmatrix}$$

$$M_S = \frac{y_S}{\Lambda}\begin{pmatrix} v_\phi & 0 & 0 \\ 0 & v_\phi & 0 \\ 0 & 0 & v_\phi \end{pmatrix}$$

$$M_{\nu N} = \frac{v}{\Lambda}\begin{pmatrix} y_\xi\xi & y_sv_s + y_av_a & y_sv_s - y_av_a \\ y_sv_s - y_av_a & y_\xi\xi & y_sv_s + y_av_a \\ y_sv_s + y_av_a & y_sv_s - y_av_a & y_\xi\xi \end{pmatrix}$$

$$m_\nu = M_{RN}(M'_{NS})^{-1}M_sM_{NS}^{-1}M_{\nu N}$$

We now define the Hermitian matrix as,

$$M_\nu = m_\nu m_\nu^\dagger$$

$$M_\nu = \begin{pmatrix} x'^2 + 2(s'^2 + a'^2) & s'^2 - a'^2 + 2s'x' & s'^2 - a'^2 + 2s'x' \\ s'^2 - a'^2 + 2s'x' & x'^2 + 2(s'^2 + a'^2) & s'^2 - a'^2 + 2s'x' \\ s'^2 - a'^2 + 2s'x' & s'^2 - a'^2 + 2s'x' & x'^2 + 2(s'^2 + a'^2) \end{pmatrix}$$

where, $x' = \lambda x, s' = \lambda s, a' = \lambda a, x = y_\xi \xi, s = y_s v_s, y_a v_a$ and $\lambda = \frac{v v_\eta v_s v_\phi^3 y_{RN} y_s}{\Lambda^4 v_\xi^2 y'_{NS} y_{NS}}$

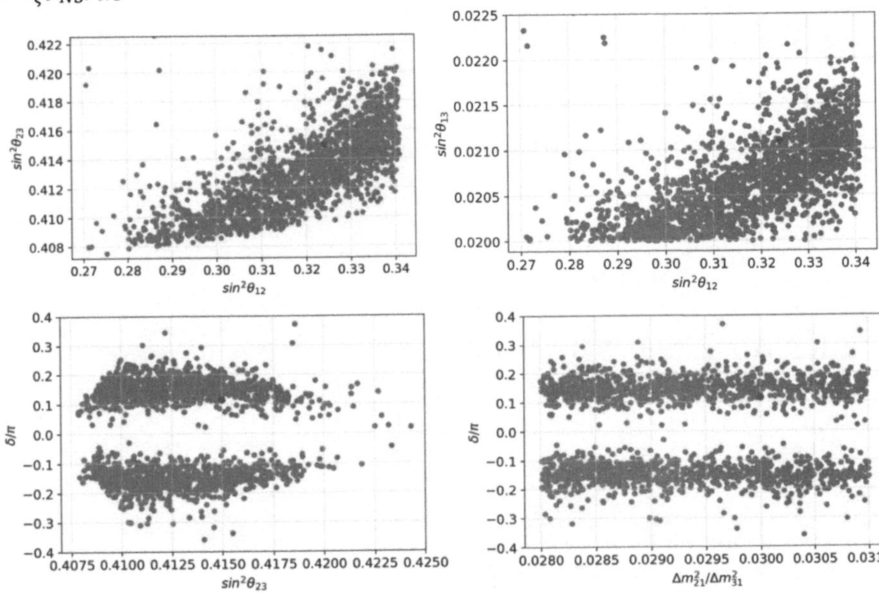

III. Conclusion

Generally, by introducing Δ54 symmetry for the model that generate TBM can explain a tiny non zero reactor angle θ_{13}. Recent data seems to be indicating towards the existence of neutrino mixing with non-zero reactor angle θ_{13}. We have constructed a flavor-symmetric approach to realize the neutrino masses and mixing. We have used our model Δ54 to fit present neutrino oscillation data, including the nonzero reactor angle θ_{13}. We use additional flavons to obtain the desired mixing pattern. The resulting mass matrix gives predictions for the neutrino oscillation parameters and their best-fit values obtained using the χ^2 analysis, are consistent with the global data from various neutrino oscillation experiments. The model cases under consideration clearly shows deviation from TBM mixing of the neutrino mixing matrix. The model do not show parameters in 3σ range under IH in terms of the prediction of the mixing angles, mass-squared differences, and the Dirac CP phase. The prediction of θ_{23} in NH case indicate that lower octant is occupied for the values of the model parameter of your work.

Acknowledgement

Hrishi Bora is grateful and acknowledges the financial support provided by the Tezpur University, India for Institutional (Non-NET) Research Fellowship (F.14-5/3/2003(Acad)/1755-A. The research of Ng. K. Francis is funded by DST-SERB, India under Grant no. EMR/2015/001683. Animesh Barman acknowledges the financial support provided by the CSIR, New Delhi, India for CSIR Senior Research Fellowship (file no 09/796(0072)/2017-EMR-1).

References

[1] Borah, D. and Biswajit, K., A4 flavour model for Dirac neutrinos: Type I and inverse seesaw, *Physics Letters B*, 780, 461–470 (2018).

[2] Kashav, M., Singh, Labh and Verma, S., Trimaximal mixing and extended magic symmetry in a model of neutrino mass matrix (2022) arXiv preprint [arXiv:2207.13328]

[3] King, S.F., Merle, A., Morisi, S., Shimizu, Y. and Tanimoto, M., *New J. Phys.* 16, 045018(2014), 1402.4271.

[4] King, S.F., Neutrino mass models, *Rept. Prog. Phys.*, 67:107–158 (2004).

[5] Arthur, B. McDonald., Nobel lecture: The sudbury neutrino observatory: observation of flavorchange for solar neutrinos, *Reviews of Modern Physics,* 88(3):030502 (2016).

Leptogenesis and Dark Matter Through Relativistic Bubble Walls with Observable Gravitational Waves

I. Saha[1,*], D. Borah[1] and A. Dasgupta[2]

[1]Department of Physics, Indian Institute of Technology Guwahati, Assam 781039, India

[2]Pittsburgh Particle Physics, Astrophysics, and Cosmology Center, Department of Physics and Astronomy, University of Pittsburgh, Pittsburgh, PA 15206, U.S.A.

*E-mail: s.indrajit@iitg.ac.in

ABSTRACT: We study a scenario where dark matter and heavy right-handed neutrino (RHN) responsible for leptogenesis acquire masses by crossing the relativistic bubble walls formed from a TeV scale supercooled first-order phase transition (FOPT). While this leads to a large out-of-equilibrium abundance of right-handed neutrino inside the bubble sufficient to produce the required lepton asymmetry, the dark matter being lighter can still remain in equilibrium with its relic being set by subsequent thermal freeze-out or freeze-in. A classical conformal symmetry ensures the origin of mass via FOPT induced by a singlet scalar while ensuring supercooling, leading to enhanced gravitational wave amplitude within the sensitivity of the LISA experiment.

Keywords: Phase Transitions in the Early Universe, Baryo-and Leptogenesis, Dark Matter.

I. Introduction

The Standard Model of particle physics has been the most successful theory to explain particle nature and its interactions. Despite its success, there are many persistent mysteries of our universe. The observed baryon asymmetry of the universe (BAU) is a persistent mystery in cosmology and particle physics. It is measured as baryon to photon ratio from Planck 2018 data, and it is,

$$\eta_B = \frac{n_B - n_{\bar{B}}}{n_\gamma} = 6.1 \times 10^{-10} \qquad \ldots (1)$$

This asymmetry is predicted from the measurement of cosmic microwave background (CMB) anisotropies and big bang nucleosynthesis. With the most common assumption, the universe starts with the balance baryon and anti-baryon, so a dynamic mechanism must lead to net baryon asymmetry called baryogenesis. There are three Sakharov conditions sufficient for BAU to have such a mechanism to generate asymmetry. The first is the baryon number violation, which can be achieved within Standard Model via chiral anomaly. Also, the B+L violating and the B-L conserving processes are controlled by the sphaleron rate depending on the strength of the phase transition. In the second condition, C and CP violation; otherwise, the asymmetry will be washed out. The third condition is out of thermal equilibrium, an assured feature of first-order phase transition. While in Standard Model, the first-order phase transition is not achieved, there is a scope of extension with a singlet, doublet, etc., which may also lead to new physics like Dark matter. There is an alternative way to generate such asymmetry, known as leptogenesis, and Fukugita and Yanagida first propose it. In this mechanism, the lepton asymmetry is generated through out-of-equilibrium decay of heavy field in CP-violating processes. In this way, the heavy field can be associated with the origin of tiny neutrino masses, which is also an unsolved puzzle of SM.

While the CMB measurements suggest that the present universe has a large amount of non-luminous, non-baryonic form of matter, known as dark matter (DM), this is also supported by other astrophysical evidence gathered over a longer period. Dark matter is around 26 % of our universe, while the ordinary baryonic matter is 5%, and the rest of the part is dark energy responsible for the universe's expansion. The present dark matter abundance is defined through density parameter Ω_{DM} and h (Hubble parameter/(100 km s^{-1} Mpc^{-1})) as,

$$\Omega_{DM}h^2 = 0.120 \pm 0.001 \quad \dots (2)$$

at 68% CL. In the Standard Model, there is no candidate to be dark matter satisfying the criteria. So, in beyond Standard Model scenarios, different proposals have been put forward, among which the weakly interacting massive particle (WIMP) is the most popular. This type of dark matter particle having interactions in order of electroweak scale can give rise to observed abundance after thermal freeze-out. However, the interaction in DM-nucleon scattering gives a tight constraint on several WIMP models through experiments like LUX, PandaX-II, and Xenon1T. Apart from this, there is another possibility of non-thermal production of DM called as feebly interacting massive particles (FIMP). In this case, it has very feeble interaction strength with an initial abundance of zero. From the decay or annihilation of other particles, DM can be produced.

In leptogenesis scenarios, the out-of-equilibrium decay can generate such asymmetries following the seesaw mechanism, but it gives a lower bound on the scale of heavy particle $M > 10^9$ GeV, which is called Davidson-Ibarra bound. For this, the direct probe of leptogenesis is beyond our present experimental limit. There are other situations where adding extra fields in the seesaw model of neutrino mass can relieve the lower bound of the scale of leptogenesis. There is another scenarios where TeV scale leptogenesis can be achieved by resonant enhancement of the CP asymmetry, known as resonant leptogenesis with the fine-tuning of mass splitting between decaying particles. In this case, the decaying right-handed neutrino also has a very tiny coupling with leptons to generate a tiny mass of light neutrino, making it difficult to probe directly. So, there are some works to probe leptogenesis via stochastic gravitational wave (GW) observation. While some of the work is based upon topological defects like cosmic strings, and domain walls formed from symmetry-breaking results, others consider first order phase transition (FOPT) responsible for the generation of GW.

II. Conformal Scotogenic Model

We study a conformal or scale-invariant version of the minimal scotogenic model where the SM is extended by three gauge singlet right-handed neutrinos N_i (i = 1, 2, 3), one extra scalar doublet η; to achieve a supercooled phase transition along with leptogenesis and dark matter. An additional Z2 symmetry is imposed under which all SM particles and $N_{2,3}$ are even while all other new particles are odd. To preserve the conformal nature and generate masses, another Z2, even singlet scalar S is added. The fermionic Yukawa Lagrangian of the model is as follows,

$$\mathcal{L} \supset \frac{1}{2} Y'_{ij} S N_i N_j + (Y_i \bar{L}_i \tilde{\eta} N_1 + h.c.) + \left(Y_{\phi i} \bar{L}_i \widetilde{\Phi}_1 N_{2,3} + h.c.\right) \quad \dots (3)$$

The scalar potential of the model is

$$V(\Phi_1 \eta, S) = \frac{\lambda_1}{4} |\Phi_1|^4 + \frac{\lambda_1}{4} |\eta|^4 + \lambda_3 |\Phi_1|^2 |\eta|^2 + \frac{1}{4} \lambda_S S^4 + \lambda_4 \left|\Phi_1^\dagger \eta\right|^2 + \left[\frac{\lambda_5}{2} \left(\Phi_1^\dagger \eta\right)^2 + h.c.\right] + \lambda_6 |\Phi_1|^2 S^2 + \lambda_7 |\eta|^2 S^2 \quad \dots (4)$$

where Φ_1 is the SM Higgs doublet.

III. First Order Phase Transition

In order to realize leptogenesis, the phase transition is propelled by singlet scalar field at a scale above the electroweak scale. To study the first order phase transition in conformal scotogenic model, the effective potential considering one loop correction can be written as,

$$V_{tot} = V_{tree} + V_{CW} + V_{th} \qquad \dots (5)$$

where V_{tree}, V_{CW} and V_{th} denote the tree level scalar potential, the one-loop Coleman-Weinberg potential, the thermal effective potential, respectively. The tree level scalar potential is given by Eq. (4). In finite temperature field theory, the effective potential, V_{CW} and V_{th}, are calculated by using standard background field method.

During the FOPT at the critical temperature, two degenerate minima are created corresponding to the scalar field, and a barrier separates it. Just below the critical temperature, the minima at the origin become a metastable vacuum, and the other minima become a true vacuum. The decay of metastable or false vacuum can happen in two ways; quantum tunneling or thermal fluctuations. The transition through fluctuation is preferred in only energetically possible states. While the rate of tunneling through the barrier is exponentially suppressed. The FOPT proceeds via tunneling, and the corresponding spherical symmetric field configurations known as bubbles are nucleated, followed by expansion and coalescence. In the early universe, it is considered to be perfectly pure. The thermal fluctuations ignite the nucleation when the bubbles expand and reach a critical volume such that it doesn't shrink back and vanish. The temperature at which the bubbles are nucleated is called the nucleation temperature T_n. We then implement the model in Phase Tracer to find the parameter space consistent with a FOPT shown in Figure 1.

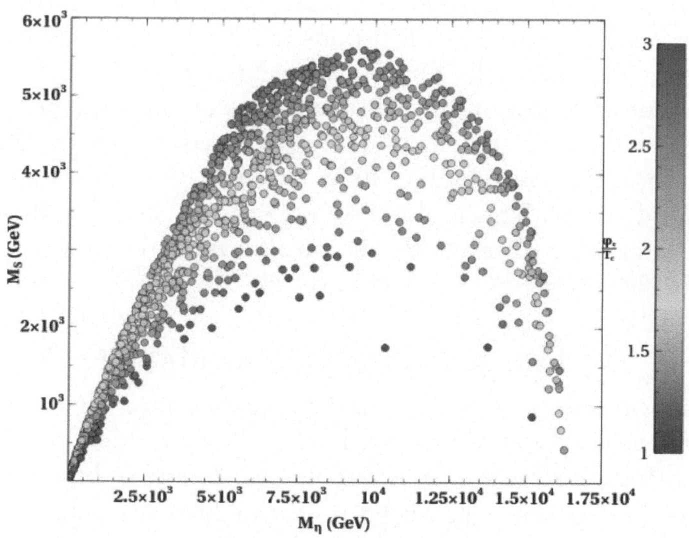

Figure 1: M_η versus M_S

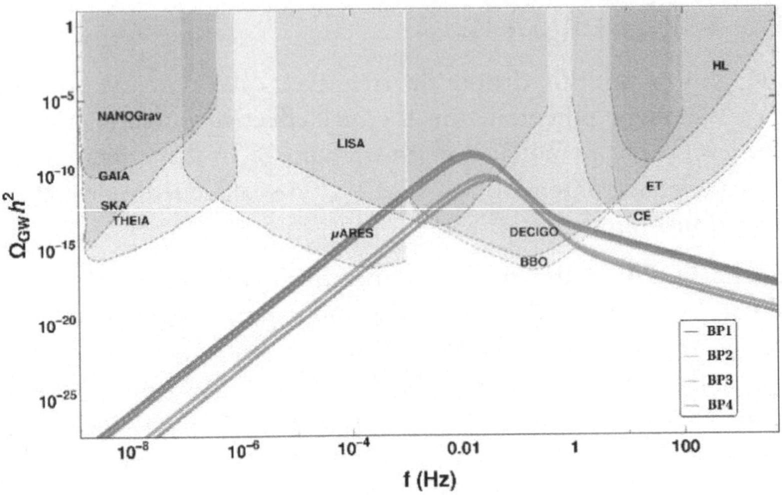

Figure 2: Gravitational Wave Spectrum from FOPT

IV. Stochastic Gravitational Waves from FOPT

During the FOPT, the formation of bubbles has a violent consequence. After the nucleation, the bubble starts to expand due to the pressure difference between the broken and unbroken phases and collides untill the whole universe goes to the broken phase. The bubble expansion into the surrounding plasma and collision leads to anisotropic stresses in the fluid, which serve as seeds of GWs. We can see a potentially observable relic signal today from FOPT in the early universe. The bubbles of true vacuum nucleate, expand, and collide, which leads to a background scalar field that varies in space and time. The Einstein equations, therefore, tell us that the energy-momentum tensor of this field is a source of curvature perturbations. Moreover, a substantial amount of energy released during the PT can be transferred into the surrounding fluid through turbulence or coherent motion (sound waves). This is also a source of GWs since the fluid couples to the metric as well. Considering all contributions, the GWs spectrum is evaluated for some benchmark points in Figure 2.

V. Leptogenesis and Dark Matter

We have shown the baryogenesis via leptogenesis through the relativistic bubble wall mechanism in our model. Due to the first-order phase transition in the scalar singlet (S) sector, some bubbles will create, and the particles like N_i,η will become massive when it go inside bubbles. After that $N_{2,3}$ decays into leptons and SM Higgs which generate the lepton asymmetry. For leptogenesis, we followed mass-gain mechanism, where we need to ensure that Lorentz boost of bubble wall should be more than the Lorentz factor of the particle in plasma frame such that all RHN goes into the bubble and maintain the equilibrium co-moving number density. The observed baryon asymmetry is generated from lepton asymmetry through sphaleron in wash-out suppressed regime.

On the other hand N_1 has small Yukawa coupling with the scalar and it is possible DM candidate. In principle, we can consider two scenarios: thermal WIMP and on-thermal DM (popularly FIMP). For DM mass below T_n, it remain in thermal equilibrium inside the bubble and freeze-out as temperature decreases. The Figure 3 shows the observed DM relic is satisfied in Higgs resonance which barely allows from direct detection bound shown in Figure 4. But this case ruled out for such singlet-SM Higgs mixing. In non-thermal FIMP case, the initial abundance of DM is zero and it produced from the decay of scalar singlet and freeze-in after some time. It can satisfy observed DM relic for DM mass to be 13.96 MeV.

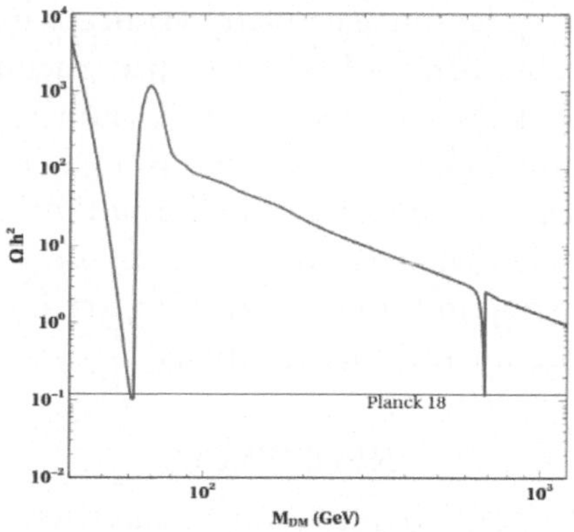

Figure 3: WIMP DM Relic vs M_{DM}

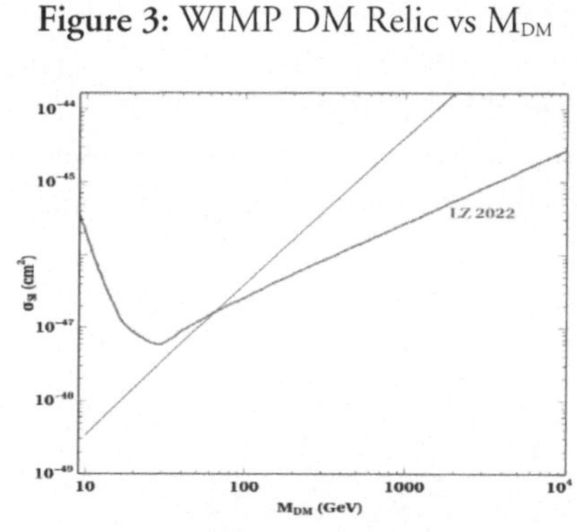

Figure 4: σ_{SI} vs M_{DM}

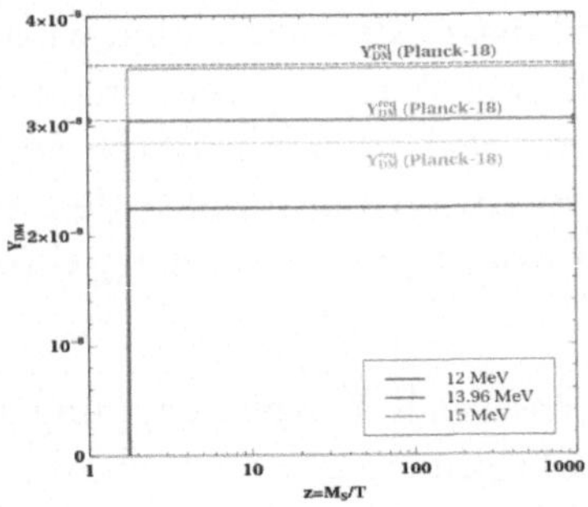

Figure 5: Comoving Density FIMP

VI. CONCLUSION

We have studied a minimal conformal scotogenic model to realize the super cooled first order phase transition along with dark matter and low scale leptogenesis. The phase transition is guided by the scalar singlet in around TeV scale. After the nucleation temperature, the right handed neutrino and dark matter are passed into the relativistic bubble and aquire masses. And those RHNs are creating lepton asymmetry via decay inside the bubble. This also leads to a large abundance of RHN in true vacuum inside the bubble sufficient for generating the required lepton asymmetry without washout or Boltzmann suppression. The dark matter is lighter than the nucleation temperature and hence can remain in equilibrium inside the bubble with its relic determined by thermal freeze-out or freeze-in at later stages. Also we have a strong supercooling to bring the resulting gravitational wave amplitude within near future experiment LISA sensitivity.

REFERENCES

[1] Baldes, I., Blasi, S., Mariotti, A., Sevrin, A. and Turbang, K., Baryogenesis via relativistic bubble expansion, *Phys. Rev. D*, 104 (2021) 115029.

[2] Dasgupta, A., Dev, P.S.B., Ghoshal, A. and Mazumdar, A., Gravitational Wave Pathway to Testable Leptogenesis, *Phys. Rev. D*, 106 (2022) 075027.

[3] Borah, D., Dasgupta A. and Saha, I., Leptogenesis and dark matter through relativistic bubble walls with observable gravitational waves, *JHEP* 11 (2022), 136.

Neutrino Masses and Mixing Using A_4 Modular Symmetry

J. Gogoi*, N. Gautam and M.K. Das
Department of Physics, Tezpur University, Tezpur 784028, Assam, India
*E-mail: jotingogoi2014@gmail.com

ABSTRACT: In this work, we have constructed a model with the help of modular symmetry in the framework of minimal inverse seesaw [ISS (2,3)]. We have used Γ(3) modular group which is isomorphic to non-Abelian discrete symmetry group A_4. *In this group there are three Yukawa modular forms of weight 2 which are functions of a single complex variable* τ, *called the modulus. Along with this,* Z_3 *is used to restrict certain interactions in the Lagrangian. Through this model, we have studied neutrino masses and mixing for both normal and inverted hierarchy. The results are found to lie within the range as predicted by experiments.*

Keywords: Neutrino Phenomenology, Neutrino Masses, Mixings, Modular Symmetry.

I. Introduction

One of the most successful models in particle physics, incorporating all the known fundamental particles in nature is the Standard Model. With the discovery of Higgs boson in 2012, both in ATLAS and CMS at the LHC, this model has been put on a firm footing both experimentally and theoretically. Inspite of its immense successes, this model is not without drawbacks. The discovery of neutrino oscillation confirms mixings between the three generation of neutrinos. The neutrinos, initially massless in the Standard Model were found to be massive. These aspects of neutrino remain unanswered and no suitable explanation is found in the model. There are some phenomena like lepton flavor violation, lepton number violation, darkmatter and dark energy etc. which require immediate attention. But these cannot be explained within the domain of Standard Model. As a result, these observations compel physicists to go for Beyond Standard Model frameworks.

Early attempts to address the masses and mixings of neutrinos came through Seesaw mechanism. It is an extension of the Standard Model. Based on the type of particles added, this mechanism is broadly classified into the following:

Type-I, Type-II, Type-III and Inverse seesaw. These extra particles include right-handed neutrinos, scalar triplets, sterile fermions etc.

Inverse seesaw is a quite realistic and experimentally probable approach that aims to explain generation of tiny neutrino masses in a much feasible way. The additional particles in this framework include three right-handed neutrinos and three gauge singlet sterile fermions. This framework can lower the mass of right-handed neutrinos to TeV, due to which there is scope of them being detected in the LHC and future oscillation experiments.

Symmetry plays an important role in model building in the field of particle physics. Discrete symmetry groups like A_4, S_4, Z_N etc. have been extensively used to build models. In recent times modular symmetry has gained much importance in this arena. Unlike discrete symmetry, modular symmetry helps to reduce the number of extra flavons and maintains minimality of the model.

A. Modular Symmetry

The modular group Γ(N) (N = 1, 2, 3,..........) can be defined as,

$$\Gamma(N) = \{\begin{pmatrix} a & b \\ c & d \end{pmatrix} \in SL(2,Z), \begin{pmatrix} a & b \\ c & d \end{pmatrix} = \begin{pmatrix} 1 & 0 \\ 0 & 1 \end{pmatrix} \ (mod\ N)\}$$

- $\Gamma(N)$acts on complex variable τ in the upper-half of the complex plane $\mathrm{Im}(\tau) > 0$ and transforms τ as,

$$\tau \rightarrow \frac{a\tau + b}{c\tau + d}$$

- $\Gamma(3)$has three modular forms of weight 2, $Y(\tau) = (Y_1, Y_2, Y_3)^T$
- The finite modular groups (N$\leq$ 5) and non-Abelian discrete groups are isomorphic to each other. As a result, $\Gamma_2 \approx S_3, \Gamma_3 \approx A_4, \Gamma_5 \approx A'_5$ etc.

B. Inverse Seesaw [ISS(2, 3)]

ISS(2, 3) apart from the Standard Model particles, contains two right-handed neutrinos (N_i) and three gauge singlet sterile fermions (S_i). The Lagrangian in the basis.

$n_L = (\nu_{L\alpha}, N_{Ri}^{\ c}, S_j)^T$ can be written as:

where M is a 8 × 8 matrix$\begin{pmatrix} 0 & M_D^T & 0 \\ M_D & 0 & M_{NS} \\ 0 & M_{NS}^T & M_S \end{pmatrix}$

The matrix M can be diagonalised with the help of a unitary matrix U as,

$$U^T MU = \text{diag}\,(m_1,\ldots, m_8),$$

where, m_i's are the masses of the eight particles. Three of them are light neutrinos and five are heavy particles. The effective neutrino mass matrix can be written in the following way,

$$m_\nu = M_D . d . M_D^T$$

where d is a 2×2 matrix, defined as,

$$M_H^{-1} = \begin{pmatrix} 0 & M_{NS} \\ M_{NS}^T & M_S \end{pmatrix}^{-1} = \begin{pmatrix} d_{2\times2} & \ldots\ldots \\ \ldots\ldots & \ldots. \end{pmatrix}$$

Diagonalising m_ν gives light neutrino masses in sub-eV range when M_D, M_{NS} and M_S are in electroweak, TeV and KeV range.

C. The Model

This section describes the interactions of the model. The tables below show the particles and their charge assignments with respect to the groups used in the model.

Table 1: Charge Assignment of Particles

Gauge Groups	L	E_i	N_1	N_2	S_i	H_u	H_d	ϕ
$SU(2)_L$	2	1	1	1	1	2	2	1
A_4	3	1,1″,1′	1′	1″	3	1	1	3
K_I	0	-2	-2	-2	0	0	0	2
Z_3	ω^2	1	ω	ω	1	Ω	1	ω

Table 2: Charge Assignments of Modular Forms

	Y (modular forms)
A_4	3
K_I	2
Z_3	ω^2

With the above description and behaviour of the particles, the relevant Lagrangian can be written as:

$$\begin{aligned}\mathcal{L} = {} & \alpha_1 E_1^c H_d (L\phi)_1 + \alpha_2 E_2^c H_d (L\phi)_{1'} + \alpha_3 E_3^c H_d (L\phi)_{1''} \\ & + [N_1 (LY)_{1''} H_u]_1 + [N_2 (LY)_{1'} H_u]_1 \\ & + \beta [\{N_1 (SY)_{1''}\} + \{N_2 (SY)_{1'}\}] + \Lambda (SS)_1\end{aligned}$$

The mass matrices for the charged leptons for the VEV, $\langle\phi\rangle = (u, 0,0)$ take the form:

$$M_L = \text{diag}\,(\alpha_1, \alpha_2, \alpha_3)vu$$

$$M_D = v\begin{pmatrix} Y_3 & Y_2 \\ Y_2 & Y_1 \\ Y_1 & Y_3 \end{pmatrix}, M_S = \Lambda\begin{pmatrix} 1 & 0 & 0 \\ 0 & 0 & 1 \\ 0 & 1 & 0 \end{pmatrix}, M_{NS} = \beta\begin{pmatrix} Y_3 & Y_2 & Y_1 \\ Y_2 & Y_1 & Y_3 \end{pmatrix}$$

The matrices M_D, M_S and M_{NS} represents Dirac, sterile fermion and mixing between sterile and right-handed neutrinos, respectively. These matrices facilitate to obtain the light active neutrino masses through the processes described in ISS (2, 3).

II. Numerical Analysis and Discussion

The neutrino mixing angles can be calculated from the elements of the unitary matrix, *U*. These relations for the three mixing angles are:

$$\sin^2\theta_{13} = |U_{e3}|^2,$$

$$\sin^2\theta_{23} = \frac{|U_{\mu 3}|^2}{1-|U_{e3}|^2},$$

$$\sin^2\theta_{12} = \frac{|U_{e2}|^2}{1-|U_{e3}|^2}$$

In order to fit the neutrino oscillation data, the following ranges of values for the model parameters are taken:

$\text{Re}(\tau) \rightarrow [0,3]$, $\text{Im}(\tau) \rightarrow [0.5, 2.8]$, $v = 125$ Gev, $\Lambda \rightarrow [10, 20]$ KeV, $\beta \rightarrow [10, 100]$ TeV

The results are displayed below:

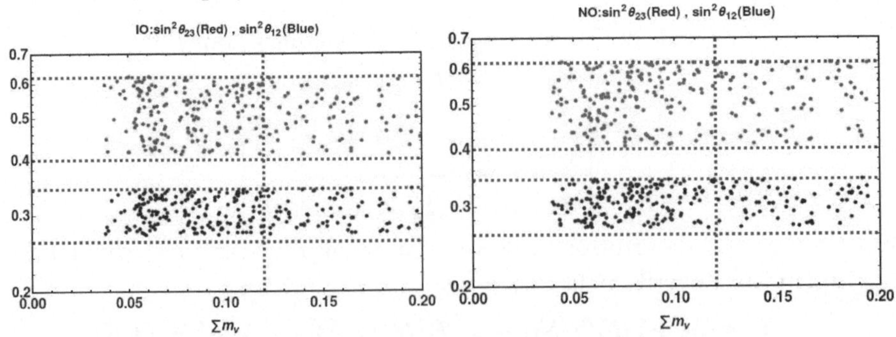

Figure 1: Shows the Relation between Mixing Angles $(\theta_{23}, \theta_{12})$ and Sum of Neutrino Mass (Σm_ν)

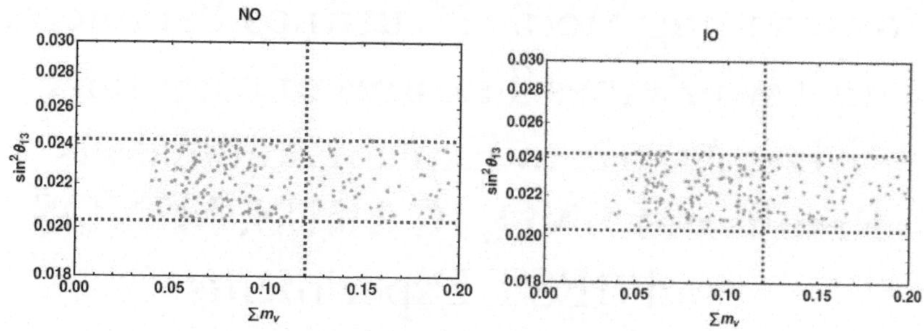

Figure 2: Shows the Relation between Mixing Angles (θ_{13}) and Sum of Neutrino Mass (Σm_ν)

III. Conclusion

ISS is anexperimentally feasible framework. Because of its potential to lower the mass scale to TeV there are chances for the particles to be detected LHC. As evident from the figures portrayed in the numerical analysis, a considerable region of parameter space of mixing angles and sum of neutrino mass lies within the allowed region. This is found to be true for both normal and inverted hierarchies. As a result this has become a popular framework in model building. This can also be used to study dark matter and related phenomena.

References

[1] Dev, P.S.B., Lee, C.H. and Mohapatra, R.N., Leptogenesis Constraints on the Mass of Right-handed Gauge Bosons, *Phys. Rev. D*, 90, 9 (2014).

[2] Abada, A. and Lucente, M., Looking for the minimal inverse seesaw realisation, *Nucl. Phys. B,* 885, 651–678 (2014).

[3] Esteban, I., Gonzalez-Garcia, M.C., *et al.*, The fate of hints: Updated global analysis of three-flavor neutrino oscillations, *JHEP*, 178, 09 (2022).

Constraining Sterile Neutrinos Parameters in Two Zeroes Textures of Majorana Mass Matrix to Explain Anomalous Results in LSND, ICARUS, MINOS and BEST Experiments

Maibam Ricky Devi

Department of Physics, Gauhati University, Guwhati-781014, Assam, India

E-mail: deviricky@gmail.com

ABSTRACT: In this work, we have analyzed different allowed two zeroes textures of low-scale Majorana mass matrix in the context of (3+1). We have constrained the neutrino oscillation parameters using the latest NuFit 5.1 data [1] and found the sterile neutrino oscillation parameters that give results consistent with the measurement of $Sin^2 2\theta$ *and* Δm^2 *as obtained in sterile neutrino experiments such as LSND [2] and ICARUS [3] regarding sterile neutrino anomaly. We also analyze the same textures to discuss the sterile neutrino parameters constrained in MINOS [4] and BEST [5] Experiments.*

Keywords: Neutrino Mixing, Sterile Neutrino, Two Zero Textures.

I. Introduction

Some the neutrinos are elusive and exotic particles that are almost imperceptible to detect in the neutrino detector due to its feeble interactions with matter. Although the Standard Model of particle physics predicts only three generations of light neutrinos, there is a possibility of an extra light neutrinos called as sterile neutrinos that can explain the recent anomalies reported in the neutrino experiments such as LSND, reactor and gallium solar neutrino experiments. In order to validate the active-sterile neutrino mixing, also called as 3+1 mixing, several experiments. Their results will either confirm or refute the existence of the sterile neutrinos.

In the LSND experiment, there was surplus of electron anti-neutrinos in the apperance channel of muon antineutrinos to electron antineutrinos which could be compensated with an additional light neutrino. Similar result was also shown by the MiniBooNE experiment. Some other reactor and solar experiments too showed such an anomalies letting the neutrino physicists to wonder if there is a sterile neutrinos in existence. A few of the anomalies have been summarized below:

- *LSND Anomaly [1]:* $\bar{\nu_\mu} \rightarrow \bar{\nu_e}$, Sin2 2θ > 10−3, Δm2 > 3 × 10−2 eV2
- *MiniBooNE [2]:* $\nu_\mu \rightarrow \nu_e$and $\bar{\nu_\mu} \rightarrow \bar{\nu_e}$
- *ICARUS [3]:* $\nu_\mu \rightarrow \nu_e$, Sin2 2θ < 10−2, Δm2> 10−2 eV2
- *GALLIUM, SAGE Experiments [4]:* deficit of neutrino
- *Reactor Anomaly:* observed neutrino fluxes are 6% smaller than the predicted.

II. Motivation and Objectives of Our Work

Following is the motivation of our work:

From both an experimental and theoretical perspective, the study of sterile neutrinos can provide valuable insights into a variety of important topics. These include:

1. **New Physics:** The discovery of sterile neutrinos could potentially lead to the discovery of new physics beyond the Standard Model.
2. **Leptogenesis and the Baryon Asymmetry of the Universe:** Sterile neutrinos could play a crucial role in explaining the observed baryon asymmetry of the universe through the process of leptogenesis.
3. **Neutrino Mass Scale and Nature:** The study of sterile neutrinos can help us better understand the nature of neutrinos, including whether they are Dirac or Majorana particles, as well as the neutrino mass scale.
4. **CP Violating Phases and Majorana Mixing Angles:** The measurement of CP violating phases and Majorana mixing angles in neutrino oscillations could provide important clues about the fundamental nature of neutrinos.

Overall, the study of sterile neutrinos is a fascinating and important area of research that has the potential to greatly enhance our understanding of the universe and the fundamental laws of physics.

III. Objectives of Our Research

- Our research aims to investigate the three zeros textures of the Majorana mass matrix, which is allowed by LSND.
- We will compare this matrix with the 4 × 4 neutrino mass matrix for parametrization to determine the unknown sterile neutrino parameters.
- Our goal is to constrain these parameters within a 3+1 global fit 3σ parameter space and determine the probability oscillation for different

experiments. This will help us to check the feasibility of our three zeros texture analysis.

Our research is crucial in understanding the properties of neutrinos and their interactions. By investigating the Majorana mass matrix and comparing it with the 4 × 4 neutrino mass matrix, we hope to uncover new information about the unknown sterile neutrino parameters. Our analysis will be conducted within a 3+1 global fit 3σ parameter space, which will allow us to constrain these parameters and determine the probability oscillation for different experiments. This will provide valuable insights into the feasibility of our three zeros texture analysis and contribute to the advancement of neutrino research.

IV. Allowed Three-Zero Textures by LSND

The 20 patterns of M_ν with all the three zeros in the active sector can also be further classified into subgroups. Out of these a few three zeros textures are allowed by LSND such as, [5]

$$A = \begin{pmatrix} 0XXX \\ X0XX \\ XX0X \\ XXXX \end{pmatrix}, \quad B_1 = \begin{pmatrix} 0X0X \\ X0XX \\ 0XXX \\ XXXX \end{pmatrix}, \quad B_2 = \begin{pmatrix} 0XXX \\ X00X \\ X0XX \\ XXXX \end{pmatrix}, \quad a_4 = \begin{pmatrix} 00XX \\ 0XXX \\ XX0X \\ XXXX \end{pmatrix}. \quad \ldots (1)$$

$$P(\nu_\alpha \to \nu_\beta) = sin^2\theta_{ij} sin^2(1.27\Delta m_{ij}^2 \, L/E)$$

In 3+1 scenario, the four flavor eigenstates is connected the mass eigenstates through a 4 × 4 lepton mixing matrix give by U,

$$U = \begin{pmatrix} c_{12}c_{13}c_{14} & c_{13}c_{14}s_{12}e^{i\alpha/2} & c_{14}s_{13}e^{i\beta/2} & s_{14}e^{i\gamma/2} \\ .. & .. & .. & c_{14}s_{24}e^{i(\delta_{14}-\delta_{24}+\gamma/2)} \\ .. & .. & .. & c_{14}c_{24}s_{34}e^{i(\delta_{14}+\gamma/2)} \\ .. & .. & .. & c_{14}c_{24}c_{34}e^{i(\delta_{14}+\gamma/2)} \end{pmatrix}. \quad \ldots (2)$$

V. Result and Discussion

To draw the correlation plots given below we have taken the following 3+1 Global fit values:

$1.3\ (0.33)\ eV^2 < \Delta m^2_{41} < 32\ (148)\ eV^2$ at 2σ (3σ) [6],

$\theta_{14} = 6.1^\circ - 15.3^\circ$, $\theta_{24} = 2.8^\circ - 12.5^\circ$, $\theta_{34} = 0^\circ - 30^\circ$ [1].

The 3σ values of neutrino oscillation parameters are taken from the following reference [7].

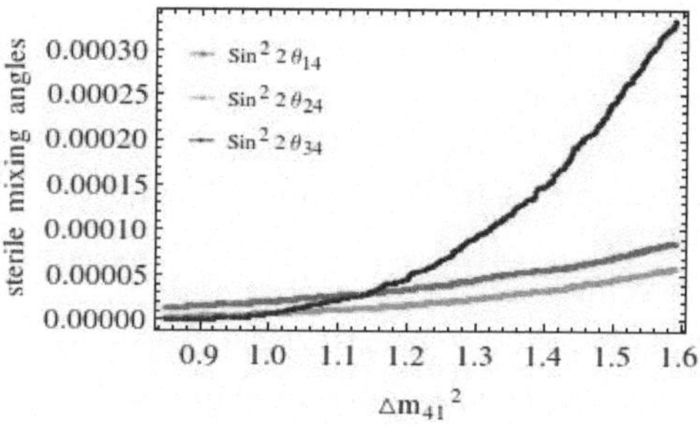

Figure 1

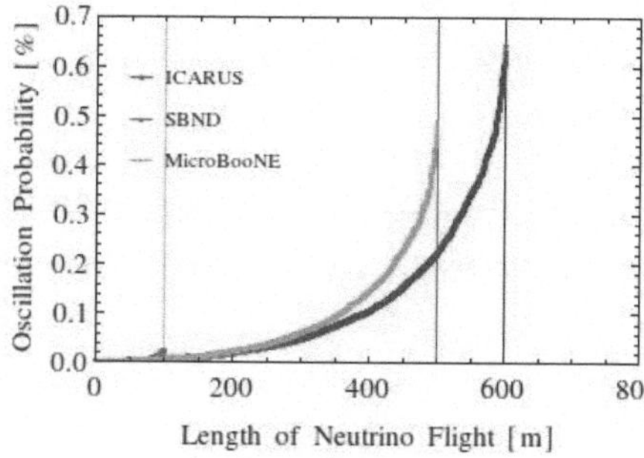

Figure 2

VI. Conclusion

In this work, we have successfully constrained the 3+1 neutrino oscillation parameters for the allowed three zeroes textures by LSND within the 3σ parameter space. By utilizing this data, we were able to determine the oscillation probability. Furthermore, this analysis provided us with the value of Δm^2_{41}, which aligns well with recent global fit values.

Our next step is to extend this analysis to include MINOS and BEST experiments. This will allow us to further refine our findings and gain a more comprehensive understanding of neutrino oscillation parameters.

It is worth noting that the discovery of a light sterile neutrino would provide a signature for new physics beyond the standard model at low energies. This exciting possibility highlights the importance of continued research in this field.

References

[1] Athanassopoulos, C., *et al. Phys. Rev. Lett.,* 77:3082 (1996), Aguilar A, *et al. Phys. Rev.* D64:112007 (2001).

[2] Aguilar-Arevalo, A.A., *et al. Phys. Rev. Lett.,* 121:221801 (2018).

[3] Antonello M, *et al. Eur. Phys. J.* C73:2599 (2013).

[4] Abdurashitov, J.N., *et al. Phys. Rev.* C73:045805 (2006), Laveder, M., *Nucl. Phys. Proc. Suppl.* 168:344 (2007), Giunti, C., Laveder, M., *Mod. Phys. Lett.* A22:2499 (2007).

[5] Yongchao, Z., *Phys. Rev.*, D 87 (5), 053020 (2013).

[6] Gariazzo, S., *et al., JHEP* 06, 135 (2017).

[7] P.F. de Salas *et al., Phys. Lett.* B 782, 633–640 (2018).

A Bottom-Up Approach to Predict the Majorana Phases

Pralay Chakraborty,* Manash Dey and Subhankar Roy
Department of Physics, Gauhati University, Guwahati-781014, India
*E-mail: pralay@gauhati.ac.in

ABSTRACT: We propose a neutrino mass matrix texture that shelters some linear correlations among its elements in light of general parametrization. The proposed texture predicts the Majorana phases and the Dirac CP phase. In addition, we try to realize the texture from A_4 symmetry.

Keywords: Majorana Neutrino Mass Matrix, Majorana Phases, Dirac CP Phase.

I. Introduction

The Standard Model (SM) [1-2] of particle physics is well-tested based on its predictions which are consistent with the experiments. But when it comes to the origin of neutrino mass, the SM keeps silent as it presents the neutrinos as a massless fermion. But the neutrino oscillation phenomenon [3] predicts the non-zero mass of the neutrino and to explain this one needs to go beyond the territory of the SM. In recent years, neutrino oscillation experiments have measured neutrino mixing angles $(\theta_{12}, \theta_{13}, \theta_{23})$ and mass squared differences $(\Delta m^2_{21}, \Delta m^2_{31})$ with much precision. However, the precise measurement of the Dirac CP phase (δ) is still in the dark. Also, the Majorana CP phases (α, β) cannot be measured in the said experiments. The neutrino mass matrix (M_ν) however, carries the information of all nine physical parameters including the Majorana Phases and it is quite relevant to constrain M_ν so that it can predict some physical parameters. In that light, we adhere to the general parametrization (GP) of the neutrino mass matrix. Under GP, we express the neutrino mass matrix elements in terms of their real and imaginary part as all the Majorana neutrino mass matrix elements are complex in general. We propose a neutrino mass matrix texture having three simple linear correlations among its elements and predicting three physical parameters $\alpha, \beta,$ and δ. This texture may become important after the experimental detection of the Majorana phases.

II. THE NEUTRINO MASS MATRIX TEXTURE

The origin of the neutrino mass matrix can be related to either the See-Saw mechanism or the dimension 5 operator. To understand the neutrino masses, we need to diagonalize the M_ν as $M^d = V^\dagger . M_\nu . V^*$, where $V = U_{lL}^\dagger . U_\nu$. The $U_{lL}^\dagger$ is the left-handed charged lepton diagonalizing matrix and U_ν is the matrix which diagonalizes the neutrino mass matrix in the symmetry basis. In the flavor basis, where the charged lepton mass matrix is already diagonal, $V = U_\nu$ and it is known as the Pontecorvo-Maki-Nakagawa-Sakata matrix (PMNS matrix) [4]. The Particle Data Group (PDG) [5] has adopted a parametrization for U_ν known as standard parametrization as shown in the following,

$$U_\nu = \begin{bmatrix} c_{12}c_{13} & s_{12}c_{13} & s_{13}e^{-i\delta} \\ -s_{12}c_{23} - c_{12}s_{23}s_{13}e^{i\delta} & c_{12}c_{23} - s_{12}s_{23}s_{13}e^{i\delta} & s_{23}c_{13} \\ s_{12}s_{23} - c_{12}c_{23}s_{13}e^{i\delta} & -c_{12}s_{23} - s_{12}c_{23}s_{13}e^{i\delta} & c_{23}c_{13} \end{bmatrix} \begin{bmatrix} e^{i\alpha} & 0 & 0 \\ 0 & e^{i\beta} & 0 \\ 0 & 0 & 1 \end{bmatrix} \quad \ldots (1)$$

where, $c_{ij} = \cos\theta_{ij}$; $s_{ij} = \sin\theta_{ij}$ and $i, j = 1, 2, 3$.

As mentioned earlier, the M_ν contains the information on all the physical parameters including the Majorana phases and we may impose certain conditions on it to make the texture predictive. In recent years, several phenomenological ideas such as texture zero [6-7], vanishing minors [8], hybrid texture [9], and $\mu - \tau$ [10] symmetry are put forward to constrain the M_ν. In the present work, we put some constraints on M_ν in terms of a few correlations among the real and imaginary parts of the mass matrix elements. The proposed neutrino mass matrix texture is shown in the following,

$$M_\nu = \begin{pmatrix} Re[m_{ee}] + i\,Im[m_{ee}] & Re[m_{e\mu}] + i\,Im[m_{e\mu}] & -2\,Re[m_{e\mu}] + 2\,i\,Im[m_{e\mu}] \\ Re[m_{e\mu}] + i\,Im[m_{e\mu}] & Re[m_{\mu\mu}] + i\,Im[m_{\mu\mu}] & -2\,Re[m_{e\tau}] + i\,Im[m_{\mu\tau}] \\ Re[m_{e\tau}] + i\,Im[m_{e\tau}] & Re[m_{\mu\tau}] + i\,Im[m_{\mu\tau}] & Re[m_{\tau\tau}] + i\,Im[m_{\tau\tau}] \end{pmatrix} \quad \ldots (2)$$

The conditions imposed are given below,

$$Re[m_{e\tau}] = -2\,Re[m_{e\mu}] \quad \ldots (3)$$

$$Im[m_{e\tau}] = 2\,Im[m_{e\mu}] \quad \ldots (4)$$

$$Re[m_{\mu\tau}] = -2\,Re[m_{e\tau}] \quad \ldots (5)$$

We solve Eq. (3)-(5) with the help of the Newton-Raphson method and predict α, β, and δ as $47.52^o, 23.02^o$ and 278.82^o respectively. To find the constrained range where α may lie, we take the scatterplots of the functions $Re[m_{e\tau}] + 2\,Re[m_{e\mu}]$, $Im[m_{e\tau}] - 2\,Im[m_{e\mu}]$ and $Re[m_{\mu\tau}] + 2\,Re[m_{e\tau}]$

against α and identify the common range for the said parameter. We see that α lies within a region $[45.83^0, 97.40^0]$ (see Figure 1). Similarly, the common range for β is obtained as $[0^0, 28.64^0]$ (see Figure 2). It is to be noted that we do not get any specific bound for δ as the three relations prescribed in Eq. (3)-(5) hold good for all values of δ (see Figure 3). For the present analysis, we stick to the normal ordering of the neutrino masses and set m_3 at 0.06 eV. The mass-squared differences are fixed at their best-fit values given by the experiments $(7.42 \times 10^{-5}$ and $2.51 \times 10^{-3} \mathrm{eV}^2)$. For the above analysis, we set sharp ranges for the mixing angles around their best-fit values.

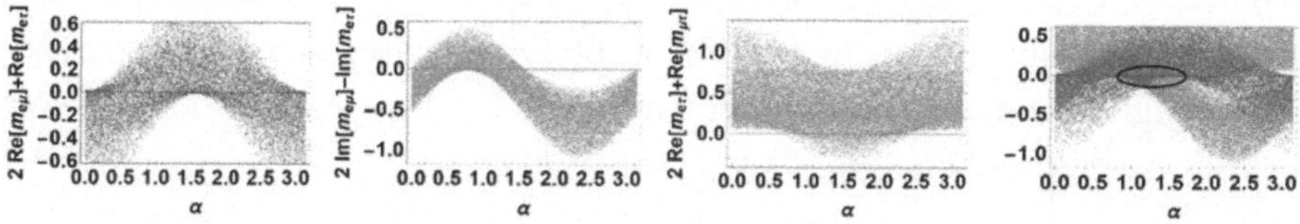

Figure 1: The Scatterplots Against α

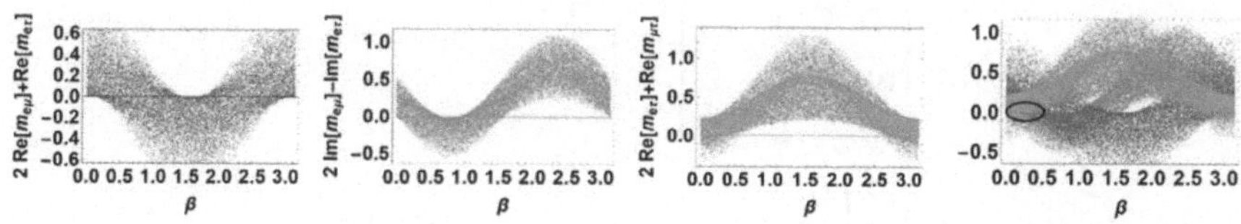

Figure 2: The Scatterplots Against β

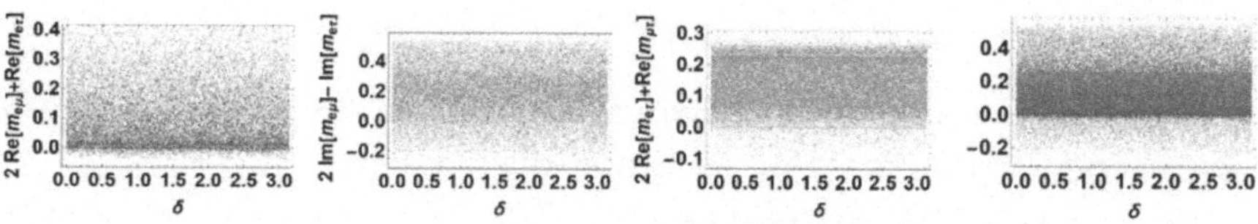

Figure 3: The Scatterplots Against δ

To understand the proposed texture from the first principle, we try to realize the proposed texture from A_4 group. In this regard, we extend the field content of the SM by adding right-handed neutrinos, scalar singlets, and scalar triplets. The transformation properties of the fields used are summarised in Table 1.

Table 1: The Transformation Properties of the Field Contents under

Fields	D_{l_L}	l_R	ϕ	ν_{l_R}	η	κ
$SU(2)_L$	2	1	2	1	1	1
A_4	3	(1, 1″, 1′)	3	3	1	3

$SU(2)_L \times A_4$

The $SU(2)_L \times A_4$ invariant Yukawa Lagrangian is constructed in the following manner,

$$\begin{aligned}\mathcal{L}_{Yukawa} = {} & y_e(\overline{D_{l_L}}\,\phi)e_R + y_\mu(\overline{D_{l_L}}\,\phi)\mu_R + y_\tau(\overline{D_{l_L}}\,\phi)\tau + \\ & y_D\left[(\overline{D_{l_R}}\tilde{\phi})_{3s}\nu_R + (\overline{D_{l_R}}\tilde{\phi})_{3a}\nu_R\right] + \tfrac{1}{2}y_{R_1}(\overline{\nu_R}\nu_R^c)\eta + \\ & \tfrac{1}{2}y_{R_2}(\overline{\nu_R}\nu_R^c)\kappa + h.c \end{aligned} \qquad \ldots (6)$$

Where, $\tilde{\phi} = i\,\tau_2\phi^*$. By assuming the specific vacuum choices for the fields ϕ, η & κ as $v(1, w^2, w)^T, v_\eta \& v_\kappa(0,1,0)^T$ respectively, we obtain the form of the charged lepton mass matrix (M_L), the Dirac neutrino mass matrix (M_D) and the right-handed Majorana neutrino mass matrix (M_R) as shown below,

$$M_L = v\begin{bmatrix} y_e & y_\mu & y_\tau \\ y_e w^2 & y_\mu\, w & y_\tau \\ y_e\, w & y_\mu w^2 & y_\tau \end{bmatrix};$$

$$M_D = 2\,y_D\,v\begin{bmatrix} 0 & 0 & w \\ w^2 & 0 & 0 \\ 0 & 1 & 0 \end{bmatrix} \text{ and}$$

$$M_R = \begin{bmatrix} A & 0 & B \\ 0 & A & 0 \\ B & 0 & A \end{bmatrix} \qquad \ldots (7)$$

in which, $A = y_{R_1}v_\eta, B = y_{R_2}v_\kappa, w = e^{\frac{2\pi i}{3}}$ and $w^2 = w^*$.

The Majorana neutrino mass matrix is constructed on the Symmetry basis (where charged lepton mass matrix is non-diagonal) as,

$$M_{\nu_s} = M_{T_I} = -M_D M_R^{-1} M_D^T.$$

The M_{T_I} is the contribution given by the Type-I Seesaw mechanism. As it is always convenient to work in a basis where the charged lepton mass matrix is diagonal (i.e. Flavour basis), we, therefore, shift from the Symmetry basis to the Flavour basis by diagonalizing M_L: $U_L^\dagger.M_L.U_R = M_L^{diag}$, where $M_L^{diag} = \sqrt{3}v\,\text{diag}(y_\tau, y_e, y_\mu)$. The U_L and U_R is written in the following way,

$$U_L = \frac{1}{\sqrt{3}}\begin{bmatrix} 1 & 1 & 1 \\ 1 & w^2 & w \\ 1 & w & w^2 \end{bmatrix}; U_R = \begin{bmatrix} 0 & 1 & 0 \\ 0 & 0 & 1 \\ 1 & 0 & 0 \end{bmatrix} \qquad \ldots (8)$$

The shifting of basis provides a charged lepton correction to the neutrino mass matrix M_{ν_S}: $U_L^{\dagger}.M_{\nu_S}.U_L^{*} = M_{\nu_f}$. For simplicity we assume $B = 2A$, and after the redefinition of parameters the final Majorana neutrino mass matrix in the flavour basis (M_{ν_f}) with simplification reduces to the following form,

$$M_{\nu_f} = \begin{bmatrix} \frac{32a}{9} & \frac{2a}{9}+\frac{2ia}{3\sqrt{3}} & -\frac{4a}{9}+\frac{4ia}{3\sqrt{3}} \\ \frac{2a}{9}+\frac{2ia}{3\sqrt{3}} & -\frac{16a}{9}+\frac{16ia}{3\sqrt{3}} & \frac{8a}{9} \\ -\frac{4a}{9}+\frac{4ia}{3\sqrt{3}} & \frac{8a}{9} & -\frac{10a}{9}-\frac{10ia}{3\sqrt{3}} \end{bmatrix} \quad \ldots (9)$$

From the matrix appearing in Eq. (9), we identify the proposed relations among the real and imaginary parts.

III. Conclusion

In the present work, we propose a neutrino mass matrix texture having some simple linear correlations among the real and imaginary parts of the mass matrix elements. We carry out the numerical analysis by considering the normal hierarchy for the neutrino masses. The proposed texture predicts three physical parameters $\alpha, \beta,$ and δ. It is to be noted that the texture imposes certain physical conditions, which are to be understood from both the first principle. So, we try to realize the proposed texture from A_4 symmetry.

It is to be highlighted that we have chosen only simple linear relations among the mass matrix elements. This is because a specific neutrino model may encompass different scalar fields in the Yukawa term and the vacuum expectation values are supposed to be assigned intelligently. So, the simple linear correlations among the parameters may help us to choose the model in a better way.

Acknowledgment

The research work of Pralay Chakraborty is supported by Innovation in Science Pursuit for

Inspired Research (INSPIRE), Department of Science and Technology, Government of India, New Delhi vide grant No. IF190651.

REFERENCES

[1] Glashow, S.L., Partial Symmetries of Weak Interactions, *Nucl. Phys*, 22, 579–588 (1961).

[2] Weinberg, S., A Model of Leptons, *Phys. Rev. Lett*, 19, 1264–1266 (1967).

[3] Pontecorvo, B., Mesonium and antimesonium, *Sov. Phys. JETP*, 6, 429, (1957).

[4] Maki, Z., Nakagawa, M., Sakata, S., Remarks on the unified model of elementary particles, *Prog. Theor. Phys.,* 28, 870–880 (1962).

[5] Zyla, P.A., *et al.,* Review of Particle Physics, *PTEP*, 2020, 083C01 (2020).

[6] Ludl, P.O. and Grimus, W., A complete survey of texture zeros in the lepton mass matrices, *JHEP*, 07, 090, (2014).

[7] Borgohain, H. and Borah, D., Survey of Texture Zeros with Light Dirac Neutrinos, *J. Phys. G*, 48, 075005, (2021).

[8] Lashin, E.I., Chamoun, N., One vanishing minor in the neutrino mass matrix, *Phys. Rev. D*, 80, 093004 (2009).

[9] Singh, M., Investigating the hybrid textures of neutrino mass matrix for near maximal atmospheric neutrino mixing, Adv. *High Energy Phys.* 2018, 5806743 (2018).

[10] King, S.F. and Nishi, C.C., Mu-tau symmetry and the Littlest Seesaw, *Phys. Lett. B*, 785, 391–398 (2018).

Explaining Dark Matter and Neutrino Mass in a A4⊗Z8 Flavour Symmetry Based ν2HDM

Lavina Sarma[1], Partha Kumar Paul[1,2,*] and Mrinal Kumar Das[1]

[1]Department of Physics, Tezpur University, Tezpur-784028, India

[2]Department of Physics, Indian Institute of Technology Hyderabad, Kandi, Sangareddy-502285, Telangana, India

*E-mail: ph22resch11012@iith.ac.in

ABSTRACT: We extend the SM with three right-handed heavy neutrinos, one scalar doublet, and a gauge singlet (sterile neutrino, which is a possible dark matter (DM) candidate in our study) to address the tiny neutrino mass and dark matter simultaneously. We calculate the active-DM mixing angle, decay rate, and relic abundance of DM considering the sterile neutrino mass in the keV range. Neutrino phenomenology like the effective mass of the neutrino is also studied in our work for low-scale right-handed neutrino masses. We validate our model with constraints from experiments such as Lyman-α, X-ray observation, and KamLAND-Zen.

Keywords: Beyond Standard Model, Neutrino Physics, Dark Matter, Neutrinoless Double Beta Decay.

I. Introduction

Though the SM of particle physics has been successful immensely in describing masses of the elementary particles and the interactions between them, it fails to account for the DM, non-zero masses of the neutrinos, baryon asymmetry of the universe, etc. In SM a fermion to have a mass term, it should have both the left and right chiral components. But the neutrinos in the SM have only the left chiral part and as a result, they are massless in the SM. But oscillation experiments suggest that they have non-zero tiny masses. Now staying in the SM, it is not possible to explain the non-zero neutrino mass. On the other hand from Planck data, we found only 4.9% of the total energy budget of the Universe consists of baryonic matter, 26.8% is some form of non-luminious, non-interacting (or very weakly interacting) matter known as Dark matter, and 68.3% is the Dark energy.

II. The Model

Neutrino two Higgs doublet model (ν2HDM) [3] is one of the natural choices for explaining tiny neutrino mass. In our work, we have studied ν2HDM in the framework of Minimal Extended Seesaw (MES) [2] where we considered a gauge singlet (S) along with 3 right-handed neutrinos (N_1, N_2, N_3), one extra Higgs doublet (η), and 3 sets of flavon fields (φ, ξ, ζ) [1].

Table 1: Particle Content and Their Charge Assignments under A_4 and Z_8 Groups

	l	e_R	μ_R	τ_R	H	η	φ	φ'	φ''	ξ	ξ'	χ	N_1	N_2	N_3	S	ζ	ζ'	ζ''
A_4	3	1	1″	1′	1	1	3	3	3	1	1′	1	1	1′	1	1	3	3	3
Z_8	ω^3	ω^4	ω^4	ω^4	ω^2	ω	ω^7	ω^2	ω^3	ω^6	ω^4	ω	ω^5	ω^2	ω	ω^2	ω^7	ω^2	ω^3

The Lagrangian of the model is given by,

$$L = \frac{y_e}{\Lambda}(\bar{l}H\varphi)e_R + \frac{y_\mu}{\Lambda}(\bar{l}H\varphi)\mu_R + \frac{y_\tau}{\Lambda}(\bar{l}H\varphi)\tau_R + \frac{y_1}{\Lambda}(\bar{l}\tilde{\eta}\varphi)N_1 + \frac{y_2}{\Lambda}(\bar{l}\tilde{\eta}\varphi')N_2 + \frac{y_3}{\Lambda}(\bar{l}\tilde{\eta}\varphi'')N_3 + \left(\frac{1}{2}\right)\lambda_1\xi\overline{N^c{}_1}N_1 + \left(\frac{1}{2}\right)\lambda_2\xi'^{\overline{N^c{}_2}N_2} + \left(\frac{1}{2}\right)\lambda_3\xi\overline{N^c{}_3}N_3 + \left(\frac{1}{2}\right)\rho\chi\overline{S^c}N_1 + \frac{y_1}{\Lambda}(\bar{l}\tilde{\eta}\zeta)N_1 \frac{+y_2}{\Lambda}(\bar{l}\tilde{\eta}\zeta')N_2 + \frac{y_3}{\Lambda}(\bar{l}\tilde{\eta}\zeta'')N_3 \quad \ldots (1)$$

Here $\tilde{\eta} = i\sigma_2\eta *$. In equation (1), the last 3 terms are the perturbations that we have included to break the μ-τ symmetry in the light neutrino mass matrix which has a crucial role to play in reproducing non-zero mixing angles. We have considered the following vacuum expectation value (vev) alignment,

$$<\varphi> = (k, 0, 0), <\varphi'> = <\varphi''> = (k,k,k), <\zeta> = <\zeta'> = <\zeta''> = (0,k,0), <\xi> = <\xi'> = k, <\chi> = u$$

Using the A_4 product rule and the above vev alignment we calculated the Dirac mass matrix (M_D), right-handed neutrino mass matrix (M_R), and sterile neutrino mass matrix (M_S) as,

$$M_D = \begin{pmatrix} a & b & c+p \\ 0 & b+p & c \\ p & b & c \end{pmatrix}, M_R = \begin{pmatrix} M_1 & 0 & 0 \\ 0 & M_2 & 0 \\ 0 & 0 & M_3 \end{pmatrix} \text{ and } M_S = (s \quad 0 \quad 0) \quad \ldots (2)$$

where $a = \frac{\langle\tilde{\eta}\rangle k}{\Lambda}y_1, b = \frac{\langle\tilde{\eta}\rangle k}{\Lambda}y_2, c = \frac{\langle\tilde{\eta}\rangle k}{\Lambda}y_3, M_1 = \lambda_1 k, M_2 = \lambda_2 k, M_3 = \lambda_3 k, s = \rho u$

III. Neutrino Mass and Sterile Neutrino Dark Matter

Now in the MES, the 4×4 active-sterile neutrino mass matrix is given by [2],

$$M_\nu^{4\times4} = -\begin{pmatrix} M_D M_R^{-1} M_D^T & M_D M_R^{-1} M_S^T \\ M_S (M_R^{-1})^T M_D^T & M_S M_R^{-1} M_S^T \end{pmatrix} \quad \ldots (3)$$

And the active-sterile mixing matrix takes the form,

$$V \simeq \begin{pmatrix} \left(1-\frac{1}{2}RR^{\dagger}\right)U_{PMNS} & R \\ -R^{\dagger}U_{PMNS} & 1-\frac{1}{2}R^{\dagger}R \end{pmatrix} \quad \ldots (4)$$

where $R = M_D M_R^{-1} M_S^T (M_S M_R^{-1} M_S^T)^{-1}$ is a 3×1 matrix that represents the strength of active-sterile mixing and U_{PMNS} is the leptonic mass matrix for active neutrino.

Now the active-sterile neutrino mass matrix can be diagonalized by this mixing matrix as,

$$M_\nu^{4\times4} = V.diag(m_1, m_2, m_3, m_4).V^T \quad \ldots (5)$$

where m_1, m_2, m_3, and m_4 are the mass eigenstates of the active and sterile neutrinos.

The effective neutrino mass can be written in the standard parametrization of (3+1) active-sterile mixing matrix as,

$$|m_{\beta\beta}| = |c_{13}^2 c_{12}^2 c_{14}^2 m_1 e^{i\phi_1} + c_{13}^2 s_{12}^2 c_{14}^2 m_2 + s_{13}^2 c_{14}^2 m_3 e^{i\phi_2} + s_{14}^2 m_4| \quad \ldots (6)$$

And the sterile neutrino mass is calculated as,

$$m_S \sim m_{DM} \simeq -M_S M_R^{-1} M_S^T \simeq \frac{s^2}{M_1} \quad \ldots (7)$$

This sterile neutrino is considered the DM candidate in our model. We considered the nonresonant production of this sterile DM. The relic abundance for this nonresonantly produced DM is given as,

$$\Omega_{DM}h^2 \simeq 0.3 \times 10^{10} sin^2(2\theta_{DM}) \left(\frac{m_{DM}\times10^{-2}}{keV}\right)^2 \quad \ldots (8)$$

where $sin^2(2\theta_{DM}) = 4(V_{14}^2 + V_{34}^2)$

The decay rate of this sterile DM to SM neutrino and photon is given as,

$$\Gamma \simeq 1.38 \times 10^{-22} sin^2(2\theta_{DM}) \left(\frac{m_{DM}}{keV}\right)^5 s^{-1} \quad \ldots (9)$$

IV. Result

We have taken the right-handed neutrino masses as $M_1 = 10^4$-10^5 GeV, $M_2 = 5*10^5$-10^6 GeV, and $M_3 = 10^7$-10^8 GeV, the lightest neutrino mass as $m_l = 10^{-10}$-10^{-9} eV.

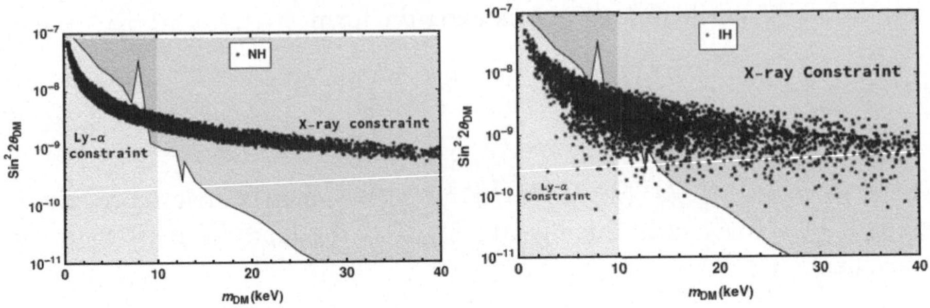

Figure 1

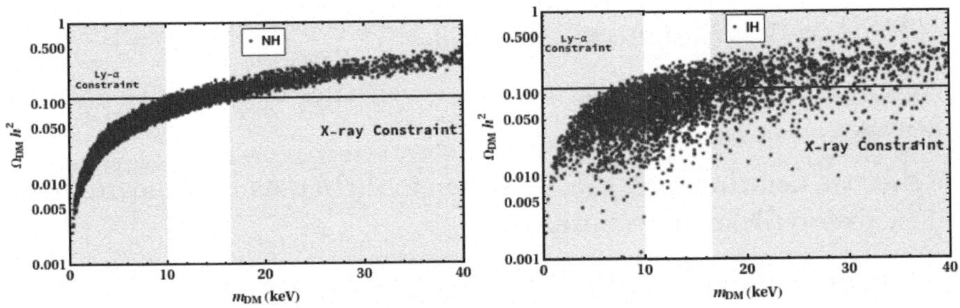

Figure 2

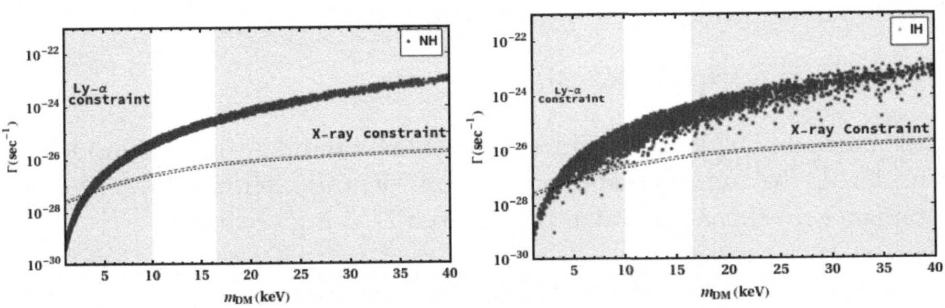

Figure 3

We have studied the active-DM mixing as a function of DM mass for both the normal (NH) and inverted hierarchies (IH), and shown in fig1. We see that NH is completely excluded by the Ly-α[5] and X-ray constraints [6] and a very small region between 10 keV and 16 keV falls in the allowed parameter space for IH. In Figure 2, we have shown the DM relic abundance vs DM mass. Though we have shown the variation of relic density vs DM mass for NH, the mass range for NH is not allowed. We got a large not of points satisfying the Planck bound [4] for IH which is 0.1187±0.0017. We have calculated the DM decay rate and plotted it against DM mass in Figure 3. We have also considered the lifetime constraint for sterile neutrino w.r.t its mass.

The red dashed line corresponds to the upper bound coming from the lifetime of sterile neutrinos for the case when energy transfer from CMB photons to gas is included and the blue dotted line is for the case when this energy is excluded in Figure 3. We found a very small number of points at 10 keV for IH.

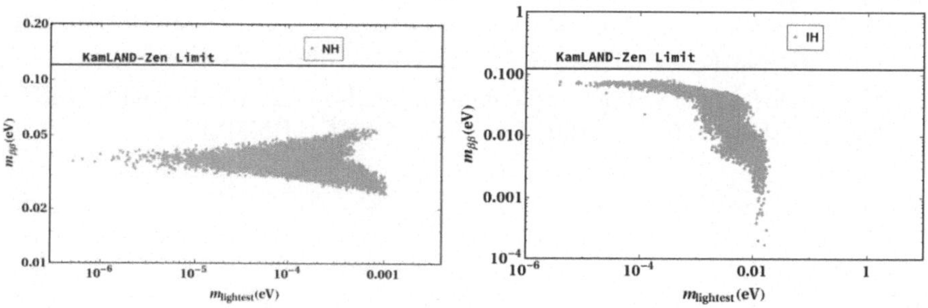

Figure 4

The lightest active neutrino mass is also constrained in our work. Both the NH and IH satisfy the KamLAND-Zen bound, which is $m_{\beta\beta} < 0.12$ eV. From Figure 4, the allowed range for NH is approximately $10^{-5} - 10^{-3}$ eV and that for IH is $10^{-4} - 10^{-2}$ eV.

V. Conclusion

We studied ν2HDM with a gauge singlet (S) in the framework of Minimal Extended Seesaw with $A_4 \otimes Z_8$ flavour symmetry. We have calculated the DM-active mixing angle, relic abundance, the decay rate of the keV scale DM, and the effective neutrino mass in the model. We observed that inverted hierarchy has shown a wider range of parameter space in the case of DM compared to normal hierarchy. On the other hand, an identical allowed parameter space in both hierarchies in the case of neutrino mass has been seen. Our model also gave a bound on the lightest active neutrino mass. Thus our model can explain the DM content of the Universe and neutrino masses and mixings simultaneously. It can also successfully explain the baryon asymmetry of the Universe which can be found in this reference [1].

References

[1] Sarma, L., Paul, P.K. and Das, M.K., Connecting dark matter, baryogenesis and neutrinoless double beta decay in a A 4 ⊗ Z 8 based ν2HDM, *Int. J. Mod. Phys.* A 37 (2022) 27, 2250157, [arXiv:2208.14764[hep-ph]]

[2] Zhang, H., Light Sterile Neutrino in the Minimal Extended Seesaw, *Phys. Lett.* B 714 262–266 (2012), arXiv:1110.6838 [hep-ph].

[3] Ma, E., Naturally small seesaw neutrino mass with no new physics beyond the TeV scale, *Phys. Rev. Lett.*, 86 (2001), 2502–2504, arXiv:hep-ph/0011121.

[4] Planck Collaboration and Aghanim, N. *et al.*, Planck 2018 results. VI. Cosmological parameters, arXiv:1807.06209 [astro-ph.CO].

[5] Baur, J. *et al.*, Constraints from Ly-α forests on non-thermal dark matter including resonantly-produced sterile neutrinos, *JCAP* 12 (2017) 013 [arXiv:1706.03118] [INSPIRE].

[6] Perez K., Ng K.C.Y., Beacom J.F., Hersh C., Horiuchi S. and Krivonos R., Almost closing the νMSM sterile neutrino dark matter window with NuSTAR, *Phys. Rev.* D 95 (2017) 123002 [arXiv:1609.00667] [INSPIRE].

Study of Two-Zero Textures of $M_\nu^{4\times4}$ in Minimal Extended Seesaw Mechanism

P. Kumar

Department of Physics, Dakshin Kamrup College, Assam-781125, India

E-mail: prianca.kumar@gmail.com

ABSTRACT: This chapter includes a phenomenological study of two-zero textures of $M_\nu^{4\times4}$ *in the context of 4 neutrino scenario considering the Minimal Extended Seesaw (MES) mechanism. MES mechanism is an extension to the canonical type-I seesaw mechanism and deals with 3 × 3 form of Dirac neutrino mass matrix* M_D *and right-handed Majorana neutrino mass matrix* M_R *along with 1 × 3 form of* M_S *which couples the right-handed neutrinos and the sterile singlet 'S'. In this work, I have realized the two-zero textures of* $M_\nu^{4\times4}$*considering MES mechanism under (5+4) scheme, that is, 5 zeros in* M_D *and 4 zeros in*M_R. *Within the given scenario, realization of the textures yields certain constraint correlations which are then examined within 3σ range of experimental values. I have considered 3σ range for Dirac CP phase* δ_{13} *while all the other five CP phases are kept unconstrained. Viability of the textures are presented in the form of scatter plots under recent neutrino oscillation data.*

Keywords: Sterile Neutrino, Minimal Extended Seesaw Mechanism, Zero-textures.

I. Introduction

Over the past decades, tremendous progress have been observed in the field of neutrino physics. Untiring efforts from physicists around the globe have been successful in providing us with precise information on neutrino parameters like mixing angles $(\theta_{12}, \theta_{13}, \theta_{23})$ and squared mass differences $(\Delta m_{21}^2, \Delta m_{31}^2)$ and also in theoretical formulation of the underlying theory in case of three neutrino paradigm. However, a number of problems like origin of sub-eV neutrino mass, CP violation, mass hierarchy still remains unsolved. Amidst these problems in three neutrino scenario, there exists a number of oscillation experiments which indicates towards the presence of an additional species of neutrino. For instance, the LSND experiment [1,2] observed the oscillation probability between $\bar{\nu}_\mu$ and $\bar{\nu}_e$ which corresponds to mass squared difference of $\sim eV^2$. This anomaly could not be explained in the context of three neutrino scenario. Similar anomalies were observed by MiniBooNE experiment [3] Gallium solar neutrino experiment [4, 5], reactor neutrino experiment [6] and cosmological observations[7]. These discrepancies indicates towards the presence of an additional species of neutrino at the eV scale.

From the theoretical view point, in order to accomodate the fourth flavour of neutrino, the Standard Model (SM) of particle physics must be extended. This can be done in a number of ways, however, the (3+1) scenario, that is, three active and one light sterile neutrino serves to be the minimal extension of SM [8]. One popular approach of including light sterile neutrino in theory is the extension of type-I seesaw mechanism, now popularly known as Minimal Extended Seesaw (MES) Mechanism [9]. Here the SM is extended by an additional gauge singlet field 'S'. In MES, even without inserting any tiny yukawa couplings or mass scale, an eV scale sterile neutrino naturally appears. The neutrino mass matrix thus takes a 4 × 4 form.

This chapter includes realization of two-zero textures of 4 × 4 neutrino mass matrix considering MES mechanism. Zero-texture model simplifies the complexities of the neutrino mass matrix by inducing relation among the different parameters of the matrix and thereby reducing the number of free parameters [10]. In this chapter, I shall consider the 4 × 4 MES neutrino mass matrix $M_\nu^{4\times4}$ which includes Dirac (M_D) and Majorana (M_R) mass matrix in its 3 × 3 form and 1 × 3 form of M_S which couples the singlet 'S' with the three right-handed neutrinos. In this work zeros in $M_\nu^{4\times4}$ is realized considering (5 + 4) scheme, that is, 5 zeros in M_D and 4 zeros in M_R. Also one-zero texture of M_S is being considered. It has been observed that under MES and (5+4) scheme only 6 out of 15 allowed two-zero textures [11] can be realized. Realization of each of the textures lead to a number of correlations which are then examined under recent neutrino oscillation data [12] and plotted against $sin\,\theta_{34}$ which is bounded by an upper limit of 0.4 [12]. Also for viability check of the textures, out of six CP phases, five of them are kept unconstrained $(\alpha, \beta, \gamma, \delta_{14}, \delta_{24})$, while 3σ range of δ_{13} is considered, that is, (120°–358°) for NH and (200°–353°) for IH [12].

II. Minimal Extended Seesaw (MES) Mechanism

In MES model one singlet '*S*' are introduced apart from the SM particle content. The Lagrangian in this case takes the form,

$$L = \overline{\nu_L} M_D \nu_R + \overline{S}^C M_S \nu_R + 1/2\, \overline{\nu}^C M_R \nu_R + h.c \qquad \dots (1)$$

Here M_S is a (1×3) matrix since only one extra singlet is considered. For $M_R > M_S > M_D$ the heavy RH ν decouple and in the basis (ν_L, S^c) and the neutrino mass matrix takes the form,

$$M_\nu^{4\times4} = \begin{pmatrix} M_D M_R^{-1} M_D^T & M_D M_R^{-1} M_S^T \\ M_S (M_R^{-1})^T M_D^T & M_S M_R^{-1} M_S^T \end{pmatrix} \quad \ldots (2)$$

In this work, two-zero textures are realized under MES considering Eq. (2).

III. Two-zero textures in $M_N^{4\times4}$

In the flavor basis the 4 × 4 Majorana neutrino mass matrix can be expressed as,

$$M_\nu^{4\times4} = \begin{pmatrix} m_{ee} & m_{e\mu} & m_{e\tau} & m_{es} \\ m_{e\mu} & m_{\mu\mu} & m_{\mu\tau} & m_{\mu s} \\ m_{e\tau} & m_{\mu\tau} & m_{\tau\tau} & m_{\tau s} \\ m_{es} & m_{\mu s} & m_{\tau s} & m_{ss} \end{pmatrix} \quad \ldots (3)$$

Out of 15 viable two zero textures [11], the 6 textures which can be realized under MES mechanism under (5+4) scheme are: Texture A_1: $m_{ee} = 0, m_{e\mu} = 0$; Texture A_2 with zeros in $m_{e\mu}, m_{e\tau}$; Texture B_3 with $m_{e\mu} = 0, m_{\mu\mu} = 0$; Texture B_4 with $m_{e\tau} = 0, m_{\tau\tau} = 0$; Texture D_1 with $m_{\mu\mu} = 0, m_{\mu\tau} = 0$ and Texture D_2 with $m_{\mu\tau} = 0$, $m_{\tau\tau} = 0$. All these 6 textures $A_1, A_2, B_3, B_4, D_1, D_2$ allows Normal Hierarchical (NH) mass spectrum, while Inverted Hierarchical (IH) mass spectrum is allowed for four textures B_3, B_4, D_1, D_2.

IV. Realization of two-zero textures under MES mechanism and (5+4) scenario

The general form of non-symmetric Dirac neutrino mass matrix M_D and symmetric right-handed Majorana mass matrix M_R and and 1 × 3 matrix M_S with three independent entries can be represented as:

$$M_D = \begin{pmatrix} a & b & c \\ d & e & f \\ g & h & i \end{pmatrix}, M_R = \begin{pmatrix} A & B & C \\ B & D & E \\ C & E & F \end{pmatrix}, M_S = \begin{pmatrix} s_1 & s_2 & s_3 \end{pmatrix} \ldots (4)$$

Realization of the two-zero textures:

1. **Texture A_1**: For realization of the texture A_1, the following combination of M_D, M_R, M_S is being considered:

$$M_D = (b, d, g, i) \neq 0,\ M_R = (A, E) \neq 0,\ M_S = (s_1 \quad 0 \quad s_3) \quad \dots (5)$$

Equation (2) then gives the following correlations:

$$\frac{m_{\mu\tau}}{m_{\mu s}} = \frac{m_{\tau\tau}}{m_{ss}} = \frac{m_{\tau s}}{m_{ss}} = \sqrt{\frac{m_{\tau\tau}}{m_{ss}}} \quad \dots (6)$$

The correlation in Eq. (6) is scanned under recent neutrino oscillation data and a scatter plot is obtained by plotting against $\sin\theta_{34}$ which is bounded by upper limit of 0.4. From Figure 1 (a), it is evident that overlapping does not exist between the different equations of the correlation. And thus this texture is not allowed.

2. **Texture B_3:** Texture B_3 can be realized with the following combination in Eq. (7).

$$M_D = (a, f, g, h) \neq 0,\ M_R = (A, E) \neq 0,$$

$$M_S = (s_1 \quad s_2 \quad 0), \frac{m_{ee}}{m_{e\tau}} = \frac{m_{e\tau}}{m_{\tau\tau}} = \sqrt{\frac{m_{ee}}{m_{\tau\tau}}} \quad \dots (7)$$

Scatter plot of correlation in Eq. (7) for Normal hierarchy is shown in Figure 1(b). It can be seen that the texture is allowed for values of $\sin\theta_{34} > 0.08$.

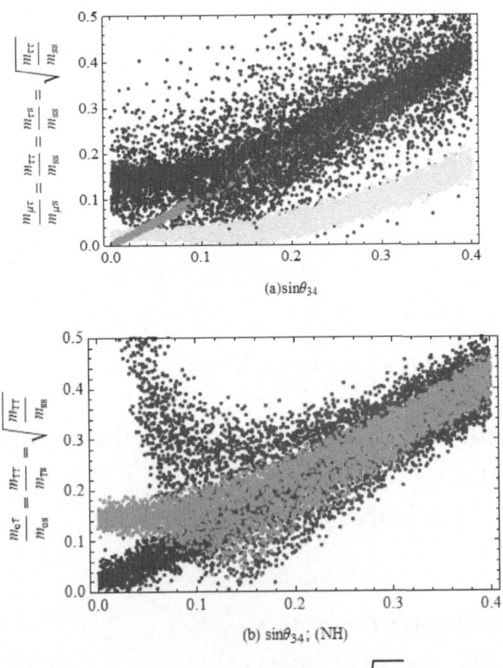

Figure 1: Scatter Plot: (a) A_1: $\frac{m_{\mu\tau}}{m_{\mu s}} = \frac{m_{\tau\tau}}{m_{ss}} = \frac{m_{\tau s}}{m_{ss}} = \sqrt{\frac{m_{\tau\tau}}{m_{ss}}}$, (b) B_3: $\frac{m_{e\tau}}{m_{es}} = \frac{m_{\tau\tau}}{m_{\tau s}} = \sqrt{\frac{m_{\tau\tau}}{m_{ss}}}$

Texture D_1 with the combination $M_D = (a,b,f,h) \neq 0,\ M_R = (C,D) \neq 0,\ M_S = (s_1 \quad s_2 \quad 0)$ yields the same correlation as in Eq. (7) and therefore shows similar phenomenology.Similarly for textures A_2, B_4, D_2,

$$\mathrm{A}_2{:}\, M_D = (b,d,f,g) \neq 0,\ M_R = (A,E) \neq 0,$$

$$M_S = (s_1 \quad 0 \quad s_3), \frac{m_{\mu\mu}}{m_{\mu\tau}} = \frac{m_{\mu\tau}}{m_{\tau\tau}} = \sqrt{\frac{m_{\mu\mu}}{m_{\tau\tau}}}$$

$$\mathrm{B}_4{:}\, M_D = (b,d,e,i) \neq 0,\ M_R = (C,D) \neq 0,$$

$$M_S = (s_1 \quad s_2 \quad 0), \frac{m_{ee}}{m_{e\mu}} = \frac{m_{e\mu}}{m_{\mu\mu}} = \sqrt{\frac{m_{ee}}{m_{\mu\mu}}}$$

$$\mathrm{D}_2{:}\, M_D = (c,d,f,h) \neq 0,\ M_R = (B,F) \neq 0,$$

$$M_S = (s_1 \quad 0 \quad s_3), \frac{m_{ee}}{m_{e\mu}} = \frac{m_{e\mu}}{m_{\mu\mu}} = \sqrt{\frac{m_{ee}}{m_{\mu\mu}}}$$

V. Conclusion

In this chapter, two-zero textures of $M_\nu^{4\times4}$ have been explored in the context of MES mechanism under (5+4) scheme. It has been found that out of 15 experimentally allowed two-zero textures only 6 can be realized under MES and (5+4) scheme. On realizing these 6 textures certain correlations among the different elements of the mass matrices have been obtained. These correlations were then examined under recent neutrino oscillation data and by considering 3σ range of Dirac CP phase δ_{13} and plotted against $\sin\theta_{34}$ which is only bounded by upper limit of 0.4. On scanning the correlations it has been observed that the texture A_1 is not allowed under the considered scenario as overlapping ceases between the different equations of the correlation. Texture A_2, B_4 is allowed for all values of $\sin\theta_{34}$. Texture B_3 is allowed for $\sin\theta_{34} > 0.08$ for NH while for IH the texture is allowed for all values of $\sin\theta_{34}$. While texture D_1, D_2 exhibits similar phenomenology to B_3, B_4 respectively.

References

[1] Athanassopoulos, C., *et al.*, Evidence for $\bar{\nu}_\mu \rightarrow \bar{\nu}_e$ oscillations from the LSND Experiment at the Los Alamos Meson Physics Facility, *Phys. Rev. Lett.*, 77, 3082 (1996).

[2] Aguilar, A., *et al.*, Evidence for neutrino oscillations from the observation of $\bar{\nu}_e$ appearance in a $\bar{\nu}_\mu$ beam, *Phys. Rev. D* 64, 112007 (2001).

[3] MiniBooNE Collaboration: Aguilar-Arevalo, A. A. *et al.*, Improved search for $\bar{\nu}_\mu \rightarrow \bar{\nu}_e$ oscillations in the Miniboone Experiment, *Phys. Rev. Lett.* 110, 161801 (2013).

[4] Acero, M. A., Giunti C. and Laveder M., Limits on ν_e and $\bar{\nu}_e$ disappearnce from Gallium and reactor experiments, *Phys. Rev.* D 78, 073009 (2008).

[5] Giunti, C. and Laveder, M., Statistical significance of the gallium anomaly, *Phys. Rev.* C 83, 065504 (2011).

[6] Mention, G. *et al.*, Reactor neutrino anomaly, *Phys. Rev.* D 83, 073006 (2011).

[7] Abazajian, K.N. *et al.*, Light Sterile neutrinos: A White Paper, FERMILAB-PUB-12-881-PPD, arXiv: 1204.5379 [hep-ph] (2012).

[8] Gomez-Cadenas, J.J and Gonzalez-Garcia, M.C., Future ν_τ oscillation experiments and present data, *Z. Phys. C* 71 443–454, (1996) [hep-ph/9504246].

[9] Barry, J., Rodejohann, W. and Zhang, H., Light sterile Neutrinos: Models and phenomenology, *JHEP* 1107 091, (2011) [arXiv: 1105.3911.

[10] Frampton, P.H. Glashow, S.L. and Marfatia, D., Zeroes of the neutrino mass matrix, Phys. Lett. B 536, 79–82 (2002).

[11] Ghosh, M., Goswami, S. and Gupta, S., Two zero mass matrices and sterile neutrinos, *JHEP* 1304103(2013); arXiv:1211.0118 [hep-ph].

[12] de Salas, P.F., *et al.*, 2020 Global reassessment of the neutrino oscillation picture, JHEP 71 (2021), arXiv: 2006.11237; Gariazzo, S. *et al.*, Light sterile neutrinos, *J. Phys. G* 43 (2016) 033001, arXiv: 1507.0820.

Section-IV

Astrophysics, Cosmology and Nuclear Physics

Study of Properties of Active Galactic Nuclei and Observational Findings

Hritwik Bora[1], Niranjan Gogoi[1], Biplob Sarkar[2*], Sree Bhattacherjee[2] and Rupjyoti Gogoi[1]

[1]Department of Physics, Tezpur University, Napaam-784028, Assam, India

[2]Department of Applied Sciences, Tezpur University, Napaam-784028, Assam, India

*E-mail: biplobs@tezu.ac.in

ABSTRACT: Active Galaxies Nuclei (AGNs) are some of the most violent sources in the universe. Several ongoing studies are trying to explain the underlying physical processes of enormous energy emission and the particle acceleration to very high energies in the jet. However, ongoing debates remain regarding its unification scheme based on the properties of radio emission and orientation. Here we present a review work that primarily includes a historical overview and discusses various AGN regions from the unification scheme. We have also incorporated some of the current observational statuses by different missions worldwide with some underlying problems and research scope in the study of AGNs.

Keywords: Active Galactic Nuclei, Radio Weak, Radio Loud, Highly Ionized Gas, Central Engine, Supermassive Black Hole, Spectrum, Broad Line Region, Accretion Disk, Starburst, Jets.

I. INTRODUCTION

Active Galactic Nuclei (AGNs) are very energetic astronomical objects which emit bright light from its nucleus. These are supermassive black holes that continuously accrete gases into themselves and grow in size, making them "active," and they are mostly found at the heart or "nucleus" of the "galaxies," hence the terminology: Active Galactic Nucleus. These astronomical sources are more powerful than the nuclei of normal galaxies and emit radiation over the entire range of electromagnetic (EM) radiation. This feature makes a zoo of AGN, of which quasars are a subset. This large number of classes and subclasses makes the AGN more fascinating but confusing at the same time. Studying the emission spectra over the EM spectrum and their evolution over time provide complementary ways to study AGN. Seyfert (1943) discovered the presence of broad and strong emission spectral lines in the nuclei of six spiral nebulae. In 1963, Schmidt discovered quasars [1], which led to the birth of an entirely new branch of astronomy. AGN have exciting properties: (a) High luminosities making them visible up to very high redshifts [2]. (b) Narrow emission region implying high energy densities [3]. (c) Strong evolution of their luminosity functions. (c) Emission spectra that cover the entire EM spectrum [4].

II. The Zoo of AGN

There are a large number of classes and sub-classes of an AGN. The Unified model, however, attempts to explain this range of AGNs under the assumption that they differ only in luminosity and viewing angle. The AGNs are primarily classified on their observed activity in the radio band: (a) weak (radio weak) and (b) strong radio emitters (radio loud). The commonly observed radio-weak AGNs are the Seyfert galaxies (Type I and II), which have a morphology of spiral galaxies. The radio-loud galaxies are further classified into Narrow Line Radio Galaxies (NLRG), Broad Line Radio Galaxies (BLRG), and Blazars. Fanaroff-Riley I (FR I) and Fanaroff-Riley II (FR II) radio galaxies are often elliptical, and their emission spectra fall in the narrow line region and, therefore, together constitute NLRG.

III. Main Components of AGN

AGN comes in various morphological structures and sizes. At one extreme, we find an unresolved compact core of the luminous sphere, and at the other extreme, we have complex structures extending up to hundreds of kiloparsecs (kpcs). However, based on typically observed characteristics, one may develop a primary morphological feature of an AGN.

Cores are compact unresolved luminous radio components that coincide with the nucleus of the galaxy associated with it (if resolved) and are found in almost all quasars and nearly 80% of all radio galaxies. The two extended radio emission regions on opposite sides of the galaxy, or the nuclei, are called the lobes. Lobes can be separated by several hundred kpcs or, in some extreme cases, even up to a few megaparsecs (Mpcs). The Broad Line Region (BLR) of an AGN is responsible for the intensive broad emission lines in the ultraviolet and optical band region of the spectrum and is considered one of the basic signatures of an AGN. The outer edge of BLR has a dusty and molecular torus. It may be due to the obscuration of the BLR clouds by this dust torus, which makes them invisible in a few AGNs. The Narrow line region (NLR) is a heated and ionized gas located further away from the central engine.

One of the prime characteristics of the NLR region is that the dust torus cannot affect its visibility as it is present outside it. So narrow lines can be seen even if the broad lines are obscured. The region between the BLR and NLR with intermediate column densities is called the Highly Ionized Gas (HIG) region. Strong emission and absorption lines are observed in the X-ray region and are often called "warm absorbers." Because of their transparency to Ultraviolet and soft X-ray radiation, intense radiation pressure leads to mass outflows, also known as HIG outflows. Torus is a thick dusty molecular structure of 1–100 parsec near the central engine. It is an important structural element because it provides the angle-dependent observational properties of AGN in the Unified theory of AGN. The orientation of the torus relative to the observer decides whether the AGN is a type I or type II. The accretion disk of an AGN is relatively larger than the X-ray binaries. It is considered that an AGN emits X-rays comprising 10% of its total energy emission. The starburst region in an AGN has a dimension of several kpcs. Supernovae explosions and fast stellar winds are the prime energy sources in these regions. The radiation bridge connecting the central compact source with the extended lobes of the AGN is called a "Jet." They are the common feature observed in most of the AGN and are either two-sided or one-sided. They are visible in radio, optical, and X-rays. They transport energetic particles at relativistic velocities from the nucleus to extended radio structures.

IV. Observational Findings

It is assumed that most galaxies in our universe have a black hole at their center. However, the active galaxies are considered to harbor a supermassive black hole [5]. The black hole in Active galaxies is a strong power source, while the black hole lies dormant in normal galaxies. The evidence of a black hole comes from observing the center of our galaxy, the Milky Way. From various telescopic observations, it has been found that our Milky Way galaxy has a compact radio source at its center and is called Sagittarius A* (Sgr A*).

In the starting 1990s, several infrared observations of our galaxy were taken in the form of images [5]. An individual star S2 has been observed, and its motions were found to be completely apparent during the comparison of subsequent frames. The star S2 is observed to have a period of 15 years around Sgr A*; along with it, other stellar objects also have been observed. It was found that the stars at the galaxy's center are not following the properly oriented orbits and are not aligned to the ecliptic plane analogous to the solar system. It was concluded that to have such random motions, there must be a presence of a body with mass $4 \times 10^6 M_\odot$ (where $M_\odot = 1.98 \times 10^{30}$ kg is the solar mass) at the center, called the black hole. But it is now known that every galaxy has an supermassive black hole at its centre. And in the case of the Milky Way, there is the presence of Sgr A*. The obvious question here is that even after harboring a supermassive black hole of the order of $10^6 M_\odot$, it is not an active galaxy. This issue has a straightforward solution: for a galaxy to be an active one there should be an infall of material onto a black hole, thus, accreting the gas and dust, which form an accretion disk.

V. Current Observational Status (Aspects of Multi-Messenger Astronomy)

The birth of radio astronomy led to a major contribution to detecting numerous bright sources in the sky. The improvement in the resolution of the telescopes has played a significant role as it successfully detected quasars in the late 1950s.

As discussed in the above sections of this paper, there are differences between the various subclasses of AGN. In the case of Seyfert galaxies, there are some observational differences in optical-UV spectra. Seyfert 1 shows strong, very broad broadening (2000–10,000 km/s), whereas Seyfert 2 does not show a broadening of more than 1200 km/s. To draw the differences between the Quasi-Stellar Object (QSO) and Seyfert galaxies, researchers have considered a bolometric luminosity of about 10^{45} erg/s; above which the object is a QSO. X-ray astronomy has a crucial role to play as X-ray originates close to the black hole, which enables the study of the immediate environment of a supermassive black hole. In soft X-ray regimes, large amplitude variations are found to have relatively shorter time scales than in hard X-ray regimes [6]. However, the correlation study still does not unveil anything between the variability amplitude and variability time scale with the frequency band. Various studies have found that multi-band light curves of many unobscured AGNs indicate variability time scales and variability amplitudes inversely correlated with source luminosity. The quest for variability has not come to an end. However, many studies state that the variability is due to the black hole mass. The low-mass AGNs show larger amplitudes variations compared to larger black hole masses [6].

AstroSat, the first multiwavelength space observatory of India has also observed blazar sources like OJ 287, Mrk 421, 4C+21.35, and 1ES 1218+304. Singh *et al.* 2022 [7] recently studied source OJ 287, where they strongly suggest that the evolution of the optical-UV synchrotron spectrum is responsible for the variation in the normal broad-band spectral state of the X-ray spectral variations. At the same time, Diwan *et al.* 2021 [8] reported a long-term study of the blazar 1ES 1218+304 using simultaneous data from various missions, including AstroSat. They reported a 20-minute time scale of minute fluctuation revealed from the light curve of the Astrosat/SXT instrument. The short-time variability of the source Mrk 421 has also been examined in detail [9] using multiwavelength observations, including soft and hard X-ray observations from Astro Sat. They computed the variability to be increasing in soft X-ray, optical, and near-infrared regions compared to hard X-rays.

VI. Some Underlying Questions

Flat-spectrum radio quasars (FSRQs) have a spectral index of less than 0.5. If it is greater than 0.5, it is called steep spectrum radio quasars (SSRQs). The spectral energy distributions of an FSRQ and a BL Lacertae (BL Lac) object, if compared, it can be found that the BL Lac objects are less luminous with a harder spectrum than the FSRQs, which are much brighter with a soft spectrum. This is a major problem that challenges AGNs' current unification scheme. So far, various observations cannot confirm the origin of the jet. It is still a matter of challenge for the research community to confirm a disk-jet connection in this aspect. With the help of X-ray observations, it might be possible to extract some information regarding the jet's geometrical configuration and the corona's energy source. Also, surveys are needed to know the entire population of AGNs. To understand the underlying science of particle acceleration, it is essential to know the connection between the emission of high-energy particles (cosmic rays, neutrinos) and gamma rays [4]. Along with this, variability in AGNs is still a mystery. Continuous efforts have been made to understand this variability in different time scales to determine whether this indicates black hole mass.

VII. Conclusion

In this paper, we have briefly described the types of AGNs, obeying the AGN unification scheme by Antonucci in 1993. There was no plausible explanation for the immense central energy output in the earlier decades. After a rigorous observational study of the galaxies, evidence of the central object being a black hole was obtained, which is briefly discussed in our review paper. The emission regions of the AGNs are broad and narrow line regions (BLR & NLR). The region closer to the central body forms the BLR, responsible for the dominating emission line, especially luminous in the ultraviolet and optical spectrum. With the increasing distance from the central region, the NLR begins. It is a heated and ionized region, providing us with knowledge about the properties of interstellar clouds. From the morphology of the AGN, one can know that they possess two lobes facing outwards. So, to connect the core and lobes of the AGN, jets are formed. Jets play a considerable role in the growth of the galaxies. The paper has concluded with the challenges that are still a topic of debate.

REFERENCES

[1] Schmidt, M., 3 C 273: A star-like object with large red-shift., *Nature,* 197 (4872), 1040–1040 (1963).

[2] Mortlock, Daniel J., *et al.*, A luminous quasar at a redshift of z = 7.085., *Nature,* 474.7353, 616–619 (2011).

[3] Ulrich, M.H., Maraschi, L. and Urry, C.M., Variability of active galactic nuclei, *Annual Review of Astronomy and Astrophysics,* 35(1), 445–502 (1997).

[4] Padovani, P., Alexander, D.M., Assef, R.J., De Marco, B., Giommi, P., Hickox, R.C., Richards, G.T., Smolčić, V., Hatziminaoglou, E., Mainieri, V. and Salvato, M., Active galactic nuclei: What's in a name? *The Astronomy and Astrophysics Review,* 25, 1–91 (2017).

[5] Bradley, M.P. and Amri, W., Evidence for supermassive black holes in active galactic nuclei from emission-line reverberation. *The Astrophysical Journal,* 540(1): L13–L16 (2000).

[6] Netzer, H., The Physics and Evolution of Active Galactic Nuclei. *Cambridge: Cambridge University Press* (2013).

[7] Singh, K.P., *et al.* Spectral States of OJ 287 blazar from Multiwavelength Observations with AstroSat. *Monthly Notices of the Royal Astronomical Society,* 509, 2: 2696–2706 (2022).

[8] Diwan, R., Prince, R., Agarwal, A., Bose, D., Majumdar, P., Özdönmez, A., Chandra, S., Khatoon, R. and Ege, E., Multi-wavelength study of TeV blazar 1ES 1218+304 using gamma-ray, X-ray and optical observations. ArXiv./abs/2301.00991 (2023).

[9] Chatterjee, R., *et al.* Short-timescale variability of the blazar Mrk 421 from AstroSat and simultaneous multi-wavelength observations., *Journal of Astrophysics and Astronomy,* 42.2, 80 (2021).

The Characteristics of Nonlinear Kinematic Parameters in *f(R)* Cosmological Models

T. Achi* and H. Shanjit Singh

Department of Physics, Rajiv Gandhi University, Arunachal Pradesh-791112, India

*E-mail: tame.achi@rgu.ac.in

ABSTRACT: In modified f(R) cosmological models, we examine the accelerated expansion of the cosmos in a vacuum. We develop a non-linear differential equation by varying the Einstein-Hilbert action with a Robertson-Walker metric. Since the differential equations are non-analytic solutions, the expansion phases of the Friedmann-Robertson Walker universe are investigated numerically. In various f(R) models, we explore the universe's acceleration phases for different eras, cosmic jerk, snap, and crackle parameters, and their reliance on the initial and coupling parameters. Using a parametric method, we also present the natures of the jerk, snap, and crackle parameters at different accelerating stages of the universe.

Keywords: Modified *f(R)* Cosmological Models, Einstein-Hilbert Action, Robertson-Walker Metric, Friedmann-Robertson Walker Universe, Cosmic Jerk, Snap, Crackle Parameters.

I. Introduction

It is both a theoretical requirement and an observationally confirmed fact that the cosmos must have accelerated recently [1]. The theoretical study is carried out to establish a plausible option that can generate this acceleration by using the *f(R)* model instead of R in Einstein-Hilbert's action [2]. In $f(R) \approx R^2$ theories, an inflationary universe scenario for the early universe is successfully constructed [3] and negative curvature powers in the *f(R)* gravity behave as possibilities for dark energy. As a result, the theory could adequately represent the recent cosmic acceleration [4, 5]. The deceleration parameter (q), which was previously thought to be constant, defines the second-order evolution of the scale factor [6]. The existing observable evidence implies that the universe recently accelerated from a decelerated phase, which immediately compels one to examine the evolution of even higher order derivative of scale factors [7, 8]. The proposed work concerns the field equations of a covariant description of *f(R)* gravity to reconstruct nonlinear kinematic parameters such as the cosmological deceleration (q), jerk (j), snap (s), and crackle (l) in a null Lagrangian density. The cosmological solutions of highly non-linear second-order differential equations in *f(R)* models depicting the time evolution of the deceleration parameter (q) are presented in [2, 5]. Similarly, we consider a *f(R)* model involving $\frac{1}{R^2}$ as an additional term to the model considered in [5] involving coupling parameters such as α, μ and λ. The characteristics of the higher-order non-linear parameters in scale factor are studied numerically by varying the Einstein-Hilbert action in Robertson-Walker Universe.

II. The Field Equations and Cosmological Solutions

The action for *f(R)* gravity with the metric $g_{\mu\vartheta}$ is given by,

$$A = -\int \left[\frac{1}{2}f(R) + L_m\right]\sqrt{-g}d^4x \qquad \dots (1)$$

Where, *f(R)* is a function of R, g is the determinant of the 4D metric $g_{\mu\vartheta}$, L_m is the Lagrangian density with $8\pi G = c = 1$. Variation of the action of Eqn. (1) with respect to $g_{\mu\vartheta}$ yields,

$$-T_{\mu\vartheta} = f'(R)R_{\mu\vartheta} - \frac{1}{2}g_{\mu\vartheta}f(R) + f''(R)\left(\nabla_\mu\nabla_\vartheta R - g_{\mu\vartheta}g_{\mu\vartheta}\nabla^\mu\nabla^\vartheta R\right) + f'''(R)\left(\nabla_\mu R\nabla_\vartheta R - g_{\mu\vartheta}\nabla^\sigma R\nabla_\sigma R\right) \qquad \dots (2)$$

where, ∇_μ is the covariant derivative operator, and prime denotes the derivative with respect to R. $T_{\mu\vartheta}$ is the energy-momentum tensor which we have considered for the field. We investigate only the role of the geometry for the cosmological evolution by putting $L_m = 0$.

Next, we consider the spatially flat Robertson-Walker metric given by,

$$ds^2 = dt^2 - a^2(t)[dr^2 + r^2(d\theta^2 + sin^2\theta d\phi^2) \qquad \dots (3)$$

where, $a(t)$ is the scale factor of the universe. We have obtained the following field equations,

$$3\frac{\dot{a}^2}{a^2} = \frac{1}{f'}\left[\frac{1}{2}(f - Rf') - 3\frac{\dot{a}}{a}\dot{R}f''\right] \qquad \dots (4)$$

and

$$2\frac{\ddot{a}}{a} + \frac{\dot{a}^2}{a^2} = -\frac{1}{f'}\left[2\frac{\dot{a}}{a}\dot{R}f'' + \ddot{R}f'' + \dot{R}^2f''' - \frac{1}{2}(f - Rf')\right]. \qquad \dots (5)$$

where over dot represents the cosmic time derivative. The scalar curvature R for the metric is obtained as $R = -6(\frac{\ddot{a}}{a} + \frac{\dot{a}^2}{a^2})$. Defining the Hubble parameter $H = \frac{\dot{a}}{a}$, deceleration parameter $q = -\frac{\ddot{a}a}{\dot{a}^2} = -\frac{\dot{H}}{H^2} - 1$ and using Eqn. (4) and Eqn. (5) with $f_1(R) = R + \alpha R^2 - \frac{\mu^4}{R} - \frac{\lambda^4}{R^2}$, we get

$$q'' - \Pi(q,H)(q')^2 - \Upsilon(q,H)q' - \zeta(q,H) = 0 \qquad \ldots (6)$$

where,

$$\Pi(q,H) = -(-432H^2\alpha + 5\lambda^6 - 8H^2\mu^4 + 2160H^8\alpha q + 3\lambda^6 q + 4H^2\mu^4 q - 4320H^8\alpha q^2 + 4H^2\mu^4 q^2 + 4H^2\mu^4 q^2 + 4320H^8\alpha q^3 - 2160H^8\alpha q^4 + 432H^8\alpha q^5)/((q^2-1)(432H^8\alpha - \lambda^6 + 2H^2\mu^4 - 1728H^8\alpha q - 2H^2\mu^4 q + 2592H^8\alpha q^2 - 1728H^8\alpha q^3 + 432H^8\alpha q^4)),$$

$$\Upsilon(q,H) = -(2160H^8\alpha + 11\lambda^6 + 8\lambda^2 q - 5184H^8\alpha q - 14H^2\mu^4 + 6H^2\mu^4 q - 864H^8\alpha q^2 + 8H^2\mu^4 q^2 + 12096H^8\alpha q^3 - 11664H^8\alpha q^4 + 3456H^8\alpha q^5)/(H(q+1)(432H^8\alpha - \lambda^6 + 2H^2\mu^4 - 1728H^8\alpha q - 2H^2\mu^4 q + 2592H^8\alpha q^2 - 1728H^8\alpha q^3 + 432H^8\alpha q^4)),$$

$$\zeta(q,H) = -(2(q-1)(108H^6 + 3888H^8\alpha - 9H^2\mu^4 + 11\lambda^6 - 324H^6 q - 11664H^8\alpha q + +15\lambda^6 q - 9H^2\mu^4 q + 324H^6 q^2 + 7776H^8\alpha q^2 + 18H^2\mu^4 q^2 - 108H^6 q^3 + 7776H^8\alpha q^3 - 11664H^8\alpha q^4 + 3888H^8\alpha q^5))/(3H^3(q+1)(432H^8\alpha - \lambda^6 + 2H^2\mu^4 - 1728H^8\alpha q - 2H^2\mu^4 q + 2592H^8\alpha q^2 - 1728H^8\alpha q^3 + 432H^8\alpha q^4)).$$

Here the prime represent a derivative of *H*. It is seen that Eqn. (6) is a highly non-linear differential equation. Hence, we utilise numerical technique to extract information about the characteristics of the expanding universe in different eras. The cosmological parameters such as jerk(*j*), Snap (s) and Crackle (l) are defined in terms of scale factor and Hubble parameter respectively as $j = \frac{\ddot{a}}{aH^3}$, $s = \frac{1}{aH^4}\frac{d^4a}{dt^4}$ and $l = \frac{1}{aH^5}\frac{d^5a}{dt^5}$. These parameters are rewritten as functions of q and *H* as,

$$j = q + 2q^2 + (q+1)Hq'[H] \qquad \ldots (7)$$

$$s = -[H^2(q+1)^2 q'' + H^2(q+1)(q')^2 + 4Hq'(2q^2+3q+1) + 6q^3 + 7q^2 + 2q] \qquad \ldots (8)$$

$$l = \frac{1}{H}[-(q^2+2q+1)(H)^2 q''' + (-Hq' + 4H^2q^2 + (7H^2-10)q + 3H^3 - 6)(q+1)Hq'' - H^2(q')^3 + (4H^2q^2 + (7H^2-18)q + 3H^2 - 14)H(q')^2 + 2(16H^2q^3 + (36H^2-13)q^2 + 13(2H^2-1) + 6H^2 - 2H - 3)q' + (24q^4 + 46q^3 + 29q^2 + 6q)H]. \qquad \ldots (9)$$

III. Characteristics of Nonlinear Kinematic Parameters

In order to find the parametric solutions for Eqn. (7), (8) and (9) with the second-order non-linear differential equation (6), we supply initial boundary conditions and coupling constants by hand for convenience and to closely match the results found by authors [2, 8], wherein the unit of *H* for the present universe is taken as $H_0 = 1$.

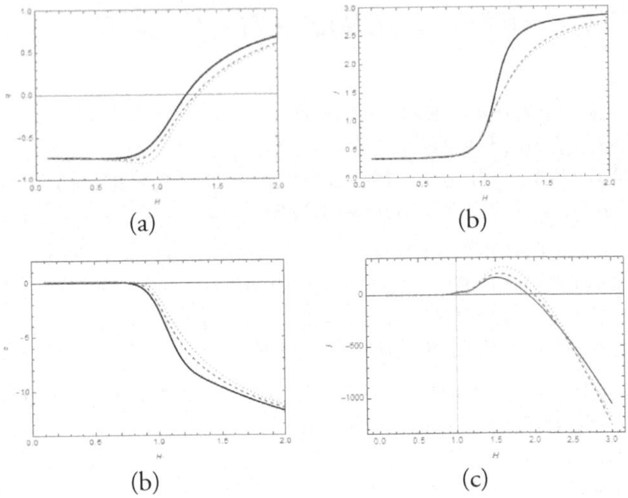

Figure 1: The plots show the characteristic of deceleration parameter q, jerk j, snap s and crackle l versus H for different initial and coupling values. Here, Thick line → ($\alpha = 2$, $\mu = 2.8$, $\lambda = 5$, $q[1] = -0.5$, $q'[1] = 1.655$, $q''[1] = 4.5$), Dashed Line→ ($\alpha = 3$, $\mu = 3$, $\lambda = 6$, $q[1] = -0.6$, $q'[1] = 1.755$, $q''[1] = 8$), Dotted Line→ ($\alpha = 4$, $\mu = 3.2$, $\lambda = 7$, $q\,[1] = -0.7$, $q'[1] = 1.855$, $q''[1] = 14$).

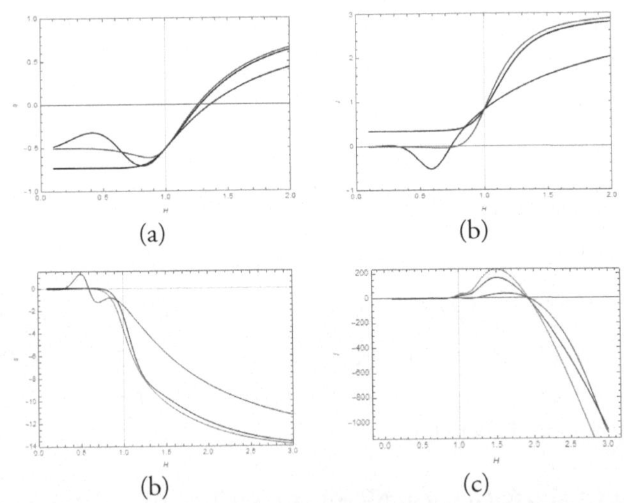

Figure 2: The plots show the characteristic of deceleration parameter q, jerk j, snap s and crackle l, respectively versus H for different $f(R)$ models. Here, Black→$f_1(R) = R + \alpha R^2 - \frac{\mu^4}{R} - \frac{\lambda^4}{R^2}$, Blue→$f_2(R) = R + \alpha R^2 - \frac{\mu^4}{R}$, Red→$f_3(R) = R - \frac{\mu^4}{R}$.

The universe's expansion is studied and understood for different ranges in the (q−H)-plane. $H < 1$ gives the future evolution of the universe and $H > 1$

describes the universe in the past. Figure (1) (Thick line), because $H_0 = 1$ represents the current universe, we observe that it is speeding, which is supported by the current observational results [2, 7, 8, 9] with $q_0 = -0.5$, $j_0 =$ 0.81, $s_0 = 3.09$, $l_0 = 18.27$. We also see from

Figure (1a) that the universe recently went through a phase transition at $H_t =$ 1.23. As a result, the phase transition points for jerk, snap, and crackle are $j_t = 2.39$, $s_t = -8.34$, and $l_t = 61.64$, respectively. When we change the initial values and coupling constants, depicted by dashed and dotted lines, we get different transition points and present values for all cosmological parameters. In figure (2), we compare the $f_1(R)$ model to other models such as $f_2(R) = R + \alpha R^2 - \frac{\mu^4}{R}$, and $f_3(R) = R - \frac{\mu^4}{R}$, but the initial values and respective coupling constants remain the same as in Figure 1, for all models, the natures of plots for all cosmological parameters are almost the same. However, the plots show distinct phase transition points and the appearance of the hump(s) between the region $H \rightarrow 1$ and $H \rightarrow 0$ for $f_2(R)$ except in Figure (1d), which is due to the absence of the λ term in the model. We also find that while q_0 and j_0 are constant among models, s_0 and $l_0$vary. Authors [10] show that the graphs between jerk (j) and snap (s) signify the behavior of dark energy aspects such as, when $j < 1$, $s > 0$ represents quintessence-type behavior of dark energy aspects, $j = 1$, $s = 0$ represents ΛCDM-type behavior, and $j > 1$, $s < 0$ suggests Chaplygin gas, from Figure 3 we can conclude that all three models are dominated towards Chaplygin Gas type region. Nevertheless, the $f_2(R)$ model also exhibits the quintessence-type behavior of dark energy parts.

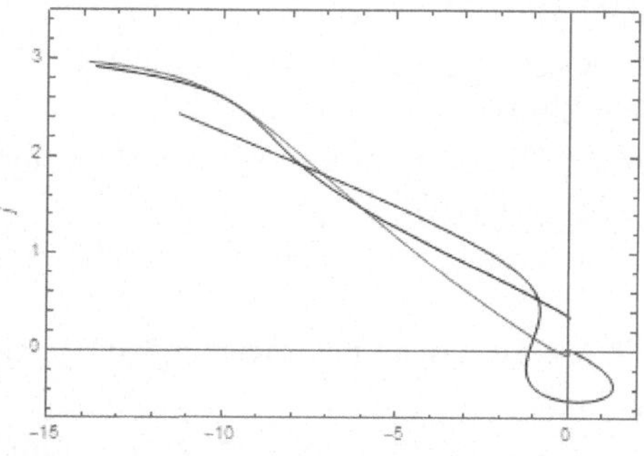

Figure 3: The Plots Show the Characteristic of Jerk j Versus Snaps for Different f(R) Models. Here, Black → $f_1(R) = R + \alpha R^2 - \frac{\mu^4}{R} - \frac{\lambda^4}{R^2}$, Blue → $f_2(R) = R + \alpha R^2 - \frac{\mu^4}{R}$, Red → $f_3(R) = R - \frac{\mu^4}{R}$

IV. Conclusion

The numerical technique is used to investigate the cosmological parameters such as q, j, s, and l of the cosmos over time for $f_1(R)$ model. We find that the current universe ($H_0 = 1$) is expanding at an accelerating rate with $q_0 = -0.5$, $j_0 = 0.81$, $s_0 = 3.09$, $l_0 = 18.27$ and has transitioned from deceleration to acceleration at $j_t = 2.39$, $s_t = -8.34$, and $l_t = 61.64$. These values are found to depend on the coupling constants used in the $f(R)$ model. Finally, we examine three models, $f_1(R)$, $f_2(R)$, and $f_3(R)$, which result in unchanged q_0 and j_0 value, while s_0 and l_0 fluctuate, and all models exhibit Chaplyg in Gas-like behaviour. On the other hand, the $f_2(R)$ model also demonstrates the quintessence-type behaviour of dark energy portions.

Acknowledgement

The authors would like to thank Rajiv Gandhi University in Arunachal Pradesh, India, for their computing resources and PANE 2022 for providing a platform for this paper.

References

[1] Riess, A.G., *et al.*, Observational Evidence from Supernovae for an Accelerating Universe and a Cosmological Constant, *Astron. J.* 116, 1009 (1998).

[2] Mukherjee, A., *et al.*, Acceleration of the universe in f(R) gravity models, AP&SS, 352, 893 (2014).

[3] Starobinski, A.A., A New Type of Isotropic Cosmological Models without Singularity, *Phys. Lett. B* 91, 99 (1980).

[4] Nojiri, S., Modified gravity with lnR terms and cosmic acceleration, *Gen. Rel. Grav.* 36, 1765 (2004).

[5] Paul, B.C., *et al.*, Accelerating Universe in Modified Theories of Gravity, *Phys Rev.* D.79, 083534 (2009).

[6] Perlmutter, S., *et al.*, Discovery of a supernova explosion at half the age of the Universe, *Nature* 51, 391 (1998).

[7] Popawski, N.J., The cosmic jerk parameter in f(R) gravity, *Phys. Lett.B* 640, 135137 (2006).

[8] Popawski, N.J., The cosmic snap parameter in f(R) gravity, *Class. Quant. Grav.* 24, 3013 (2007).

[9] Rapetti, *et al.*, Constraints on modified gravity from the observed X-ray luminosity function of galaxy clusters *H, Mon. Not. R. Astron. Soc.* 375, 1510 (2007).

[10] Sahni, V., *et al.*, State finder A new geometrical diagnostic of dark energy, *Exp. Theor. Phys. Lett.* 77, 201 (2003)

Eigen Mode Analysis of Discrete Slow Mode $m = 1$ Warp of Self-Gravitating Galactic Disc

Sagar Sing Goyary* and H. Shanjit Singh

Department of Physics, Rajiv Gandhi University, Arunachal Pradesh–791112, India

*E-mail: sagar.singhgoyary@rgu.ac.in

ABSTRACT: Recent observations have revealed that outer edges of most of the disc galaxies are warped and quite common in the disc galaxies. In the present paper we study analytically the most common observed stellar disc integral sign warp (S - shape) of self-gravitating disc galaxies represented by discrete eigen mode (m = 1) by eigen mode analysis. We also analyse the morphology of the warp and so far, the observation of warping radius span which lies in between 4–6-disc scale-lengths. The present analysis based on an eigen mode approach brings out clearly that the disc shows a net warp only in between the radius span of 4–6-disc scale-lengths and the warp geometry in the outer disc is not only very sensitive to softening parameter of the disk but also significantly sensitive to the contribution from the dark matter halo.

I. Introduction

The outer parts of most of the spiral galaxies are distorted in vertical direction from the mid-plane of disc galaxy and it is commonly known as galactic warp and now considered as common feature of the disc galaxies. Approximately 50% of all disk galaxies have stellar warps [1, 2]. Recent observation of various stellar surveys also revealed the distinct warp in the stellar disc of the Milky Way [3]. Warp in stellar disc can easily be observed in an edge-on view disc galaxies by radio and optical band. Stellar warps have smaller amplitudes as compared to HI-warps [4]. The most common shape of the galactic warp is integral S-shaped warp [1, 2, 5]. Based on 21 cm hydrogen (HI)-line observations it is found that 70% of the stellar warps are (integral) S-shaped [6]. The integral S-shaped warp in the Milky Way was first discovered in [6, 7]. The possible mechanism of generation of such type of warp is studied extensively in past by many authors. Some of mechanisms till put forwarded are: tidal interaction, bending instabilities, halo axis misalignment, gas accretion etc. We present eigen modes analysis of global bending modes to explore the properties of most commonly observed S-shaped galactic warp such as bending radius described by Fourier component $m = 1$. This approach can describe the observed S-shaped warp only from the internal dynamics of disk without any external perturbation.

This paper is organised as follows: In section 2, the dynamical model of the galactic disk and dark matter halo for calculating related physical quantities is formulated. In Section 3, numerical analysis and result are summarized. In Section 4; we conclude the work.

II. Mathematical Modelling and Formulation

We consider an axisymmetric disk having surface density $\Sigma(R)$ embedded in a rigid or non-responsive dark matter halo. For describing dynamics in the disc we use the cylindrical polar coordinate system (R, Φ, z). The disk is assumed to rotate in the equatorial plane ($z = 0$) of the spheroidal dark matter halo with angular speed $\Omega(R)$ about the halo's symmetry axis ($z = 0$). When the disc is displaced vertically by a small amount $Z(R,\phi,t)$, the subsequent dynamical equation of a small bending of the disc perpendicular to its unperturbed plane ($z = 0$) at any time t is given by,

$$\left(\frac{\partial}{\partial t} + \Omega(r)\frac{\partial}{\partial \phi}\right)^2 Z = F_{\mathrm{d}} + F_{\mathrm{h}} \qquad \ldots (1)$$

where F_d and F_h are forces due to the disc and the halo respectively and are given by,

$$F_d = -G\int_0^\infty \Sigma_d(R')R'dR' \times \int_0^{2\pi} \frac{Z(R,\phi,t) - Z(R',\phi',t)}{[R^2 + R'^2 - 2RR'\cos(\phi - \phi') + z_0^2]^{\frac{3}{2}}} d\phi' \qquad \ldots (2)$$

$$F_h = -\nu_{\mathrm{h}}^2(R)Z(R,\phi,t) \qquad \ldots (3)$$

Here $\nu_{\cdot}^2$ is the vertical frequency due to the dark matter halo. In the above equation, z_0 represents the softening parameter used to make the integral regular at $R = R'$. When equations (2) and (3) are substituted into equation (1), a linear equation for $Z(R,\phi,t)$ results. We seek solutions in the form of Fourier component m as: $Z_m(R,\phi,t) = \mathrm{Re}\{h_m(R)e^{i(\omega_m t - m\phi)}\}$ in which m represents the bending mode of the disc. Substituting the expression for $Z_m(R,\phi,t)$ into equation (1) we can obtain the dynamic equation of bending of $m = 1$ mode and can be recast it into a matrix-eigen value problem on a uniform grid with radial points that takes the quadratic eigen value problem (QEP) (for more details see [8]) of the form:

$$\left(\omega^2 I + \omega R_{ij} + D_{ij}\right)h = 0 \qquad \ldots (4)$$

Where, mass matrix I is an identity matrix, damping matrix$R_{ij} = -2\Omega_{ij}\delta_{ij}$, Ω is the angular frequency contributed from the disc and the dark matter halo and D_{ij} is known as stiffness matrix. h represents the eigenvector corresponding to the eigen value ω. The eigen equation is a nonlinear eigen value problem which describes the global behaviour of the bending mode in a self-gravitating disc representing stable S-shaped warp for $m = 1$. For solving such eigen value problem (4) we discuss the mechanism in Section 3.

A. Disk and Dark Matter Profiles

For simplicity we consider the density distribution of the galaxy disc in our model and the galactic disc that follows an exponential fall-off with central surface density Σ_0 and a scale-length R_d as $\Sigma(R) = \Sigma_0\, exp(-R/R_d)$ [9], where Σ_0 is the central surface density of disc. The disk is assumed to be embedded in the field of external potential of the dark matter halo having in the form as (Binney).

$$V(R,z) = \frac{v_0^2}{2}\, ln\left(R^2 + R_c^2 + \frac{z^2}{q^2}\right)$$

Here v_o, R_c and q are velocity at large radius, core radius and flattening parameter of the halo.

III. Numerical Solution and Result

We solve QEP numerically by using standard technique for diagonalization. We treat the galactic disc with a finite radius having a system of *N* uniformly spaced concentric rings system. On solving the *N* -dimensional QEP, we obtain *2N* real eigen values. As all the eigen values in our analysis are real, the eigen modes obtained are stable. Such stable global mode oscillates like a simple harmonic oscillator with frequency ω.

A. Input Parameters

In order to explore the $m = 1$ warp mode in the self-gravitating disk, we introduce a system for numerical calculations such that $G = M_d = R_d = 1$ and all the frequencies are normalised in the unit of$\sqrt{GM_d/R_d}$. He input parameters for the dark matter halo are M_h-mass of the dark matter halo; R_c-the core radius and q-the flattening parameter.

B. Eigen Modes and Disk Warp

In this subsection we present the result of m = 1 eigen mode. The eigen spectrum for the disk with parameters $R_c = 3[R_d]$, $q = 0.78$, and $M_h = 3[M_d]$ giving flat rotation curve is shown in the left panel of Figure 1.We obtain two continuum branches in an Argand diagram with $(\Omega - \nu)$ and $(\Omega + \nu)$ representing retrograde and prograde modes of the disk respectively. In between these two eigen modes a continuum gap known principal gap or P-gap exists. In the P-gap we obtain one isolated eigen mode separated from the boundary of slow and fast modes continuum which is clearly visible in inset figure of the Argand diagram. The isolated eigen mode in the P-gap is the main characteristic of most commonly occurrence S-shaped bending mode in the disk galaxies. The eigen modes are stable neither growing nor decaying in cosmic time. The right panel of Figure 1 shows the surface face map of the disk displacement, $Z(R, \phi, t)$ of this stable mode at epoch$t = \tau/8$, where $\tau = \frac{2\pi}{\omega}$ is characteristic time scale. The face map shows clearly S-shaped warp in the disk galaxies.

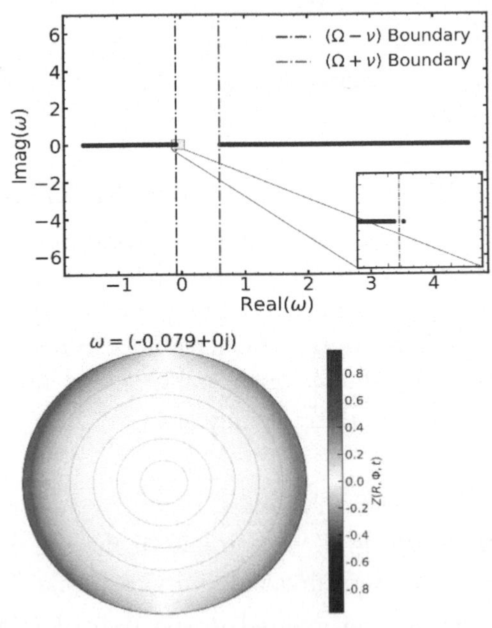

Figure 1: Left: Nature of complete eigen spectrum clearly representing slow and fast continuum modes. Right: Face map eigen modes of isolated eigen value in between the slow and fast mode continuum in eigen spectrum representing m = 1 S-shaped bending mode in the disc. The concentric rings in the plot represent the increment of $1R/R_d$ along radial axis in the disk.

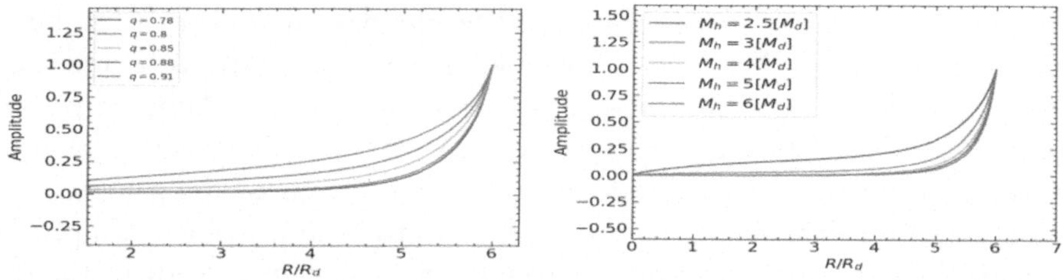

Figure 2: Plot showing vertical displacement of the galactic disk for various dark halo parameters. In left panel, amplitude of disk bending for various shapes of dark matter is shown. In right panel, amplitude of vertical bending of the disk at various dark matter halo mass is shown.

We present the role of the dark matter halo on the onset radius of disk warp. In Figure 2 we show the two plots of vertical displacement of the disk along the radial axis. The left panel shows the dependence of warp amplitude on halo flattening parameters. We vary the halo flattening parameters q from 0.78 to more spherical 0.91. It is observed that when the dark matter halo is made flat, the warp bending in the outer galaxy is much higher than the spherical dark matter halo. Next, we vary the halo to disk mass ratio from 2.5 times mass disk to 6 times of mass of disk to analyse the role of the halo mass on bending mode. The higher dark matter halo mass makes the disk bend very sharp near the edge. This is explained as follows, when the disk is embedded in the high halo mass, the torque exerted by the dark matter halo towards the external radius increases resulting sharper bending towards the outer edge. We examine the disk bending radius for different cases of halo parameters. From the Figure 2 it is seen that the bending radius of the disk is generally not much sensitive to the flattening parameter and mass of the dark matter halo. The bending radius of the disc is constrained in for both cases of the halo parameters i.e., whether the dark matter halo is more flatten or for the disk is embedded in high halo masses. It is seen prominently in the region starting from $4R_d$ up-to $6R_d$ disk scale lengths. This result of our model confirms the finding of Briggs rule that the disk bend typically at 4–5 disk scale lengths [10].

IV. Conclusion

We propose and study the disk warp using eigen mode analysis considering the dynamical equation of small bending from internal disk dynamics only. Our analysis confirms the finding of Briggs (1990) [10] that galactic disk starts to bend typically at 4–6 disk scale lengths. It is found that as we make the halo more spherical, the disc edge is wrapped with lesssharp. We show that when the galactic disk is dominated by the dark matter, the disk bends sharper at the edge. With our model we have studied the observed warp structure without need of external perturbation. This explains why some isolated galaxies show warp in outer portions of the disk galaxies. For more realistic warp onset and amplitude in disk galaxies, we need the parameters of the disk and the dark matter halo from recent observations which will provide us a much better and accurate understandings of S-shaped warp in the disk galaxies. A detailed analytical and numerical exploration by using realistic-galactic disk and dark matter halo parameters is a subject of further study.

ACKNOWLEDGMENT

The first author, Sagar Sing Goyary would like to thank PANE organisers, UGC, India for Junior research fellowship and thanks to the authority of Rajiv Gandhi University, Arunachal Pradesh and IUCAA, Pune for providing the necessary facilities during the preparation of this research work.

REFERENCES

[1] Sanchez-Saavedra, M.L., Battaner, E. and Florido, E., Statistical Study of Optical Galactic WARPS, *Astrophysics and Space Science*, vol. 171, 239–241, Sept. 1990.

[2] Sanchez-Saavedra, M.L., Battaner, E., Guijarro, A., L ′ opez-Corredoira, M., and Castro-Rodrıguez, N., A catalog of warps in spiral and lenticular galaxies in the Southern hemisphere, *Astronomy & Astrophysics,* vol. 399, 457–467, Feb. 2003.

[3] Romero-Gomez, M., Aguilar, C.L., Figueras, F. and Castro-Ginard, A., Gaia kinematics reveal a complex ′lopsided and twisted Galactic disc warp, *Astronomy & Astrophysics*, vol. 627, A150, July 2019.

[4] van der Kruit, P.C., Optical surface photometry of eight spiral galaxies studied in Westerbork., *Astronomy & Astrophysics*, vol. 38, 15–38, Oct. 1979.

[5] Ann, H.B. and Park, J.C., "Warped disks in spiral galaxies," *New Astronomy*, vol. 11, pp. 293–305, Jan. 2006.

[6] Burke, B.F., Systematic distortion of the outer regions of the galaxy, *Astronomical Journal,* vol. 62, 90, May 1957.

[7] Kerr, F.J., A Magellanic effect on the galaxy., *Astronomical Journal,* vol. 62, 93–93, May 1957.

[8] Saha, K. and Jog, C.J., On the generation of asymmetric warps in disk galaxies, *Astronomy & Astrophysics,* vol. 446, no. 3, 897–904, 2006.

[9] Freeman, K.C., On the disks of spiral and s_0 galaxies, *The Astrophysical Journal,* vol. 160, 811, 1970.

[10] Briggs, F., Rules of behavior for galactic warps, *The Astrophysical Journal,* Vol. 352, 15–29, 1990.

Evolutionary Path for Molecular Cloud under the Influence of Magnetic Field: A Numerical Approach

D. Gogoi and E. Saikia*

Department of Applied Sciences, Gauhati University, Guwahati-781014, Assam, India

*E-mail: eeshankur@gauhati.ac.in

ABSTRACT: The Interstellar Medium (ISM) is the site for various physical processes, such as star formation, supernova explosion, etc. Molecular clouds present in ISM play an important role in the star formation process when influenced by external perturbations like shock or winds etc. The importance of such a process under the influence of a ubiquitous magnetic field is probed in this work. Using the numerical simulation method we allowed two clouds placed in juxtaposition to the shock path to get destroyed. Results obtained highlight that the magnetic field influences the evolution of such a system. The presence of a magnetic field resulted in the generation of clumpy regions in contrast to a case where it was absent which generated vortices. Simulated results indicate two evolutionary paths such a system can take depending on the magnetic field strength and orientation, thus laying the importance of the present work, and extending it further involving realistic constraints and thermal properties.

Keywords: ISM-Molecular Clouds-MHD-Star Formation.

I. INTRODUCTION

Stars are formed in Nebulae and are influenced by several physical and chemical processes. Such Nebular regions are seen to host molecular clouds comprising cold gases. Physical processes, such as shock, stellar winds, etc. are known to play key roles in the destruction of such clouds [1, 2]. Shock plays an important role in material movements through hydrodynamic ablation in an astrophysical system such as discs, clouds, etc. that are gravitationally bounded and can result in a different evolutionary state [3]. The studies done so far are focused on the effect of the shock speed and length and time scale of the destruction of the molecular cloud [1, 4, 5]. Reported results suggested that the magnetic field's role was also limited to inhibitions of both instabilities and cloud destruction [6]. The work here attempts to look at the importance of the field by considering a two-cloud system placed in the shock path which was allowed to evolve for a longer period of time beyond its cloud-crushing time scale in an attempt to understand a nebular region which consists of a large number of such clouds. The following subsections highlight the methodology and results obtained in this study.

II. Methodology

We use the publicly available numerical code "PLUTO" [7] for simulating the two cloud systems. We model the shock to approach two cylindrical clouds from the left with its initial position in the domain at x = 0.6. The cloud's initial radius before interaction was $0.1r_{cc}$ (r_{cc} computer unit) and the density contrast was 10. The domain of simulation was $2r_{cc}$ with the clouds placed at the position (1.0, 1.0) & (1.5, 1.5) with a separation of $0.3r_{cc}$. Following the work of Dai [2], we divided the interaction region as Pre- and Post- shock regions. We also defined their initial physical parameters (listed in Table 1) following the Rankine-Hugoniot (RH) conditions.

Table 1: Table for Initial Parameters

Sl No.	*Parameters*	*Symbol*	*Pre-Shock State*	*Post-shock State*
1.	Density	ρ	3.86859	1.0
2.	Pressure	p	167.345	1.0
3.	Velocity along x	v_x	0	-11.2536
4.	Magnetic Field along y	B_y	0.56418958	2.1826182
5.	Magnetic Field along z	B_z	-0.56418958	-2.1826182

The Navier-Stokes equations were handled numerically by the code, with the additional equation of state defined in terms of the adiabatic index 'γ'. For our simulation, γ was taken to be 1.67. The grid resolution for simulation was (1028×1028). Additional parameters that were calculated, for inference purposes were the vorticity 'ω' and Plasma Beta 'β'. The z-component of vorticity is defined as the curl of the fluid elements ($\omega = \partial_x v_y - \partial_y v_x$) in terms of the velocity along the *x* and *y* directions. β is defined as the ratio of the thermal pressure 'p' to magnetic pressure ($B^2/2\mu_0$) where 'μ_0' is the magnetic permeability and 'B' is the magnetic field. For comparative purposes, we construct two models, one with B applied obliquely along y and *z* direction (Table 1) referred as the 'MHD model' and the other with the absence of B referred to as the 'HD model'. The magnetic field has a magnitude of 6μgauss [8]. The energy equation with gravitational potential 'Φ', thermal pressure 'p', and velocity 'v' is given by,

$$\frac{\partial(E_t + \rho\phi)}{\partial t} + \nabla.[(\frac{\rho \mathbf{v}^2}{2} + \rho e + p + \rho\phi)\mathbf{v} + c(\mathbf{E} \times \mathbf{B})] = \mathbf{m}.\mathbf{g}$$

where, g is the force vector, E is the electric field and 'E_t' is the total energy density. In our case, the right-hand side of equation (1) is zero. The timescale for the crushing of cloud 't_{cc}' is ($t_{cc} = \chi^{1/2}\, a_0/v$) where χ is the ratio of the density of cloud materials to ambient medium taken as 10, a_0 is the initial cloud radius and v is the initial shock speed. For our calculation, $t_{cc} = 0.028t_0$ where t_0 is the computer time scale. The simulation was run to observe the long-term evolution of the system for a total time of about $3.18t_{cc}$.

III. Results and Conclusion

The results presented are simulated using the PLUTO code for a two-cloud system. The clouds were allowed to evolve for a time period of $3.18t_{cc}$. The simulated density morphology is shown in Figure 1 plotted for both HD (right) and MHD (left) models. Sub-panel(a) shows the shock encountering the two clouds and subsequently, it completely destroys the clouds. In the process, it also generates secondary shock fronts. Generally, materials on facing the shock tend to move laterally due to the pressure difference created by the shock front. The pressure at the middle of the cloud face is higher in comparison to the face perpendicular to the shock's propagation direction resulting in material movements to the side. This creates a wing-like structure in the compressed clouds as seen in both models. In the case of the HD model, this pressure difference and the lateral movement of the materials are higher in comparison to the MHD model which leads to the generation of vortices. Figure 2 shows the vorticity peaks at regions where materials are moving away from the shock front. Figure 2 sub-panel (7) and (8), highlights that in the HD model, the vortex formed is sustained over a longer period of time and results in the creation of more number of instabilities in the post-shock region.

The magnetic field too plays an important role in the evolution of such a system but evidently, its role in the star formation process has been considered to be nominal as already stated. We found that the presence of the field in the system promotes the clumping of materials in the post-shock region (reference Figure 1(d)&1(e)). Magnetic fields are known to inhibit the formation of instabilities and the destruction of clouds [6]. In our case, it is evident that it prevents the cloud materials to spread laterally. At a later evolutionary time period, it is seen to generate long filamentary structures that can be formidable places for the formation of stars if enough gravitationally bound materials are accumulated. As compared to the HD case where there are generations of vortices that are the known to promote material mixing between the ISM and crushed cloud materials.

As seen in our results, we can conclude the following points:

- The two clouds kept at a distance of $0.3r_0$ separation when allowed to evolve for a longer period of time $\sim 3.18t_{cc}$ showcase the different state the two models (HD and MHD) reaches owing to the generated instabilities and vortex.
- HD model forms and sustains the vortex formed in the post-shock region which is also known to promote material mixing.
- MHD model has clump formation after the destroyed cloud's material movements are confined by the field. Such clumps can be probable future star-forming regions. The magnetic field strength changes in the post-shock region which results in the inhibition of instabilities and movement of crushed cloud material.

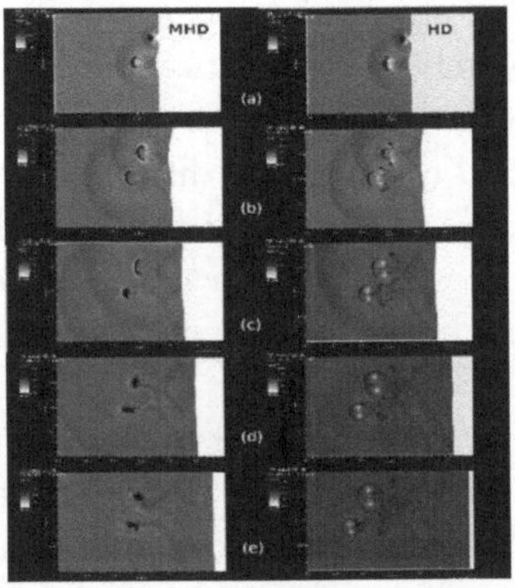

Figure 1: The simulated density of the two cloud systems considered under **B** field MHD (Left) and without it HD (Right) showed at time step sub-panels (a) to (e) ~ 10-50. The contrast in the evolutionary path is observed in both models. MHD models show crushed materials clumping together after the shock passes by while the HD model has a distinct vortex formed as the materials spread laterally (refer (b)-(e)).

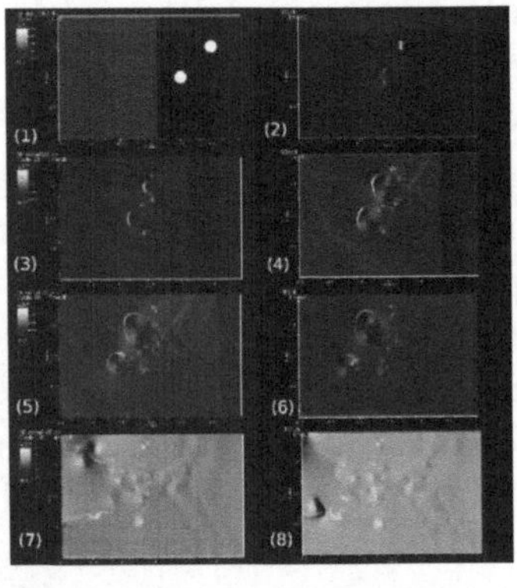

Figure 2: The morphological output of the z-component of vorticity calculated for the HD model. Sub-panels (1)-(6) are shown for the time-step 0 to 50 in steps of $t = 10\ t_0$ while sub-panels (7) and (8) correspond to time-step 100 and 131 respectively. We observed unstable regions generated in the post-shock region which can be considered a feasible place for the mixing of materials of crushed cloud and ISM.

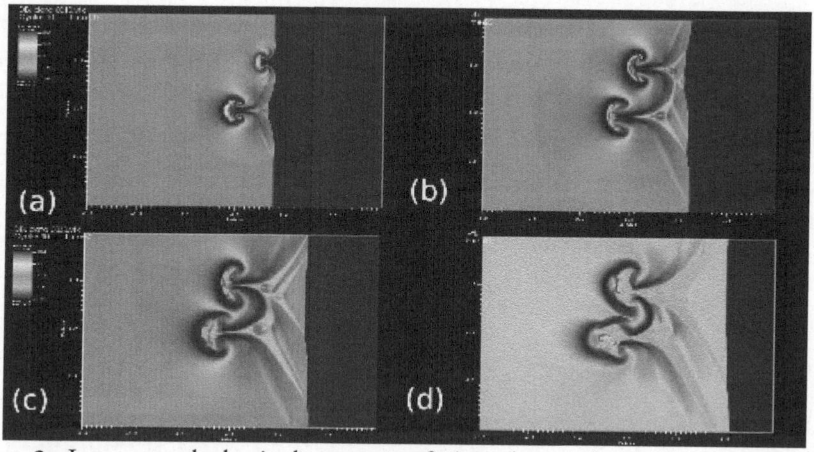

Figure 3: Is a morphological output of the plasma beta (β) calculated for time-step 10–40 shown in sub-panels (a)-(d) respectively. Regions lower in magnitude has the highest magnetic pressure. It is seen that the field bends around the compressed crushed materials in the post-shock region.

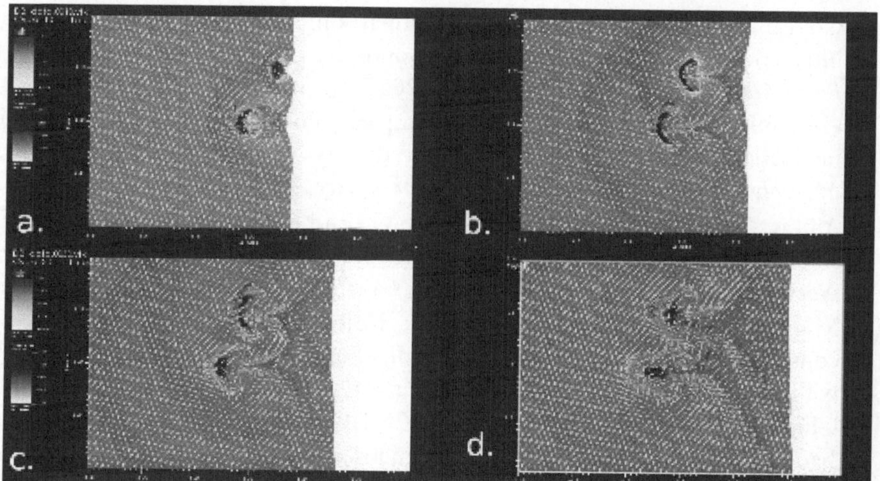

Figure 4: The vector plots of the **B** field is shown plotted over the density morphology where the color depicts magnitude & the arrowhead its direction. The magnetic field strength is maximum in the region where density clumping is high resulting in decrease of β (refer Figure 3).

Figure (3) and (4) are plots of β and B's vector field both of which highlights the increase in the field strength along the crushed materials that results in inhibiting material movement and generates density clump in the post-shock regions.

- Thus, this work highlights that the magnetic field plays an important role in the molecular cloud system which when perturbed by a shock can result in a different evolutionary path for the system resulting either in the mixing of materials or forming a new star.

Future work includes scaling the process in terms of density contrast and magnetic field strength and orientation while incorporating realistic physical and thermal conditions.

REFERENCES

[1] Mac Low, M.-M., McKee, C.F., Klein, R.I., Stone, J.M. and Norman, M.L., Shock interactions with magnetized interstellar clouds. 1: Steady shocks hitting nonradiative clouds, *The Astrophysical Journal,* 433, 757 (1994).

[2] Dai, W. and Woodward, P. R., Interactions between magneto hydrodynamical shocks and denser clouds, *The Astrophysical Journal*, 436, 776 (1994).

[3] Gogoi, D., Saleh, P.R., Borah, S.M. and Saikia, E., A new insight into changes in protoplanetary disk structure caused by change in stellar mass, New Astronomy, 93, 101767 (2022).

[4] Goldsmith, K.J. and Pittard, J.M., A comparison of shock-cloud and wind-cloud interactions: The longer survival of clouds in winds, *Monthly Notices of the Royal Astronomical Society,* 470, 2427 (2017).

[5] Goldsmith, K.J. and Pittard, J.M., A comparison of shock-cloud and wind-cloud interactions: Effect of increased cloud density contrast on cloud evolution, *Monthly Notices of the Royal Astronomical Society,* 476, 2209 (2018).

[6] Alūzas, R., Pittard, J.M., Falle, S.A.E.G. and Hartquist, T.W., Numerical simulations of a shock interacting with multiple magnetized clouds, *Monthly Notices of the Royal Astronomical Society,* 444, 971–993 (2014).

[7] Mignone, A., Flock, M., Stute, M., Kolb, S.M. and Muscianisi, G., A conservative orbital advection scheme for simulations of magnetized shear flows with the PLUTO code, *Astronomy and Astrophysics*, 545, 1 (2012).

[8] Ching, T.C., Li, D., Heiles, C., Li, Z.Y., Qian, L., Yue, Y.L., Tang, J. and Jiao, S. H., An early transition to magnetic super criticality in star formation, *Nature,* 601, 49 (2022).

Study of Fission Fragment Angular Anisotropy in Heavy-Ion Induced Reactions

Lakhyajit Sarma[1,*], Amar Das[1,2], Harun Al Rashid[1,3], N.K. Dev[1] and K. Kalita[1]

[1]Department of Physics, Gauhati University, Guwahati-781014, India
[2]Department of Physics, Suren Das College, Hajo-781102, India
[3]Department of Physics, BBK College, Nagaon, Barpeta-781311, India
*E-mail: Lakhyajitsarma0@gmail.com

ABSTRACT: Fission fragment angular distribution measurements play an important role to understand the fusion-fission dynamics in heavy-ion induced fission reactions. Fission fragment angular distribution (characterized by anisotropy) is a strong probe to differentiate fusion-fission and quasifission reactions at near barrier energies. The commonly used transition-state model (TSM) to analyze the fragment angular distribution which includes standard saddle-point statistical model (SSPSM) and the scission-point statistical (SPS) model are not able to reproduce the experimental anisotropies for many heavy-ion induced reactions because of the admixture of noncompound nuclear events with compound nuclear processes. In this work we have adopted the upgraded saddle-point statistical (USPS) model to explain fission fragment anisotropies for the systems $^{28,30}Si + ^{180}Hf$, $^{16}O + ^{194}Pt$, $^{11}B + ^{232}Th$ *and* $^{13}C + ^{207}Pb$ *and compared the results with TSM model calculations. For reaction systems with anomalous fragment anisotropies, the prediction of the USPS model is found in better agreement with the experimental anisotropies than the TSM prediction.*

Keywords: Fusion-Fission Dynamics, Anisotropy, Transition State Model, Saddle-Point Statistical Model, Upgraded Saddle-Point Statistical Model.

I. Introduction

One of the most popular ways to distinguish between fusion-fission and quasifission processes at low barrier energies is to look at the angle distribution of fission products.

The entrance channel characteristics of the colliding nuclei, such as charge product ($Z_P.Z_T$), mass asymmetry (α), nuclei deformation (β), collision energy, etc., have a significant impact on the fusion dynamics in heavy ion collision reactions [1]. The reaction dynamics in the nuclear system are also heavily influenced by the direction of the mass flow. The formation of a compound nucleus (CN), which may later decay via fission or particle evaporation, is said to occur in systems with entrance channel mass asymmetry (α) greater than the Businaro-Gallone mass asymmetry (α_{BG}). For $\alpha < \alpha_{BG}$, on the other hand, mass flow happens in the opposite direction, resulting in the formation of a dinuclear system that will decay before reaching equilibrium in all degrees of freedom, resulting in quasifission [2]. The commonly used transition-state model (TSM) to analyze the fission fragment angular distribution which includes standard saddle-point statistical model (SSPSM) and the scission-point statistical (SPS) model are not able to reproduce the experimental anisotropies for many heavy-ion induced reactions. In this work we have adopted the upgraded saddle-point statistical (USPS) [3] model to explain fission fragment anisotropies for the systems $^{28,30}Si + {}^{180}Hf$ [4], $^{11}B + {}^{232}Th$ [1] and $^{13}C + {}^{207}Pb$ and compared the results with experimental and TSM model calculations.

II. SSPM Formalism for Angular Distribution

According to the standard transition state model (TSM) or standard saddle point statistical model (SSPM), fission fragment angular distributions are characterized by angular anisotropy (A). The angular anisotropy is defined as the ratio of yield at 0° or 180° to that at 90° and it is given by,

$$A = 1 + \frac{<l^2>}{4K_o{}^2} \qquad \dots (1)$$

where $< l^2 >$ is the second moment of spin distribution of the compound nucleus and $k_0{}^2$ is the variance of *k* distribution (*k* is the projection of *I* on the nuclear symmetry axis). The quantity $k_0{}^2$ is given by,

$$k_0{}^2 = \frac{I_{eff}T}{\hbar^2} \qquad \dots (2)$$

here, $I_{\{eff\}}$ is the effective moment of inertia and *T* is the *CN* temperature at saddle point. The expression for *T* is,

$$T = \sqrt{\frac{E_{cm} - Q - B_f - E_{rot} - E_n}{a}}$$

$$= \sqrt{\frac{E_{ex}}{a}} \qquad \dots (3)$$

III. UPGRADED SADDLE-POINT STATISTICAL MODEL (USSPM)

In the present study, SSPM calculation is modified by USPSM by correcting the effective moment of inertia of the compound nucleus at the saddle point, considering the calculated nuclear temperature and the second moment of spin distribution of the compound nucleus based on the SSPS model at this point are correct. Since the motion after scission is governed simply by long-range Coulomb forces, the change can be estimated in a classical calculation of the post-scission trajectories.

The modified $k_0{}^2$ can be written as [2],

$$k_0{}^2 = (K_0)'^2 \left(1 + 1.245\frac{y}{\chi}\right) \qquad \dots (4)$$

y is the measure of angular momentum and expression is given by,

$$y = \frac{1.9249\, l^2}{\left(1 - 1.7826\frac{(N-z)^2}{A^2}\right)A^{\frac{7}{3}}} \qquad \dots (5)$$

Applying this correction and the assumption that standard deviation of $k_0{}^2$ distribution at the scission point is identical to the saddle point k distribution, the corrected expression for effective moment of inertia can be written as,

$$\left(\frac{I_{eff}}{I_{sph}}\right)^2 = \left(\frac{I_{eff}}{I_{sph}}\right)^2 \left(1 + 1.245\,\frac{y}{\chi}\right) \qquad \dots (6)$$

IV. RESULT AND DISCUSSION

For the studied heavy-ion reaction systems, the derived effective moments of inertia at the saddle point based on the USPS model are in agreement with the experimental values of the effective moments of inertia. USPSM calculations can reproduce the experimental data more closely as compared to SSPM calculation (as shown in figures 1–4) for the reactions under study. In the figures we have shown the experimental anisotropy with the SSPM and USPSM calculated anisotropies. The entrance channel mass asymmetry parameter α with respect to the Businaro-Gallone critical mass asymmetry parameter α_{BG} plays a very dominant role in the reaction dynamics. Effect of entrance channel is observed for the reactions studied. It is observed that the fission fragment angular distributions calculated based upon the USPS model are able to explain consistently the energy dependence of the angular anisotropies. The observed deviation in figures is due to the presence of non-compound nucleus fission events. The experimental fission fragment anisotropies have been observed to be anomalous in comparison with the prediction of the SSPS model.

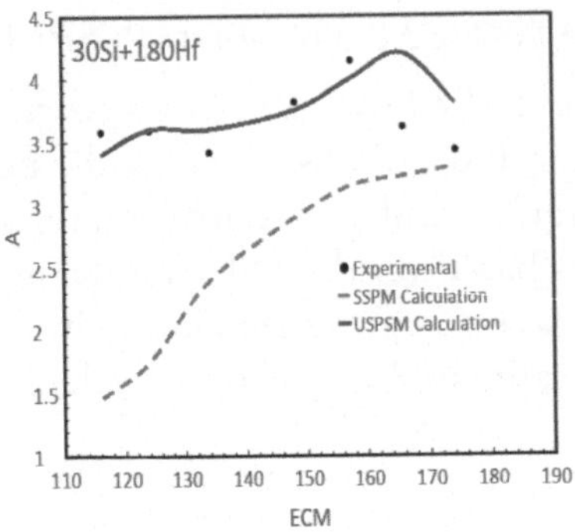

Figure 1

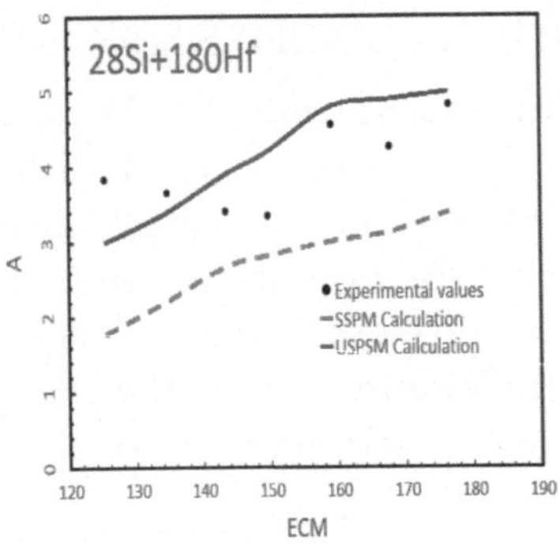

Figure 2

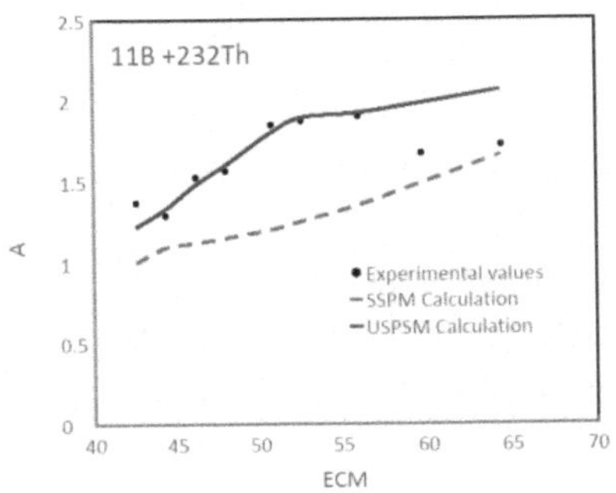

Figure 3

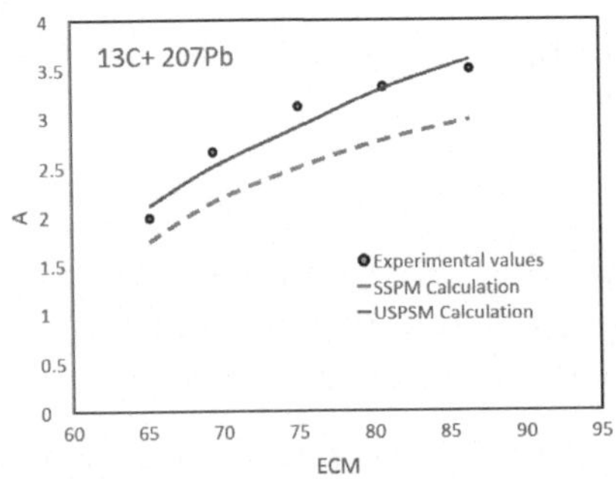

Figure 4

V. Conclusion

Fission fragment angular distributions for the four reaction systems with an anomalous behavior within the framework of the standard saddle-point statistical and the upgraded saddle-point statistical are calculated. The prediction of the upgraded saddle-point statistical model for the studied systems are found in better agreement with the experimental data of angular anisotropies than the predictions of the standard saddle-point statistical and the scission-point statistical models. Therefore, we can use the upgraded saddle-point statistical model to predict the anomalous angular fragment distributions for most reaction systems where non-compound nuclear fission probability is high.

Acknowledgement

The authors Lakhyajit Sarma and K. Kalita thanked IUAC, New Delhi for providing financial assistance through a major research project (IUAC/ XIII.7/UFR-67302).

REFERENCES

[1] Majumdar, N., *et al.*, Anomalous Peak like Structure in the Fission Fragment Anisotropies at Sub-barrier Energies in ^{11}B, ^{12}C, ^{16}O, ^{19}F + ^{232}Th Reactions, *Phys. Rev. Lett.* 77, 5027 (1996).

[2] Back, B.B., Complete fusion and quasifission in reaction between heavy ions, *Phys. Rev. C* 31, 2104 (1985).

[3] Soheyli, S., *et al.*, Fission fragment angular distribution in heavy-ion-induced fission with anomalous behavior, *Phys. Rev. C* 90, 024611 (2014)

[4] Murali, R., *et al.*, Anomalous fragment angular distributions in the fission of composite systems formed iN ^{28,30}Si+ ^{180}Hf Reactions, *Proceedings of the DAE Symp. on Nucl. Phys.* 65 (2021).

Enhancement in Neutron Production Rate via Adaptation of Different Confinement Mechanisms in a Cylindrical Inertial Electrostatic Confinement Fusion Device

L. Saikia[1*], D. Bhattacharjee[1], S. Kalita[1] and S.R. Mohanty[1,2]

[1]Centre of Plasma Physics-Institute for Plasma Research, Nazirakhat, Sonapur, Assam-782402, India

[2]Homi Bhabha National Institute, Anushakti Nagar, Mumbai, Maharashtra-400094, India

*E-mail: lucky.saikia@cppipr.res.in

ABSTRACT: The research on the development of a table top neutron source working on the principle of an inertial electrostatic confinement fusion (IECF) device has been drawing considerable attention because of its simple design and low cost. It offers either a 2D or 3D compression of oscillating fuel ions in a negative potential well depending upon electrode geometry and thereby producing a fusion reaction at the central part of the device. The wide range of applications of the device demands an enhancement in the neutron production rate. Therefore, the device needs modification to generate highly energetic ion beams with a strong plasma confinement mechanism to improve neutron yield to a large extent. With the aid of additional confinement mechanisms either by electrostatic or magnetic field, it may be possible to enhance the neutron yield because of better ion number density, resulting in more ion-ion and ion-neutral fusion reactions. In this work, we have modified the existing IECF device with a multi-gridded configuration to have a strong electrostatic confinement. Moreover, a cusp magnetic field is also used with the existing electrostatic field to improve the plasma confinement. These new configurations of the device may also be able to produce more neutrons in comparison to the single-gridded IECF system which we have partially verified. An attempt is also made to improve neutron yield by using a smaller IECF chamber with a solenoidal magnetic field arrangement. The recent findings will be presented in this paper.

Keywords: Inertial Electrostatic Confinement, Triple-gridded Arrangement, Cusp Magnetic Field, Electromagnet.

I. Introduction

The inertial electrostatic confinement fusion (IECF) device is primarily designed to produce thermonuclear energy. However, the device's poor efficiency for thermonuclear power generation prompted the researchers to look into alternative applications for the device. In 1950, P.T. Farnsworth and Lavrent'ev proposed the IECF concept, which was later modified by R.L. Hirsch [1]. Ions recirculate and confine the plasma in this IECF device via electrostatic fields, and then fuse them to produce neutrons, protons, X-rays, and so on. For neutron production, deuterium gas is used as the background gas. The primary fusion reactions that basically take place in such a device are,

$$D + D \rightarrow {}^{3}He\ (0.82\ MeV) + N\ (2.45\ MeV) \quad \ldots (1)$$

$$\rightarrow T\ (1.01\ MeV) + P\ (3.02\ MeV) \quad \ldots (2)$$

$$D + T \rightarrow {}^{4}He\ (3.5\ MeV) + N\ (14.1\ MeV) \quad \ldots (3)$$

In this IECF device (50 cm in diameter and 30 cm in height), a transparent cylindrical central grid is used which acts as a cathode and chamber wall as an anode, (see Figure 1). Plasma is created in between cathode and anode via filamentary discharge or cold cathode discharge mode. Ions are attracted towards the cathode grid and oscillate within the grid, finally, they fuse together at higher cathode potential and produce neutrons. In comparison to other fusion reactors, the IECF device stands out due to its simple construction and capacity to offer a reasonably high fusion rate in a small volume; as a result, researchers are focusing on using the device in a variety of near-term applications, such as coal and gold mining, cancer therapy (BNCT), fusion material study, explosive materials detection, radioisotope production, neutron/x-ray radiography, and so on [1,2]. This single gridded cylindrical IECF device is able to produce 10^6 n/s at -90 kV cathode voltage. But wide range of application of the device needs enhancement in neutron production rate [3].

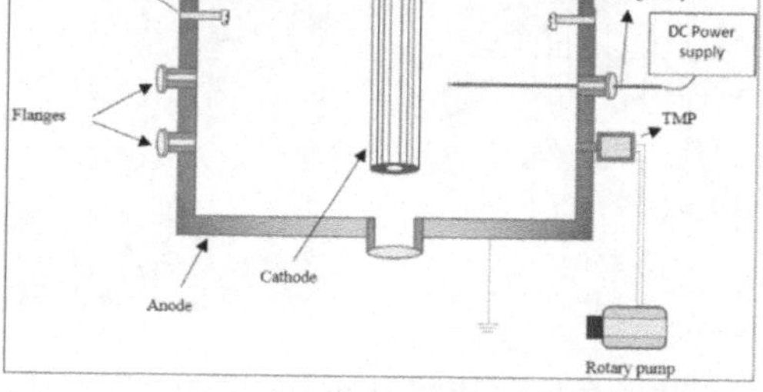

Figure 1: Schematic Diagram of Single Grid Cylindrical IECF Device

As a consequence, alternative configurations must be used to improve ion confinement approaches [1,4]. In order to explore the confinement mechanism, we first added two external grids in addition tothe central cathode grid and applied various potentials to the external grids.The plasma parameters and the neutron yield have been measured in this configuration. The plasma can also be confined using a magnetic field. Therefore, we have coupled a magnet assembly consisting of Samarium Cobalt permanent magnets so as to create a cusp magnetic field. Moreover, we recently installed a compact IECF device with a solenoid magnetic field to enhance the neutron yield in such a small device. We optimised the required magnetic field strength to effectively confine the plasma and to fuse it in a way that increases the neutron yield.

II. Different Plasma Confinement Mechanism

A. Triple-gridded Cylindrical IECF Device

In this work we have modified our device by adding two external grids (Figure 2) along with the existing IECF device to improve ion flow dynamics. This new configuration modifies electrostatic fields and channalize ions by small angle scattering mechanism with the help of independently bias external grids. Hence, ion confinement mechanism is now changed. This change has been characterised with the help of Langmuir probe I-V measurements in the filamentary discharge mode. Potential and ion density profiles are studied by varying intermediate grid biasing, applying –1kV on central grid, and keeping the outer grid in floating condition.Potential profile observations show a decrease in well depth and the formation of a hill-like

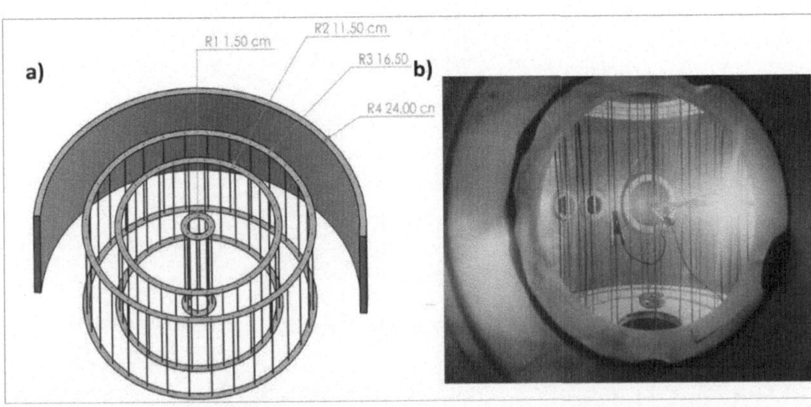

Figure 2: (a) Triple Grid Arrangement and
b) Filamentary Discharge Mode with Langmuir Probe

area near the cathode grid, as shown in Figure 3(a). This could be owing to the intermediate positive field, which allowed most of the primary electrons to flow towards the central cathode. Ion density is almost identical to that of the original single-gridded IECF (10^{15} m^{-3}), but when one of the external grids is positively biased, it intensifies between the central cathode and intermediate grid [Figure 3(b)]. Ions failed to cross the potential barrier produced near the cathode edge due to insufficient recirculation energy. These patterns have been observed due to different discharge mechanisms and plasma diffusion.

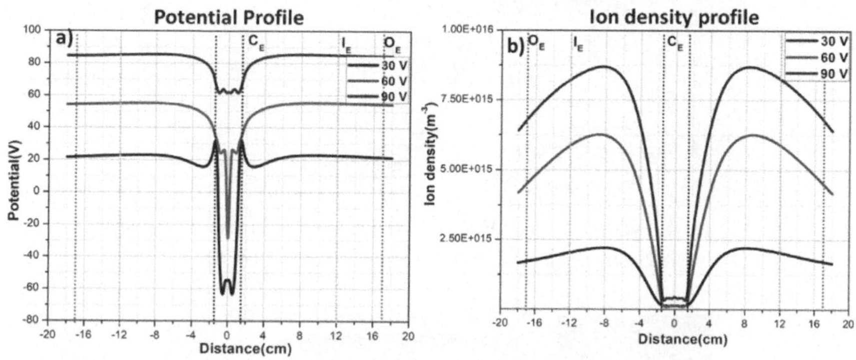

Figure 3: a) Potential and, b) Ion Density Profiles by Varying Intermediate Grid with Positively Biased, -1 kV on Central Grid and Outer Grid is Floating

This new configuration is also able to produce neutrons at higher cathode potential. By using neutron area monitor (NAM), we have measured the neutron dose rate at -40 kV cathode voltage and is found to be 1102 $\mu Svh^{-1,}$ which is equivalent to 10^7 ns^{-1} (Figure 4). Hence, we can say that the triple-gridded configuration with a modified electrostatic field can enhance the neutron yield approximately one order higer than the single-grid IECF device.

B. Effect of Electro-Magnetic Field in an IECF Device

For better ion confinement, we have modified the device by imposing a magnetic cage inside the IECF device, as shown in CAD diagram [Figure 5 (a)]. We believe that the effect of electromagnetic field can enhance the ion density with strong confinement mechanism. Samarium Cobalt (SmCo) is used as a permanent magnet to provide 0.1 to 0.3 T field strengths. Figure 5(b) depicts the simulated cusp magnetic field pattern that shows a field-free zone at the central area. The pattern of the glow discharge in the star mode [Figure 6(a)] reveals that the particles are following the magnetic field lines and are highly confined within the cathode grids. Hence, ion density in the core increased when the magnetic effect came into the picture compared to the initial single-gridded scenario. It implies that the neutron yield has similarly risen. According to observations of the neutron dose rate, this configuration produces nearly a 10^7 n/s neutron yield at our working limitations as seen from Figure 6(b).

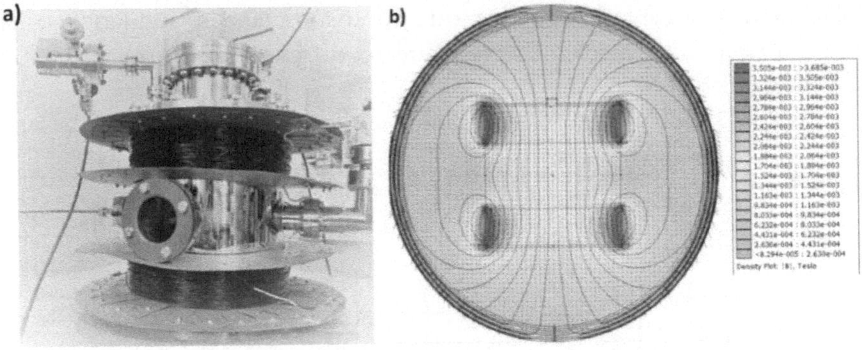

Figure 4: Neutron Dose Rate Recorded by the NAM at -40 kV Discharge in the Triple-Gridded IECF Device

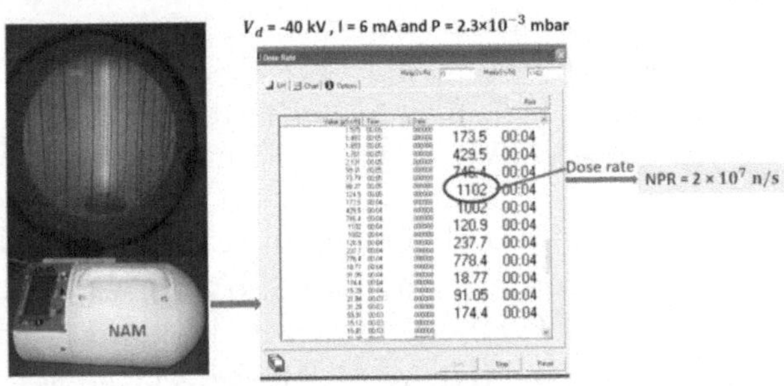

Figure 5: a) CAD Diagram of Magnetic Cage with Chamber, and b) Simulated (FEMM) Magnetic Field Patterns

C. Compact IECF Device with Electromagnet

A compact version of the cylindrical IECF device having a diameter of 25 cm and a height of 30 cm has been designed and installed with the goal of achieving more NPR using solenoid magnetic field, as shown in Figure 7(a). The pattern [Figure 7(b)] shows that the simulated magnetic field strength behaves similarly to bar magnets. In order to generate the required magnetic field strength that will strongly confine the plasma and fuse them in a way that will increase the neutron yield, the length, coil currents, and number of turns of the solenoid are fixed. By varying coil currents from 2 to 6A, we are able to produce magnetic field strengths 0.1 to 0.4 T.

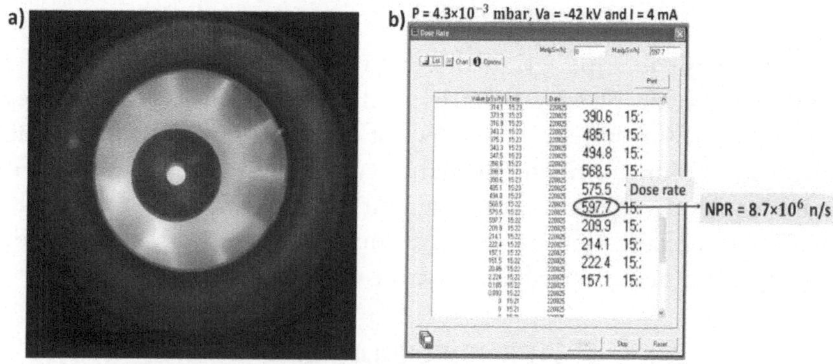

Figure 6: a) Glow Discharge Observation (star mode) from Bottom View and, b) Screenshot of Neutron Dose Rate

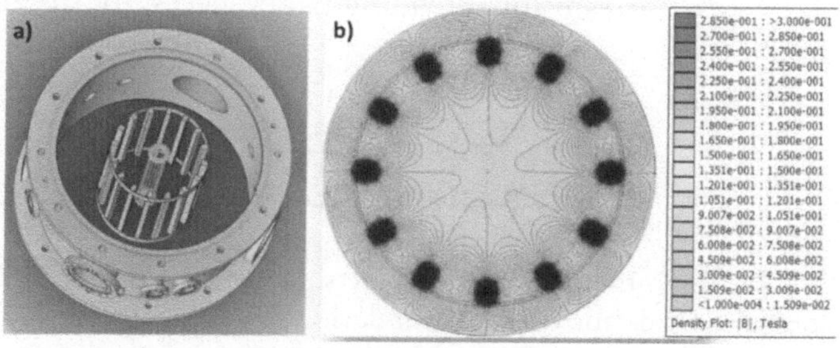

Figure 7: a) Compact IECF with Electromagnetic Coils and, b) Simulated (FEMM) Magnetic Field Patterns

The glow discharge observation [Figure 8] shows that increasing the solenoid current results in a strong ion confinement near the cathode. It means that for such a reasonable coil current, an increase in neutron production is achievable, which we shall verify in later period.

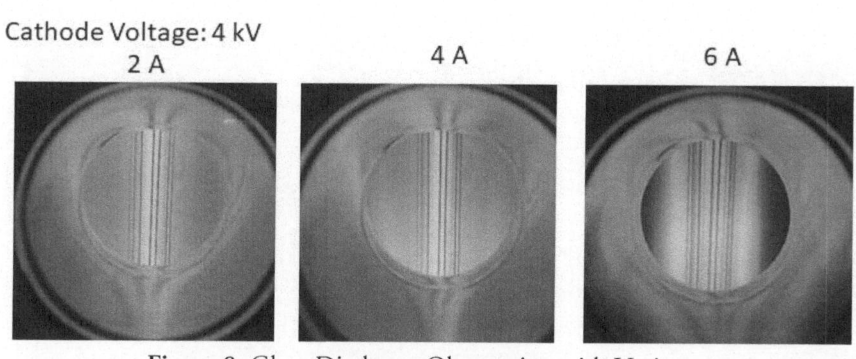

Figure 8: Glow Discharge Observation with Various Coil Currents at Cathode Voltage -4 kV

III. Conclusion

I-V curve analysis in triple-gridded arrangement demonstrates the difference in ion flow behavior, plasma potential, and core ion density as compared to a single-gridded IECF system. The effect of electromagnetic fields also able to provide a strong ion confinement at the central region of the system. Finally, we must conclude that suitable magnetic field strength and external grid potentials can enhance ion density and ion beam energy in an IECF device, which improves the rate at which neutrons are produced.

Acknowledgement

We are grateful to the Director of Institute for Plasma Research (IPR), Gujrat and the Centre Director of Centre of Plasma Physics –Institute for Plasma Research (CPP-IPR), Assam for providing us the necessary supports to carry out this work.

References

[1] Miley, G.H. and Murali, S.K., Inertial electrostatic confinement (IEC) fusion, Springer (2014).

[2] Murali, S.K., *et al.*, Effect of chamber pressure variation on the grid temperature in an inertial electrostatic confinement device, *Phys. Plasmas* 17, 102701 (2010).

[3] Noborio, K., *et al.,* Confinement of ions in an inertial electrostatic confinement fusion (IECF) device and its influence on neutron production rate, Fusion engineering and design 81, 8 (2006): [1701-1705].

[4] Thorson, T.A., *et al.,* Convergence, electrostatic potential, and density measurements in a spherically convergent ion focus. *Phys. Plasmas* 4, 4 (1997): 4-15.

SECTION-V

Communication and Instrumentation

A Comparative Study of Fastest Settling PI and PID Controlled Phase-Locked Loop for Communication Systems

Geetamoni Konwar* and Tulshi Bezboruah

Department of Electronics & Communication Technology,
Gauhati University, Guwahati, India
*E-mail: geetak222@gmail.com

ABSTRACT: We have introduced a new higher order fast settling proportional-integral-derivative controlled Phase-locked loop and compared it with two other most viable models in this area viz., the proportional-integral controlled phase locked loop and the conventional 5th order phase locked loop. We have constrained the model using MATLAB simulation. We have studied the Settling time, Phase margin and Bandwidth properties of the proposed models and it is found that the proportional-integral-derivative controlled phase-locked loop is significantly more stable and also provides too much fast settling time with an improvement of 24.43% (maximum) as compared to the conventional 5th order phase-locked loop as well as the proportional-integral controlled phase-looked loop.

Keywords: Phase-Locked Loop, Proportional-Integral Controller, Proportional-Integral-Derivative Controller, Settling Time, Phase Margin.

I. Introduction

The phase-locked loop (PLL) is a very interesting and useful building block available as single integrated circuits in modern communication system design. The PLL concept as we know it today was originally described in a published work by de Bellescize in 1932 but did not fall into widespread use until the era of television where it was used to synchronized horizontal and vertical video scans [1-3]. The first PLL ICs were available at around 1965; it was built using purely analog component [4,5]. Recent advances in IC circuit design techniques have led to the development of high performance PLL which has become more economical and reliable.

Settling time (ST) is a very important property of PLL for applications which need fast frequency switching. Due to the rapid growth of high speed data communication, many communication systems such as Frequency Hopping Spread Spectrum (FHSS), Step Frequency Radars (SFR) and Wireless Local

Area Networks (WLANs) use frequency hopping to avoid multipath fading and interference by protecting data security [6-10]. However, currently, there is no PLL available that can hop the frequency quickly enough to meet the timing requirements of the Global System of Mobile system (GSM) [11, 12]. So, there is a need to switch from one frequency to another in a fast way. But, the switching from one frequency to another causes a ripple due to the sudden change in frequency. The fast ST has therefore become a very important challenge to achieve stable PLL system. Consequently, to improve the ST of the proposed system, the conventional filter of the PLL system is replace by a proportional-integral (PI) controller and a proportional-integral-derivative (PID) controller in the loop.

II. Proposed PLL System

Figure 1 shows the basic block diagram of the proposed PLL system. The system consists of a Phase frequency detector (PFD), a 4th order passive low pass filter (PLPF) and a voltage controlled oscillator (VCO) along with optional PI controller and PID controller within its loop.

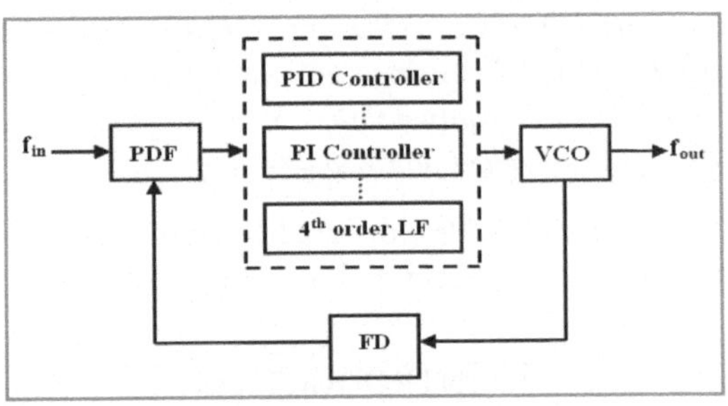

Figure 1: Functional Block Diagram of the Proposed PLL

The TF of 5th order PLL can be derived as:

$$H_{5th}(s) = \frac{K_d K_o (1 + sC_2R_2)}{C_1C_2C_3C_4R_2R_3R_4s^5 + \{C_1C_2R_2R_3(C_3+C_4) + C_4R_4(C_2C_3R_3 + C_1C_3R_3 + C_1C_2R_2 + C_2C_3R_2)\}s^4 + \{C_2R_2(C_1+C_3+C_4) + R_3(C_1+C_2)(C_3+C_4) + C_4R_4(C_1+C_2+C_3\}s^3 + (C_1+C_2+C_3+C_4)s^2 + KC_2R_2S + K} \quad \dots (1)$$

Where, K_d = PDF gain, $F(s)$ = TF of corresponding filter, $\frac{K_o}{s}$ = VCO sensitivity, $\frac{1}{N}$ = Fractional counter value and $K = \frac{K_d K_o}{N}$ = Loop Gain constant [3, 16].

The TF of the PI controller PLL can be derived as:

$$H_{PI}(s) = \frac{K_d K_o (K_p s + K_i)}{s^2 + K_p K s + K_i K} \qquad \dots (2)$$

The TF of the PID controller PLL can be derived as:

$$H_{PID}(s) = \frac{K_d K_o (K_D s^2 + K_P s + K_I)}{s^2(1 + K_D K) + K_P K s + K_I K} \qquad \dots (3)$$

III. The Simulation

Step responses of simulation for PLL with 4th order LF, with PI controller and with PID controller are shown in Figures 2, 4 & 6 respectively. Figures 3, 5 & 7 shows Bode plots of simulation of PLL with 4th order LF, PI controller and PID controller for four different samples. It is observed that the ST for different test cases of the PID controlled PLL is faster than that of the 5th order PLL as well as the PI controlled PLL.

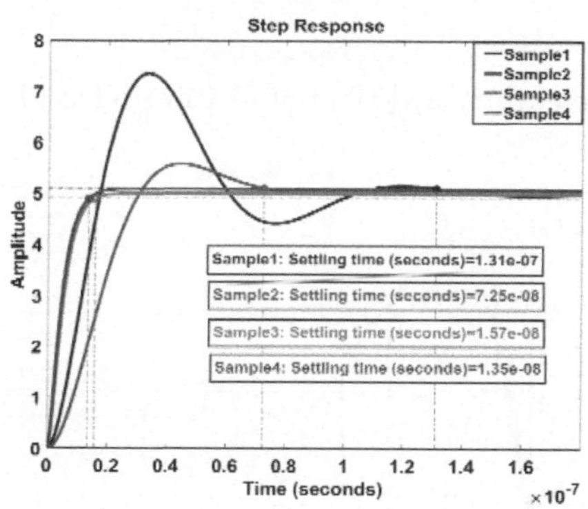

Figure 2: Step Response for 5th Order PLL

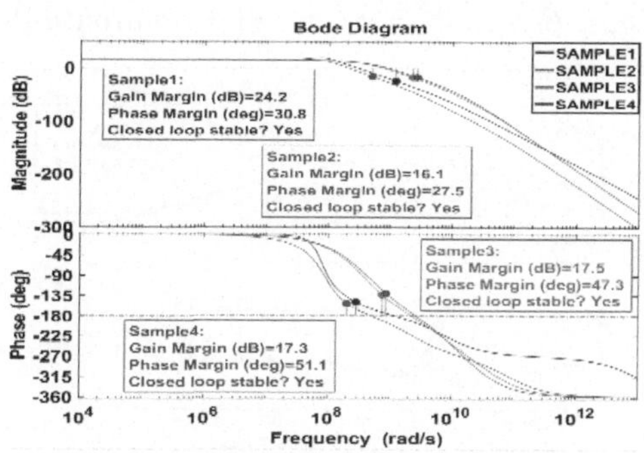

Figure 3: Bode Plot of 5th Order PLL

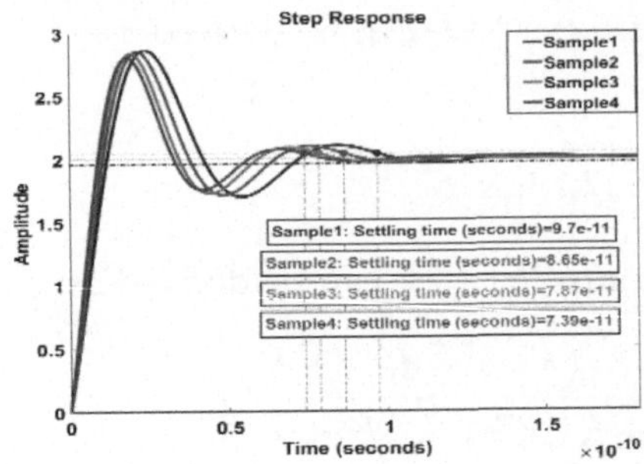

Figure 4: Step Response for PI Controlled PLL

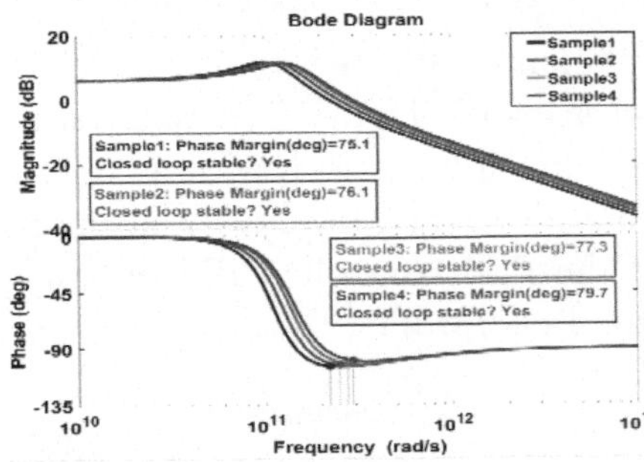

Figure 5: Bode Plot of PI Controlled PLL

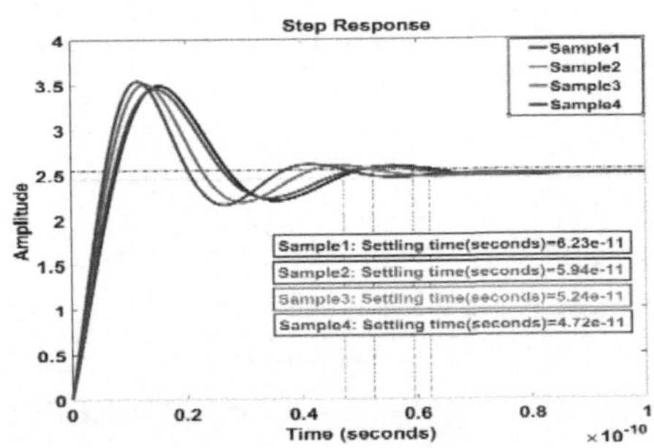

Figure 6: Step Response for PID Controlled PLL

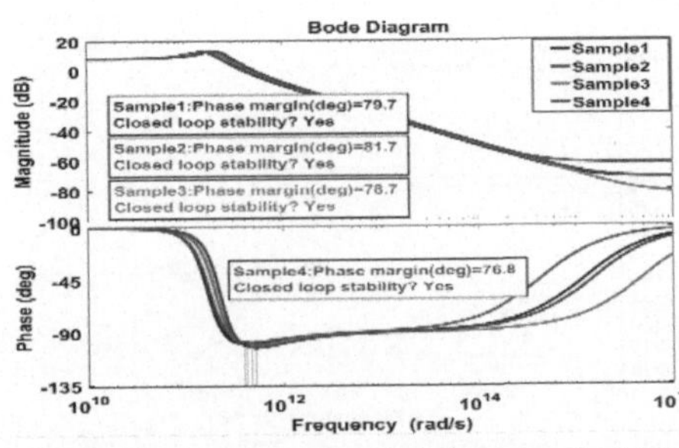

Figure 7: Bode Plot of PID Controlled PLL

IV. Results and Discussion

The overall simulation results of the 5th order PLL, PI controlled PLL and PID controlled PLL is summarized in Table 1. It is observed that the 5th order PLL provides *ST* in the range of 13.1 nS to 72.5 nS while the PI controlled and PID controlled PLL provides very fast ST in the range of 0.0739 nS to 0.097 nS and 0.0595 nS to 0.0733 nS respectively. The resultant PMs are varies in a range of 27.5° to 79.7° which satisfies the limit of stability. The variation of loop BW is from 0.231 GHz to 52.527 GHz. The percentage (%) of improvement in the ST of PID controlled PLL with respect to the PI controlled PLL is also summarized in Table 1. It is observed that the ST of the PID controlled PLL system improves up to 24.43% in comparison to the PI controlled PLL.

Table 1: Simulation Results

PM (degree)			*BW (GHz)*			*ST(Nanoseconds)*			*% of Improvement in ST*
5th order PLL	*PI Controlled PLL*	*PID Controlled PLL*	*5th Order PLL*	*PI Controlled PLL*	*PID Controlled PLL*	*5th Order PLL*	*PI Controlled PLL*	*PID Controlled PLL*	
30.8	75.1	76.5	0.0231	28.996	38.383	13.1	0.097	0.0733	24.43%
27.5	76.1	76.9	0.0164	32.513	41.272	72.5	0.0865	0.0681	21.27%
47.3	77.3	73.8	0.0384	35.749	47.227	15.7	0.0787	0.0595	24.39%
51.1	79.7	71.8	0.0409	37.979	52.527	13.5	0.0739	0.0655	11.36%

V. Conclusion

This work will provide the overall results of simulation for 5th order PLL along with PI controlled as well as PID controllered PLL. The result of the proposed model increases the stability with fast ST PID controlled PLL. Also, it is observed that the PM values for each cases of PLL with LF, PI controller and PID controlled PLL are well within the limit of stability. So, the model may be applied as a trade – off for research and industrial applications for acquiring better ST in communication system. This section concludes a summery; a comparison of fast settling time higher order PID controlled PLL with conventional PLL and also PI controller. Future work includes hardware implementation of the proposed model.

References

[1] Daniels, B., Baldwin, G., Farrell, R. and Loone, Mc S., Modelling and Design of High-Order Phase Locked Loops, *Proceedings on VLSI Circuits and Systems II,* Vol. 5837 (2005).

[2] Kuang, X. and Wu, N., AFast-Settling PLL Frequency Synthesizer with Direct Frequency Presetting, *IEEE International Solid State Circuits Conference,* pp. 741–750 (2006).

[3] Uyanik Ugur, H. and Tarim, N., PID-Controlled PLL for Fast Frequency-Hopped Systems, *Workshop on 6th IEEE Dallas Circuits and Systems*, pp. 1–3 (2007).

[4] Patel Janakkumar, B., Anand, R.S. and Patel Nilam, J., Implementation of frequency demodulator using the PLL demodulation method, *National Conference on Computational Instrumentation*, pp. 13–16 (2010).

[5] Karuppanan, P. and Mahapatra, K.K., PLL Synchronization with PID Controller Based Shunt Active Power Line Conditioners, *International Journal of Computer and Electrical Engineering,* Vol. 3(1), pp. 42–47 (2011).

[6] Santos, L.A. d., Kaster, M. d. S. and Silva, S.A.O. da., Applying a nonlinear PID in asingle phase PLL control, *IEEE International Conference on Power Electronics, Drivesand Energy Systems*, pp. 1–4 (2012).

[7] Paliwal, P., Fadadu, J., Chawda, A. and Gupta, S., A Fast Settling 4.7–5 GHz Fractional-N Digital Phase Locked Loop, 2016. *29th International Conference on VLSI Design and 2016 15th International Conference on Embedded Systems,* pp. 553–554 (2016).

[8] Kong, L. and Razavi, B., A 2.4-GHz RF Fractional-N Synthesizer with BW = 0.25 fREF, *IEEE Journal of Solid-State Circuits*, pp. 1–12 (2018).

[9] Yang, W.-B., Wang, H.-H., Chang, H.-I. and Lo, Y.-L., A Fast-locking all-digital PLL with dynamic loop gain control and phase self-alignment mechanism for sub-GHz IoT applications, *Japanese Journal of Applied Physics,* Vol. 59(SGGL08), (2020).

[10] Beiraghdar, F., Ghobadi Rad, A., Sheikhaei, S. and Tohidian, M.A., Fast settling frequency synthesizer with switched-bandwidth loop filter, *International Journal of Circuit Theoy and Applications*, Vol. 49, pp. 2021–2031 (2021).

[11] Bertulessi, L., Cherniak, D., Mercandelli, M., Samori, C., Lacaita, A.L. and Levantino, S., Novel Feed-Forward Technique for Digital Bang-Bang PLL to Achieve Fast Lock and Low Phase Noise, *IEEE Transactions on Circuits and Systems I: Regular Papers*, Vol. 69(5), pp. 1858–1870 (2022).

[12] Pan, Z., Wang, X., Hoang, TGT., Tian, L., An enhanced phase-locked loop for non-ideal grids combining linear active disturbance controller with moving average filter, *International Journal of Electrical Power & Energy Systems*, Vol. 149, pp. 1–13 (2023).

Design of a PWM Based 50 Hz, 12 V DC/ 230 V AC, 2.5 KVA Solar Power Inverter

H.C. Medhi

Department of Electronics, St. Edmund's College, Shillong-793003, Meghalaya, India
E-mail: hemenmedhi1984@gmail.com

ABSTRACT: The epileptic property of power generation via traditional methods, lead us to shift our focus to alternate forms of power generation. On account of erratic nature of power, some communities are deprived for days, weeks or months from the nation's grid. This gap period of no supply or cut off from the grid, can be solved by designing and constructing an alternate source using solar power inverter suitable for house hold applications. The aim of this paper is to design and construct a 50 Hz, 12 V DC/230 V AC 2.5 KVA solar power inverter. The inverter circuit consists of four important stages which include the transformation stage, oscillator stage (implemented with IC SG3524 pulse width Modulator (PWM), driver stage (using MOSFET IRFP 260) which controls the switching and feed back stage.

Keywords: Inverter, Pulse Width Modulator, MOSFET, Feedback.

I. Introduction

The erratic nature of power supply has translated into high cost of production of goods and services. Power electronic solutions such as inverters convert direct current to alternating current for the purpose of domestic, commercial and industrial uses are increasing attention [1]. An inverter is an electrical device that converts direct current (DC) into alternating current (AC) which is possible with the appropriate design of transformer, oscillator, MOSFET driving stage, feedback section and associated electronic circuits. The inverter which are available in the market to drive light loads in the range of 500 to 1000 watts. This paper introduces a method to calculate total household electrical loads for realizing the higher wattages and sizing the entire system (inverter, PV array and battery) to meet such loads.

II. Electrical Load Information

The size of the inverter depends upon the loador energy consumption. This can be estimated from the knowledge of power consumption for a specified load. In this paper the size of the solar inverter is estimated by following some important steps: (i) list all the electrical appliances, (ii) record the required operating wattage for each item, (iii) specify the number of hours per day each item to be powered, (iv) calculate watt hour uses per day by multiplying the first three columns and (v) determine the total watt hour per week for each item is obtained from the number of days per week that each item to be used.

From the individual load information, it is estimated as 1980 Watt but a sudden addition of loads hence a 2.5 KVA unit has designed to meet these needs. The circuit is designed so that if a power spike occurred, or something malfunctioned then the key components of the system would be saved, as the breaker opens. The loads include one washing machine, one TV set, a DVD player, one refrigerator, one blender, bread toaster, electric iron which uses are determined according to time of need is depicted in Table 1.

Table 1: Estimation of Power Consumption for Different Loads

Individual Loads Qty × Volts × Amps = Watts Watts × Use × Use ÷ 7 = Watt Hours								
AC	Hrs/Day		Days/Week		Days AC			
Electric Iron	1	220	2.5	550	.5	3	7	117.85
Washing Machine	1	220	1.5	330	2	2	7	188.57
TV	1	220	0.5	110	3	7	7	330.00
DVD Player	1	220	0.5	110	2	5	7	157.14
Refrigerator	1	220	1.0	220	7	7	7	1540.00
Blender	1	220	1.5	330	.5	2	7	47.14
Bread Toaster	1	220	1.5	330	.5	2	7	47.14
AC Total Connected Watts 1980 AC Average Daily Load								2,427.84

III. Inverter

Inverter is a device that converts Direct current (or DC) to Alternating Current (or AC) using step up transformers, switching and control circuits. Domestic inverters take the DC power from the batteries and converts into AC power used by the load connected at home. Similarly, an off grid solar inverter operates. In case of solar inverter [2] the DC power is generated from the solar panels and AC power is given to the grid. In case of grid connected solar inverter when the power is coming from the grid, the Inverter system charges the batteries using the power coming from the grid. When the power coming from the grid is off, then the inverter takes the DC power from the batteries and converts it into AC to supply it to appliances.

In this regard an automatic switch is used to sense whether the power is coming from the grid and if not then it switches the UPS into battery mode.

IV. Working of Inverter Circuit

The inverter system consists of an oscillator unit, MOSFET driver unit, Centre–tap transformer and a feedback unit. These units are shown in the design and construction circuit diagram of Figure 1. The method of operation of each unit is explained thus.

A. Oscillation Unit

To generate the required frequency 50 Hz, an oscillator is designed using IC SG3524, a dedicated pulse width modulator as depicted infigure1 known as free- running relaxation oscillator and it has no stable state, but two half-stable states between which it keeps oscillating continuously on its own without any external excitation. This versatile PWM controller can be used in a variety of isolated and non-isolated switching power supplies such as inverters. By supplying a constant 12 V DC, through frequency of the oscillating signal is determined using (130 KΩ), comprising a 100 KΩ resistor, connected in series with a 47 KΩ resistor and both connected in parallel with a capacitor 0.2 μF to form the RC time constant network.

The SG3524N consists of an on-board 5 V regulator facility that serves as a reference, as well as supplying power to the SG3524N internal regulator control circuitry. The pulse width modulation is used to keep the inverter output constant at 230VAC. To ensure constant output, the SG3524N always receives a feedback output of AC supply generated by the inverter [3]. If the IC does not receive the feedback, the PWM output from pins 11 and 14 will also change; as a result the inverter output will fluctuate. This dedicated PWM IC SG3524N operates best between 5.40 volts DC and at temperatures 0°C –70°C.

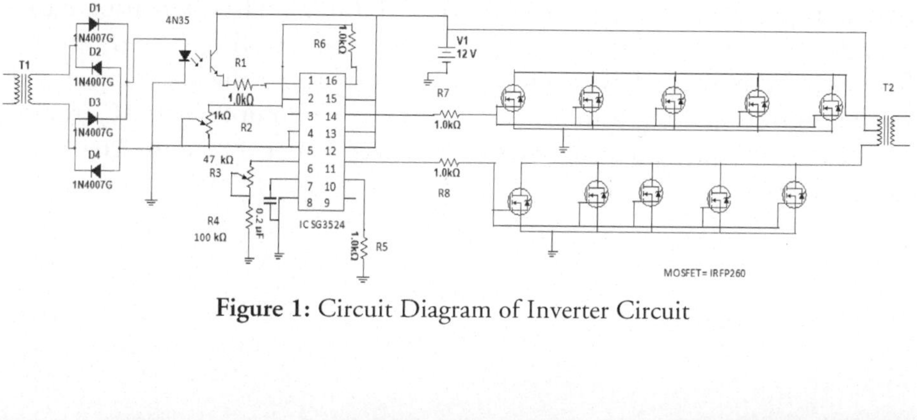

Figure 1: Circuit Diagram of Inverter Circuit

B. MOSFET Driver Unit

In the forward direction output is available at pin number 14 MOSFET M1 gets triggered; it will now act as closed switch and will get completed due to which voltages would be induced in the secondary of the transformer. When MOSFET M1 acts as closed switch at that time the direction of flow of current in the primary winding will be clockwise and at the same time the direction of flow of current in the secondary will be in anticlockwise. The output obtained from the secondary winding of the transformer would be square wave and positive in nature.

When we get an output at pin 11 at that time MOSFETM2 gets triggered. It will now act as a closed switch. Thus the primary circuit of the transformer will become closed switch and will get completed due to which the voltages would be induced in the secondary of the transformer. When MOSFET M2 acts as closed switch at that time the direction of flow of current in the primary will be anticlockwise and in the secondary will be clockwise. The output obtained from the secondary winding of the transformer would be a square wave. This wave will be negative in nature. The supply from the battery to the load receives the oscillated DC voltage signal; boost the power and conducts in the positive and negative half cycles thereby producing an AC output which is stepped up by the transformer to the required AC voltage of 230 V.

V. Transformation Stage

A centre tapped transformer, steps–up from 12V-0-12V AC to 230VAC is used to increase supply voltage for the circuit. It consists of two coils called the primary and the secondary, wound round a soft iron core which made of sheets of soft iron. The number of turns for primary is 21 and for secondary is calculated as 383.28 which is approximated as 400 turns. The value of SWG for primary winding is 10 and SWG is 18 respectively for secondary winding, is used for the voltages using the calculated turn's ratio and the coil thickness provides a power transformer of 3500 W. The alternating current is applied at the primary winding induced an alternating current at 50Hz frequency in the secondary winding of the transformer which is stepped up by the transformer causing it to become 230 V. The output 230 V, AC from the secondary winding is transferred to the socket outlet of the inverter system for appliances.

VI. Optocoupler Feed Back Unit

The feedback unit, made this design unique, strong in terms of instantaneous loading, and reliability. It consists of transformer T2, a bridge rectifier, resistorand an optocoupler which is shown in the Figure 1. The optocoupler used in this design is 4N35, the LED has a forward current of 10 mA, isolation voltage 5300 V_{RMS} and input–output coupling capacitance < 0. 5 pF. The bridge rectifier provides DC input to the LED of the optocoupler. The feedback unit stabilizes the output power in the event of sudden rise in load voltage, monitors the output and prevents voltage fluctuation keeping the output steady at 230 V based on the internal circuitry of the SG3524N fixed variable regulator.

VII. Conclusion

The solar inverter circuit completed successfully and gives voltage and current stability on load, providing an alternate source of power to charge batteries. This prototype system is sensitive to sinusoidal inputs can easily be implemented in houses for saving electricity.

References

[1] Kapadia, N., Patel, A. and Kapadia, D., Simulation and design of low cost single phase solar inverter, *International Journal of Emerging Technology and Advanced Engineering*, Volume 2, pp 158-159 (2012).

[2] Kulkarani, V. and Nehete, R., Simulation and analysis of photovoltaic (PV) based solar inverter system", *International Journal of Soft computing and Engineering*, ISSN: 2231-2307, Volume 3, pp. 114 (2014).

[3] Najafi, E. and Yatim, A.H.M., Design and Implementation of a New Multilevel Inverter Topology, *IEEE Transaction on Industrial Electronics*, Vol. 59, No.11, pp-4148-4154 (2012).

Smart Instrumentation System for Medical Application

Jayshri Narzary[1] and Hidam Kumarjit Singh[2]
[1]Department of Physics, Bijni College, Bijni, Assam
[2]Department of Electronics & Communication Technology, Gauhati University, Assam
E-mail: [1]ijayshri@gmail.com; [2]kumarjit_hidam@gauhati.ac.in

ABSTRACT: We have designed a novel IoT-based system that will be used in the monitoring of saline water infusion to patients in the hospital. The system includes a load cell sensor and HX711 ADC integrated with the raspberry pi circuit. The load cell sensor senses the weight of the saline water through the piezo-resistivity principle and then the HX711 ADC amplifies and converts the output electrical signal to the digital form. Python program has been developed for controlling the raspberry pi hardware. The system continuously monitors the saline level and sends warning messages to the mobile phone of on-duty medical staff when the saline level is about to cross a predefined threshold value. The experimental results show that the design system can measure the saline volume with a sensitivity of 0.00477 mV/mL, which shows the promising potential of the proposed system for use in hospitals. The novelty of the proposed system includes good repeatability and accuracy in addition to being inexpensive to develop.

Keywords: Load Cell Sensor, Saline, Raspberry pi.

I. Introduction

In health care centers, continuous liquid level monitoring is also one of the important parameters. Liquid monitoring may be saline water level or blood level monitoring. A liquid volume monitoring system in hospitals is a process where a doctor or a nurse can continuously monitor many patients at a time with less human intervention [1-3]. In traditional liquid level monitoring of a saline bottle, the nurses use to monitor the saline level from time to time until the saline finishes or once the saline liquid level is about to run out, a relative of the patient informs the indication signal of the saline level to the nurses or doctors. Consistent monitoring of saline is important because there is the possibility of blood backward flow from the patient's body due to low pressure in the saline bottle and high pressure of blood in the patient's body [4-6]. It is also risky of entering an air bubble from a saline bottle into the bloodstream of the patient stopping the blood flow of the patient [7-8]. But this constant monitoring system is laborious, time-consuming and the doctors or nurses cannot monitor many patients at the same time. So, we need innovative methods for health monitoring systems where a few people can manage the monitoring in low-cost design with highly accurate data. The rapid growth in medical technologies along with engineering technologies has solved this problem to a large extent. Over the decades, many researchers have explored the smart sensing techniques and monitoring of intravenous saline infusion with the common aim to develop a nonintrusive sensor system.

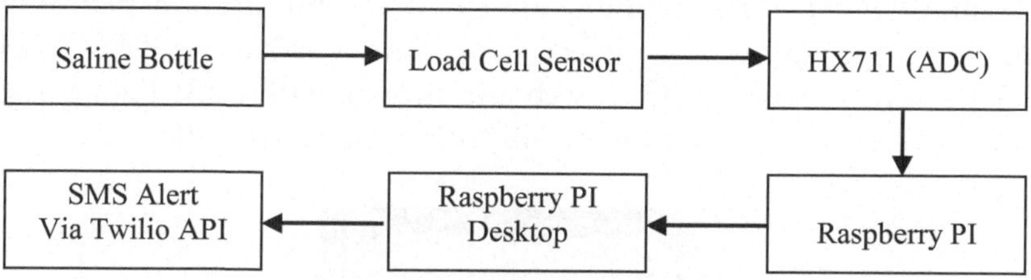

Figure 1: Block Diagram of the Developed Instrumentation System

II. METHODOLOGY

A. Instrumentation of the System

Figure 1 shows the block diagram describing the instrumentation of the system. The methodology followed in instrumenting the system is as follows:

The load cell was fixed with a hanger where a saline bottle was suspended to allow the flow of saline under gravity. The flow of saline reduces the weight of the saline bottle which was enabled to be detected by the load cell through strain gauge deflection. Using the Wheatstone bridge configuration, the amount of strain gauge deflection was then converted into an electrical signal in a millivolt (mV) unit. The electrical signal generated was amplified and then converted to digital form using HX711. This amplified output is finally connected to the Raspberry Pi. This amplified output was interfaced to the GPIO pins of the Raspberry Pi and then the GPIO pins were controlled using Python programming. Python programming was finally used to connect with Twilio Communication API to send the SMS alert to the cell phone of the responder once the saline level is about to cross the set threshold value.

The various components used in the developed instrumentation system are described below:

- **Load Cell:** The system used the load cell of model YZC-133 having a rated load of 1 kg as the weight sensor. A load cell is a transducer that measures weight or pressure and converts it into a measurable electrical signal in the output.

- **HX711:** The HX711 is a 24-bit analog-to-digital converter (ADC) specially designed for weight measurement. The differential analog input channel A can be programmed with a gain of either 168 or 64 and can be directly interfaced with the differential sensor output.
- **Raspberry Pi:** The system used Raspberry Pi 3 B+ model. Its various interfaces can interact with small electronic devices such as sensors and actuators
- **Twilio:** Twilio is a programmable cloud communication platform that provides *Application Programming Interfaces* (API) to send SMS, MMS, IP messages worldwide from web and mobile APP [10]. Figure 2 shows the SMS alert received from Twilio Communication API.

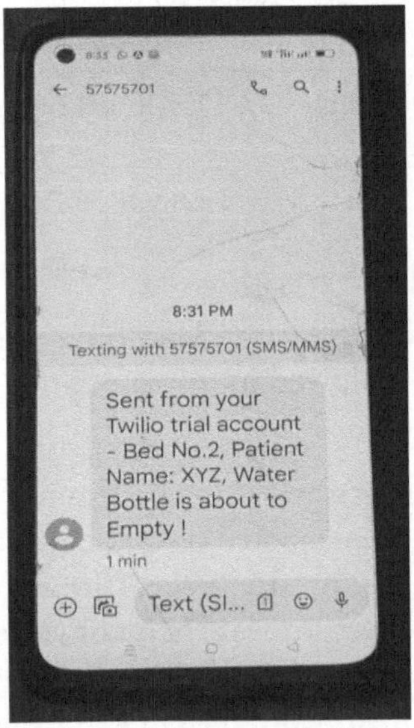

Figure 2: SMS Alert Using Twilio

B. Experiments

Experiment 1: Validation of the instrumentation system with standard weights

The standard weights of 5, 10, 20, 30, 40, 50, 60, 80, 100, 120, 140, 160, 180, 200, 220, 240, 260, 280, 300, and so on were put on the load cell module and then the corresponding output was measured and recorded. The measurement was repeated for six times as a trial to investigate the repeatability of the instrumentation system.

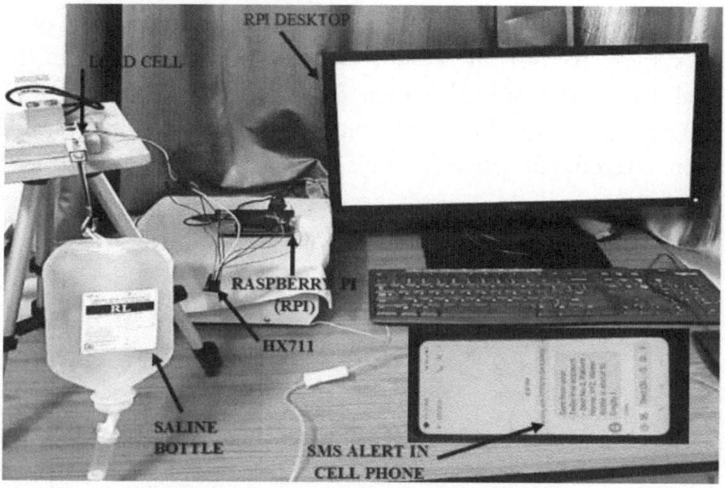

Figure 3: Experimental Setup of Proposed System

Experiment II: Measuring standard weight in unit of mV

Here same experimental procedure and setup was used as in experiment I. The output was only measured in the mV unit. The measurement was repeated for six times as a trial to investigate the repeatability of the instrumentation system.

Experiment III: Validation of the instrumentation system with saline water

In the third experiment, a beaker was put on the load cell and then with the help of a 10 ml syringe pump saline water was added to the beaker upto 500 ml. For every trial, the output of the system was recorded in mV unit. This experiment was done to establish the relationship between the saline water volume measured by the developed system and the input saline volume.

Finally, the 500 ml saline bottle was hang on the load cell and then the continuous variation of saline water was observed every 15 seconds, until it reaches the predefined threshold level set in the python programming. The experimental set up is shown in Figure 3.

III. Results and Discussion

A. Experiment I

Figure 4 shows the graph plotted between input weight and sensor output weight in gm. It shows the strong correlation between the input and output weight having coefficient of determination (R^2) of 0.9999 in all 6 nos. of trial. In all the trails having the high R^2 represents the ability of the instrumentation system to output weight with minimum error. Further, R^2 of 0.9999 in all the 6 nos. of trial depicts the high repeatability of the system in proving the output result. Therefore, based on the result, it is established that the developed system can provided measurement of weight with least error.

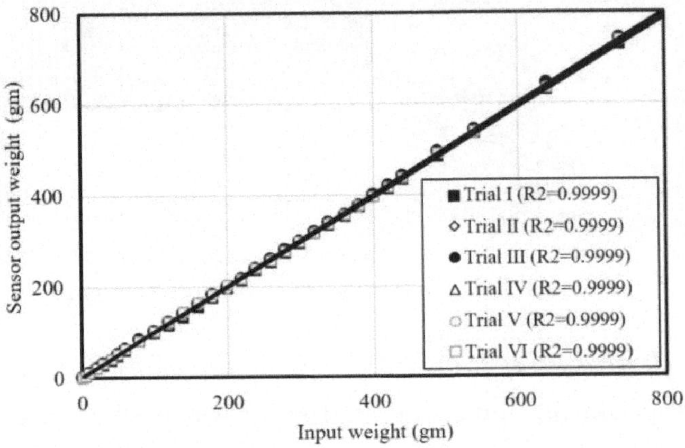

Figure 4: Input Weight vs. Sensor Output in gm

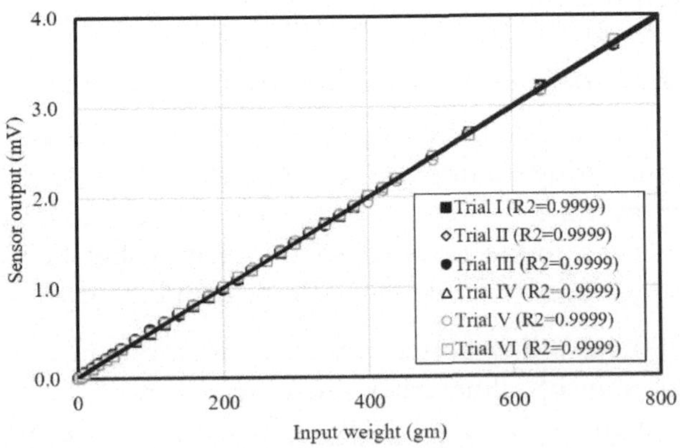

Figure 5: Input Weight in gm vs. Sensor Output in mV

B. Experiment II

Figure 5 shows the graph plotted between input weight in gm and sensor weight in mV. It shows the strong correlation between the input and output having coefficient of determination (R^2) of 0.9999 in all 6 nos. of trial.

C. Experiment III

Figure 6 shows the relationship between the average weight measured by the developed system and the volume of the saline water infused with the syringe pump. The coefficient of determination (R^2) of 0.9999 in all six trial shows that the result obtained has strong agreement with the first and second experiments.

The results obtained in all three experiments show that the designed system can be used for the monitoring of the saline volume.

For the smart monitoring of the system, python program was then developed for the sending SMS alert using Twilio API. The threshold value was set for alert system. The flowchart of it is shown in Figure 7. The system allows the real time monitoring of the saline water volume on the Raspberry Pi desktop.

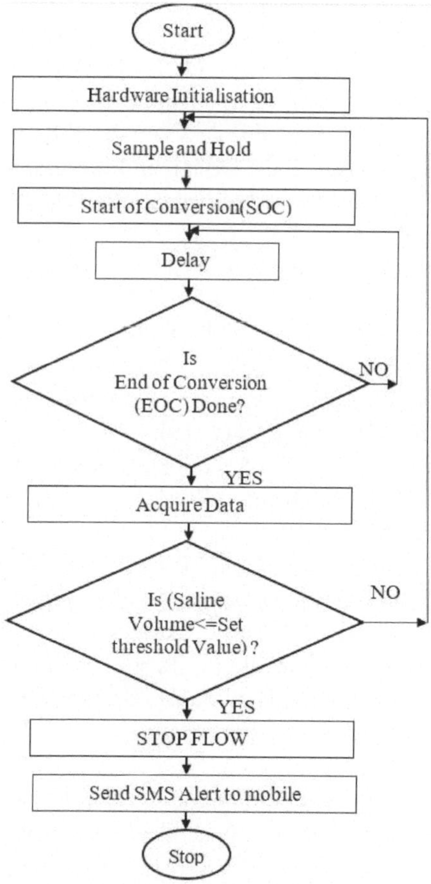

Figure 7: Flowchart of Proposed System

IV. Conclusion

The experimental results show that the design system can measure the saline volume with a sensitivity of 0.00477 mV/mL, which shows the promising potential of the proposed system for use in hospitals. The novelty of the proposed system includes good repeatability and accuracy in addition to being inexpensive to develop. The system allows the real time monitoring of the saline water volume and can send SMS alert to the phone of the nurse or doctor. With the proposed system, the nurse can monitor multiple patients at the same time.

References

[1] Ajibola, O.O., Sunday, O.O. and Eyehorua, D.O., Development of automated intravenous blood infusion monitoring system using load cell sensor, *Journal of Applied Sciences and Environmental Management*, 22(10), 1557–1561 (2018).

[2] Ghosh, D., Agrawal, A., Prakash, N. and Goyal, P., Smart saline level monitoring system using ESP32 and MQTT-S, *IEEE 20th International Conference on e-Health Networking,* Applications and Services, Healthcom (2018). https://doi.org/10.1109/HealthCom.2018.853117.

[3] Gaikwad, K.D. and Dahikar, P.B., Design and Development of Novel Weighing Scale System, *International Journal of Engineering Research & Technology (IJERT)*, 2(5), 1668–1671(2013).

[4] Adesina, O., Alamuoye, H. and Familusi, T., Development of Microcontroller Milligram Weighing System, *International Research Journal of Engineering and Technology (IRJET),* 8(8), 1796–1803 (2021).

[5] Yahaya, F.R., Ahmad, R.L. and Bakar, S.A., Dorper BSI Monitoring with Load Cells and Raspberry PI, *IEEE Symposium on Computer Applications and Industrial Electronics,* ISCAIE 2019, 208–212 (2019).

[6] Naik, K.S. and Sudarshan, E., Smart Healthcare Monitoring System using Raspberry Pi on IOT Platform, *ARPN Journal of Engineering and Applied Sciences,* 14(4), 872–876, 2019.

[7] Raghavendra, K.K., Sharanya, P.S. and Patil, S., An IoT Based Smart Healthcare System Using Raspberry Pi, *International Journal of Research and Scientific Innovation (IJRSI),* V(VI), 103–106 (2018).

[8] Jain, S., Vaibhav, A. and Goyal, L., Raspberry Pi based Interactive Home Automation System through E-mail, *International Conference on Reliability, Optimization and Information Technology - ICROIT 2014,* India, 227–280, (2014).

[9] Raspberry Pi Official Website (Online). Available: https://www.raspberrypi.org/

[10] Twilio official Website (Online). Available: https://www.twilio.com

A Comparative Study of Hollow Core Photonic Crystal Fiber and Conventional Optical Fiber for Femtosecond Optical Pulse Transmission

P. Talukdar* and D. Phukan

Department of Physics, Royal Global University, Guwahati-781035, Assam, India

*E-mail: priyankatalukdar17031995@gmail.com

ABSTRACT: The purpose of the current study is to investigate optical fiber dispersion compensation by designing a hollow core photonic crystal optical fiber. In order to study a Gaussian pulse with wavelength centered at 0.55 m, both a hollow core and a conventional solid core optical fiber are used. According to the study, hollow core fibers can effectively reduce the effects of dispersion. Both the hollow core photonic crystal fiber and the conventional optical fiber's individual optical pulse output spectra and transmission graphs are examined. In comparison to traditional optical fiber, the designed hollow core photonic crystal fiber's maximum transmission was determined to be about 20% higher.

Keywords: Dispersion, Hollow Core Optical Fibers, Gaussian Pulse, Conventional Solid Core Optical Fiber, Transmission.

I. Introduction

In optical communication systems, hollow core photonic crystal fibers (HCPCFs) are prepared to outperform the traditional solid core optical fibers now in use. A typical HCPCF consists of a 2-D array of air holes with a very high air filling factor along the entire fiber length. Rayleigh scattering and the nonlinear Shannon limit prevent conventional solid core optical fibers from performing any better than they already do [1]. Hollow core photonic crystal fibers are the best choice for high-speed data transmission and the delivery of powerful pulses because they have extremely low nonlinearity, a high damage threshold, very low latency, and are temperature insensitive [2]. The primary benefit of this fiber is its ability to sustain single mode wave guiding over a wide spectral range [3]. A single mode conventional optical fiber becomes multimode as the frequency of optical radiation rises, while optical losses rise as the frequency of optical radiation decreases [4]. In the case of hollow core photonic crystal fibers, the circumstance is different.

The two-dimensional structured cladding reduces the overall fiber nonlinearity by more than 1000 times by creating a photonic band gap that allows light to be guided in a low-refractive zone in hollow core fibers [5]. This study simulates femtosecond pulses in order to examine their spectral broadening and dispersion properties in both conventional and hollow core photonic crystal fibers.

II. Optical Simulations

We have modeled pulse propagation inside the fiber using the non linear envelope equation, from which the propagation equation can be written as [6],

$$\frac{\partial \mu}{\partial \in} = \sum_{n=2}^{4} -(i)^{n-1} \frac{L_{ds}}{n! L_{ds}^{(n)}} \frac{\partial^n \mu}{\partial \tau^n} + i\left(1 + \frac{i}{\omega_o \tau_p} \frac{\partial}{\partial \tau}\right) p^{nl} \quad \text{... (1)}$$

Where $\tau = \frac{t - \frac{z}{v_g}}{\tau_p}$ is the retarded frame time normalized to the input pulse width τ_p. τ is the normalized time, ω_0 is the central frequency, v_g is the group velocity, n is the order number, p^{nl} is the non linear polarization and L_{ds} is the dispersion length.

Finite Difference Time Domain (FDTD) method is one of the most popular simulation techniques for solving electromagnetic equations. The FDTD approach can simulate dispersive and nonlinear electromagnetic processes [7]. The designed fiber is a silicon photonic crystal fiber with a hexagonal lattice and five air hole rings in the cladding. The centre air hole, which serves as the central high index defect and is essentially the fiber's core, is eliminated. The pitch, written as "Λ" is the assumed hole-to-hole distance that is taken as 2 μm.

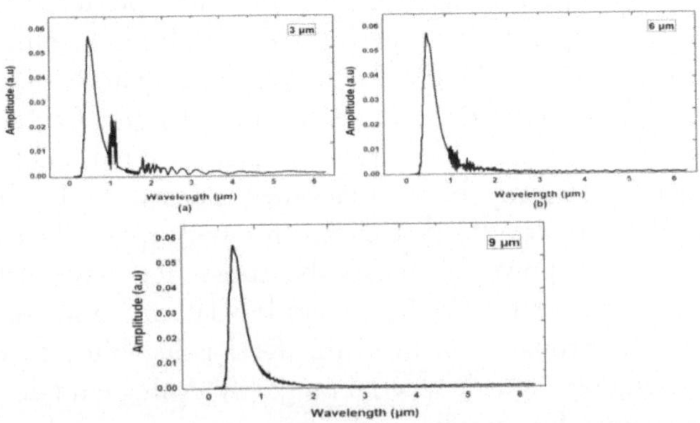

Figure 1: The Broadened Spectrum of a 2.6 Femtosecond Laser Pulse at the Output of a 1 Metre Long Conventional Optical Fiber for (a) 3 μm (b) 6 μm (c) 9 μm Core Radius

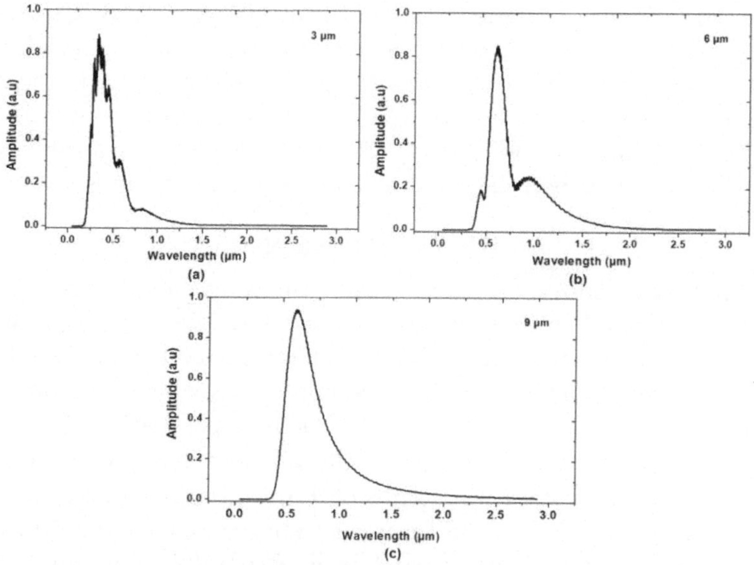

Figure 2: Output Spectrum for (a) 3 µm (b) 6 µm (c) 9 µm Core Radius of the Designed Hollow Core Photonic Crystal Fiber

While a significant portion of light energy was scattered and underwent dispersion when going through a 1 meter long conventional optical fiber, only a small portion of light energy was visible at the output spectrum. The femtosecond pulse at the fiber's output exhibits significant broadening as well. The oscillating structure at the pulse's tail that reduces optical compression is a result of third order dispersion (Figure 1). For a core radius of 9 µm, the fundamental mode is more tightly constrained within the core as

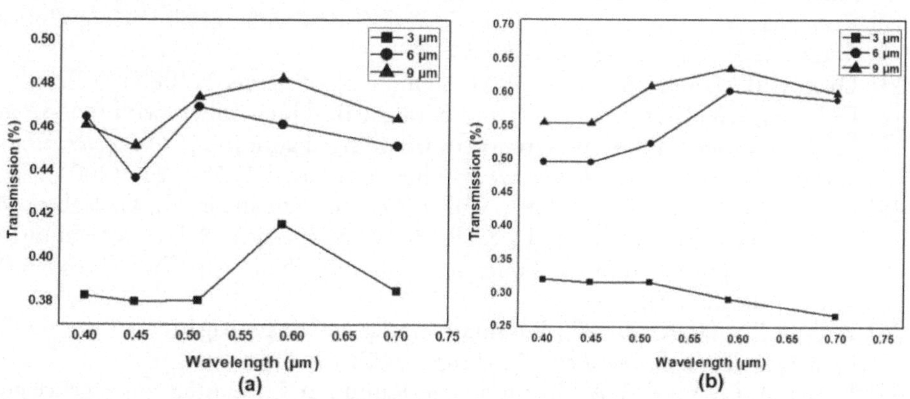

Figure 3: Transmission (%) vs. Wavelength Plot for (a) Conventional Optical Fiber and (b) Designed Hollow Core Photonic Crystal Fiber

the core size is increased, and as a result, an efficient output spectrum is detected at the fiber end (Figure 2). For various core radius considered, the transmission graph in relation to wavelength is presented. In both fibers, it was found that the transmission rises as the size of the core is increased. The designed hollow core photonic crystal fiber has a maximum transmission that is around 20% higher than that of conventional solid core optical fiber.

III. CONCLUSION

The spectral broadening of femtosecond pulses in hollow core photonic crystal fibers is found to be significantly more effective than in an ordinary optical fiber. The ability to adapt the dispersion properties of these fibres to the needs is made possible by the flexibility with which photonic crystal fibres can be designed by altering their structural parameters. Femtosecond optical pulses with a wavelength span of 0.3 μm were safely coupled into the photonic crystal fiber with a core radius of 9 μm. Photonic crystal fibres have a wide range of applications in lasers, sensors, switches and optical fibre communication systems due to their flexible structural designing, ability to fill the air holes with gas or liquid as per requirement, and variety in their operating principles.

ACKNOWLEDGEMENT

Authors acknowledge Royal Global University for providing the required software for this study.

REFERENCES

[1] Ellis, A.D., McCarthy, M.E., Al Khateeb, M.A.Z., Sorokina, M. and Doran, N.J., Performance limits in optical communications due to fiber nonlinearity, *Advances in Optics and Photonics,* 9(3), 429–503 (2017).

[2] Komanec, M., Dousek, D., Suslov, D. and Zvanovec, S., Hollow Core Optical fibers, *Radio Engineering,* 3, 417–430 (2020).

[3] Dianov, E.M. and Prokhorov, A.M., Handbook of Lasers 2, 108 (1972).

[4] Fedotov, A.B., Tarasishin, A.V. and Kirillov, B., Holey Fibers with 0.4–32 m-Lattice-Constant Photonic Band-Gap Cladding: Fabrication, *Characterization, and Nonlinear-Optical Measurements*, Physics of Lasers 1, 138–145 (2001).

[5] Ouzounov, D.G., Ahmad, F.R., Muller, D., Venkataraman, N., Gallagher, M. T., Thomas, M.G., Silcox, J., Koch, K.W. and Gaeta, A.L., Generation of megawatt optical solitons in hollow-core photonic band-gap fiber, *Science* 301, 1702–1704 (2003).

[6] Brabec, T. and Krausz, F., Nonlinear optical pulse propagation in the single-cycle regime, *Phys. Rev. Lett.,* 78, 3283 (1997).

[7] Lesina, A. C., Vaccari, A., Berini, P. and Ramunno, L., On the convergence and accuracy of the FDTD method for nanoplasmonics, *Optics Express* 8, 10481–10497 (2015).

Measurement of Small Change in Angle Using Optical Moiré Technique and Computer Generated Hologram

S. Buragohain* **and A. Das**

Department of Physics, Gauhati University, Guwahati-781014, Assam, India

*E-mail: buragohainswarnav@gmail.com

ABSTRACT: In this work we propose to use optical Moiré pattern technique to measure small angle disturbances precisely. The small angle disturbances are produced by applying a vertical load to a rod attached to another rod placed vertically over a firm base. The Moire pattern is generated with the help of two computer generated holograms having parallel opaque and transparent regions and having slightly different spatial frequencies. One of the holograms is attached to the vertical rod and another is placed very close to the first one so that the Moire pattern is formed. When a load is applied vertically the rod is bent and hence the angle between two holograms changes. This change in angle changes the Moire pattern. By comparing the two Moire patterns before and after the application of load the small change is angle is measured.

Keywords: Moire Pattern, Computer Generated Holography, Small Angle Measurement.

I. Introduction

Optical moiré pattern[1, 2] is formed when two amplitude grating of slightly different periodicity overlap with each other. The amplitude grating is formed with consequent opaque and transparent lines each with same width. Moiré pattern is common in our daily lives. Two combs when placed over one other forms moiré lines, shirts with regular spaced lines when viewed also forms moiré pattern. The moiré lines formed is also in the form of consecutive opaque and transparent lines. This moiré effect is very effective for different metrology applications, strain analysis etc. because of its high sensitivity. The sensitivity depends on the density of the lines on the two gratings [1]. Historically, moiré effect was first used in 1971 by Dan Post along with his co-worker Maclaughlin. From then it has been widely used as highly sensitive optical metrological methods for deformation, rotation, displacement measurement etc.

In this work we have applied optical moiré method to calculate very small change in angle of a steel rod when it is loaded with different loads. Two gratings of different periodicity are used to obtain the moiré lines when the rod gets bent. The gratings used in the experiment are obtained using computer generated holography (CGH)[3,4] technique with suitable frequency. The methods presented in this work has the advantage over other methods as in this methods sophisticated optical components such as lasers, lens, beam splitter etc. are not used. Therefore, in addition to being highly sensitive, the method is also very simple to implement.

II. Results and Discussions

The method to calculate the small change in angle using line moiré pattern is explained in Figure 1 (a) and (b). This moiré pattern is formed when two rulings having slightly different periodicity one attached to the reference rod and the other attached to the test rod overlaps as shown in Figure 2. When different load is applied the ruling attached to the test rod gets rotated by a small amount and the moiré angle α_m also gets changed. From this we can calculate the change in angle by which the test rod gets bent upon different applied loads. Let us assume P_r as the periodicity of the revealing layer and P_b be the periodicity of the base layer. Let P_m be the periodicity of the moiré pattern formed when the two layers overlap each other making an angle. If n number of base layer equals n+1 number of revealing layer then from Figure 1 (a) and (b) we can write [5],

$$\tan\alpha_m = \frac{P_2(\tan\theta)}{P_2 - P_1^{/}} = \frac{P_2 \sin\theta}{P_2\cos\theta - P_1} \quad \ldots (1)$$

Where θ is the angle made by the base layer with the revealing layer and α is the moiré angle formed. When θ is very small we can write,

$$\tan\alpha_m = \frac{P_2\theta}{P_2 - P_1} \quad \ldots (2)$$

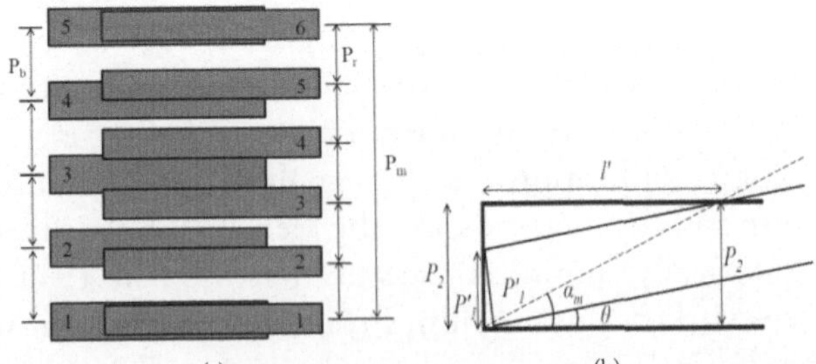

Figure 1: (a) Overlapping of Base Layer and Revealing Layer and (b) Formation of Moiré Angle

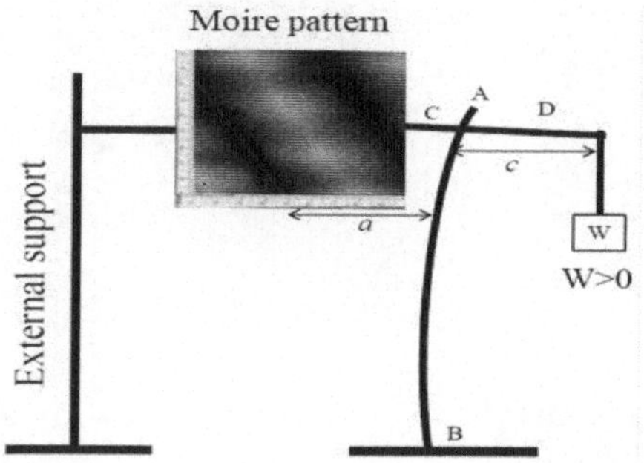

Figure 2: Schematics of Experimental Setup

From the above equation (eq. 2) we can calculate the small change in angle with different applied loads. Equation 2 also shows us that even if the relative angle between the two layers change by a small amount the moiré angle change by a large amount. So we can easily observe change in moiré angle and calculate the small change in angle of the steel rod upon subsequent applied loads. The formula to calculate the small change in angle is,

$$\Delta\theta = \frac{|\tan\alpha_{i+1} - \tan\alpha_i|}{m} \quad \text{.... (3)}$$

Here $m = \frac{P_2}{P_2 - P_1}$ represents the magnification factor and in our case its value is 9.199.

Table 1: Change in Angle for Different Applied Loads

Load Applied (gm)	$tan\alpha_m$	*Moire angle (α_m) (degrees)*	$\Delta\theta$ ($\times 10^{-3}$) *(degrees)*
0	1.1886	49.9268	
50	1.2142	50.5274	2.7836
100	1.2333	50.9644	4.8543
150	1.2786	51.9725	9.7845
200	1.2903	52.2238	11.0491

The experimental arrangement is shown in Figure 2. Four different loads 50 gm, 100 gm, 150 gm and 200 gm is applied at point c subsequently and then small change in angle is measured using the equation 3.

In our experiment l = 0.18 m, a = 0.25 m, c = 0.14 m and $r = 5 \times 10^{-3}$ m. The experimentally obtained moiré patterns for four different applied loads is shown in Figure 3. From the different moiré lines we choose a suitable moiré line and calculate the change in angle for different applied loads.

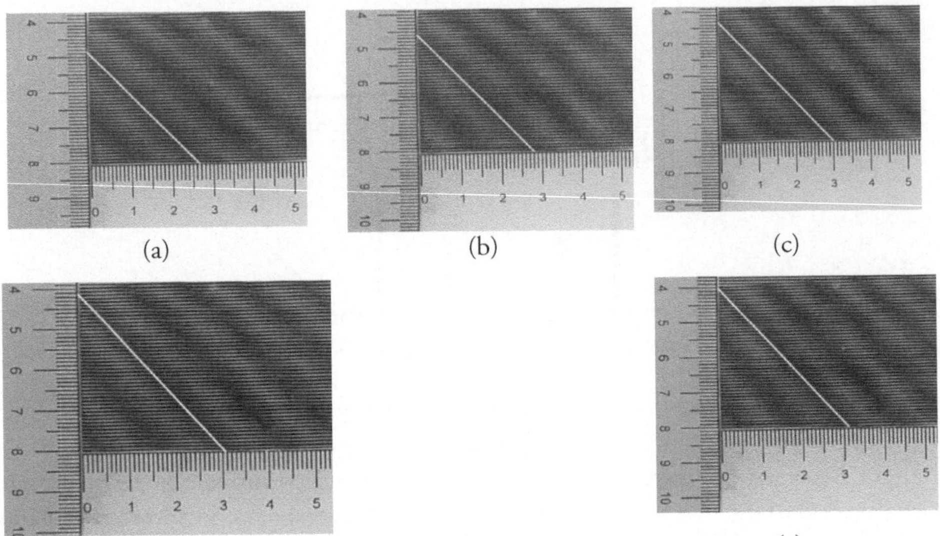

(a) (b) (c) (d) (e)

Figure 3: Experimentally Obtained Moiré Patterns for (a) 0 gm, (b) 50 gm, (c) 100 gm, (d)150 gm and (e) 200 gm

The calculated values are presented in the Table 1. From the table it is seen that a very small change in angle by the experimental rod makes a large moiré angle. Therefore, we can measure very small changes in angle precisely using the geometry of the moiré pattern. The change in angle is plotted with respect to the applied load and the variation is coming out to be almost linear. The moiré angle formed by the two rulings is also plotted against different applied loads and the variation is linear. The plots are shown in Figure 4 (a)&(b).

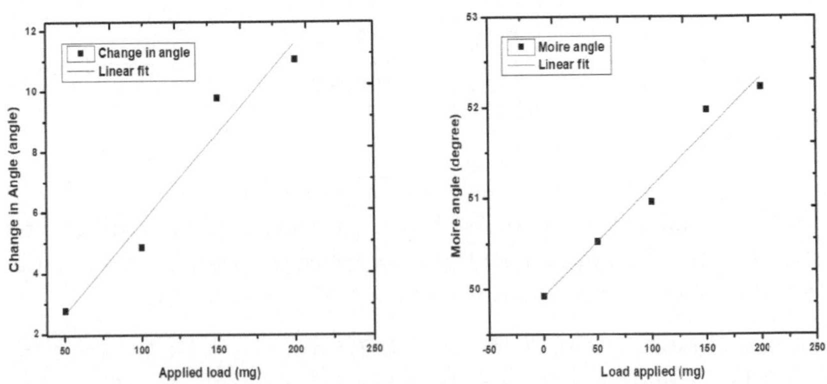

(a) (b)

Figure 4: (a) Change in Angle vs. Applied Loads and (b) Moiré Angle vs. Applied Loads

III. Conclusion

We have used Optical Moiré technique and computer generated hologram to measure small change in angle of a steel bar. We are able to measure change in angle of the order of 10^{-3} degree. The method can further be used in strain analysis and deformation analysis etc. Further, this method is relatively simple than other such methods because it does not use any optical components such as lens, mirror, laser etc.

References

[1] Walker, C.A., Moiré Interferometry for Strain Analysis, *Opt. Laser Engg.* 8, 213–262 (1998).

[2] Saveljev, V. and Kim, S.K., Simulation and measurement of moiré patterns at finite distance, *Opt. Express,* 20, 2163–2177 (2012).

[3] Boruah, B.R., Dynamic manipulation of a laser beam using a liquid crystal spatial light modulator, *Am. J. Phys.*, 77, 331 (2009).

[4] Kumar, D., Das, A. and Boruah, B.R., A Simple Experimental Arrangement to Generate Optical Vortex Beams, *Rev. Sci. Instrum.,* 84, 026103 (2013).

[5] Saveljev, V., Kim, S.K. and Kim, J., Moiré effect in displays: A tutorial, *Opt. Eng.,* 57, 030803 (2018).

A Flexible Metamaterial-Based Microwave Absorbing Tape at X-band

Sanghamitra Saikia and Nidhi Saxena Bhattacharyya*
Department of Physics, Tezpur University, Tezpur-784028, India
*E-mail:nidhisbhatta@gmail.com

ABSTRACT: This work presents a metamaterial absorber using single hexagonal closed ring resonator in the unit cell. The resonator is developed using aluminium on top of urethane rubber substrate backed by aluminium for applications in X-band. Aluminium is light, more flexible and less expensive metal [1] and on the other hand urethane rubber is a flexible and easy to processpolymer [2], hence considered to be suitable for absorbance study and developing the substrate. Despite of being thin with thickness of ~0.027λ, the absorber shows 98% absorbance at resonant frequency 10.13 GHz. Thus, can be used as absorbing tape. The measured results show good concurrence with simulation results. In addition, the structural symmetry of the unit cell enables polarization insensitive absorption.

Keywords: Flexible, Urethane Rubber, Aluminium, Microwave Absorber.

I. INTRODUCTION

Constant improvisation of electromagnetic (EM) absorbers for adequate suppression of the unwanted environmental electromagnetic interference (EMI) has resulted from rising expectations for the efficacy of electronic devices and systems. Despite the growing demand, the development of thin and flexible broadband absorbers has been an uphill battle.

Unique properties of metamaterial have extensively been researched to develop electromagnetic (EM) absorbers since its first demonstration by Landy *et al.*[3]. Generally, metamaterial absorbers (MMAs) are flat dielectric sheets with periodic arrangement of conducting resonant structures[4]. Material parameters such as effective permittivity ($\varepsilon_{eff} = \varepsilon' - j\varepsilon''$) and permeability ($\mu_{eff} = \mu' - j\mu''$) in MMAs can be tailored to desired value by optimizing structural dimensions of the unit cell. The artificially engineered unit cells with sub-wavelength resonators induce perfect absorption at microwave frequencies due to the matched impedance to free space. Ultra-thin thickness, light weight, and design flexibility in the unit cell set metamaterial absorbers apart from traditional EMI absorbers [5]. In the majority of published studies, copper and gold metallic layers are used to fabricate metamaterials on rigid substrates such as FR4. Inflexible absorbers rendered from such substrates are not ideal for use in devices with non-planar surfaces [5-7]. In addition, copper and gold unit cells make the absorber expensive. Flexible MMAs absorbers have recently attracted huge research interest due to their ability to conform to desired shape [8]. Fabrication of thin and flexible substrates however, requires costly equipments and the quest to obtain better absorption limits the flexibility while absorption is challenged when flexibility is sustained for a certain bending curvature.

This work is an attempt to design and develop a flexible metamaterial absorber (MMA) operating at X-band using urethane rubber as the substrate and array of hexagonal ring resonators pasted on the substrate. Aluminium tape is used for developing the resonators while urethane rubberis chosen for developing the substrate. Simulations are carried out on commercial CST microwave studio suite prior to fabrication for optimizing the structural design. Measurements are taken using waveguide technique for experimental verification.

II. Design and Simulation of the MMA Unit Cell

Unit cell of the proposed MMA is designed using metallic hexagonal ring resonator on top of a polymer substrate with metal back. Hexagonalring shape is chosen for the resonator as it shows six-fold symmetry unlike four-fold symmetry in square rings, thus enhancing polarization insensitive absorption. Hexagon shows better fit to curved surfaces and in addition the perimeter of a hexagon is lesser than an equal sized radius circle, hence decreasing the amount of material used reducing the overall cost. Both the hexagonal ring resonator and the metal back are developed using aluminium. Aluminum, in contrast to copper and gold, is inexpensive, lightweight, and malleable [1]. Moreover, aluminum's electrical conductivity, $\sigma_{Al} = 3.56 \times 10^7 \mathrm{S/m}$, is comparable to copper's, $\sigma_{Cu} = 5.96 \times 10^7 \mathrm{S/m}$, hence can be used instead. The substrate is developed utilizing urethane rubber. The dielectric constant of urethane rubber is found to be $\varepsilon_r = 2.8 \pm j0.2$ at X-band. The polymer is found to be flexible polymer with tensile strength of 1 MPa and it is observed to resist heat upto ~300°C. Besides urethane rubber is customizable, lightweight and shows good adhesion property [2]. Figure 1(a) shows the proposed absorber and the unit cell with structural parameters is represented in Figure 1(b). Structural optimization and simulation are carried out using CST software. Prior to simulation permittivity values of urethane rubber at X-band is imported to CST software. The simulation setup with along electric, magnetic fields and direction of propagation of the incident EM wave on the absorber are shown in Figure 1(c). The optimized values of unit cell parameters are tabulated in Table 1. The total thickness of the absorber is $8\ \text{mm} \cong 0.027\lambda$.

Table 1: Optimized Structural Parammets of the Proposed MMA

Dimension	*Value (mm)*
l	10.0
t	0.7
r_1	4.0
r_2	3.0

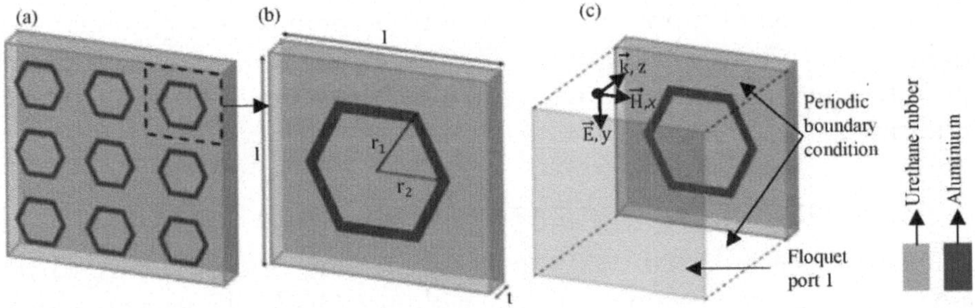

Figure 1: (a) Schematic of Proposed MMA, (b) MMA Unit Cell and (c) Simulation Setup

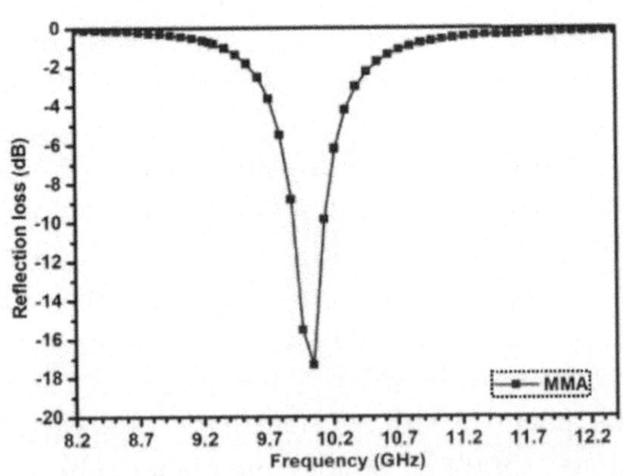

Figure 2: Simulated Reflection Loss Curve of the MMA at X-band

The absorbance of an absorber (as a function of frequency(ν)) is mathematically given as [3],

$$A_\nu = 1 - R_\nu - T_\nu \qquad \dots (1)$$

Where, $R_{\nu} = |S_{11}|^2$ and $T_{\nu} = |S_{21}|^2$arethe reflectance and transmittance respectively. Since the proposed MMA is a metal backed absorber, hence $T_{\nu} = 0$ reducing absorbance to $A_{\nu} = 1 - R_{\nu}$. Therefore, absorbance can be maximized by minimizing reflection loss solely. The MMA is found to resonate at 10.05 GHz with a reflection loss (S_{11}) of -17.3 dB as shown in Figure 2. The simulated with incident wave at different polarization angles and the reflection curves obtained are plotted in Figure 3(a). It can be observed from the plot that the curves at different polarization angles overlaps with each other. This implies that the structure is polarization insensitive attributing to the six-fold symmetrical hexagonal ring resonator.

To comprehend the absorption mechanism of the MMA simulations of the electric field and magnetic field at the resonant frequency are performed. Figure 3(b) demonstrates that the electric field concentration is greatest along the y-axis, whereas Figure 3(c) shows the magnetic field concentration is maxiumum along the x-axis. This is because, electric and magnetic fields of incident wave are parallel to the y and x-axis respectively [8].

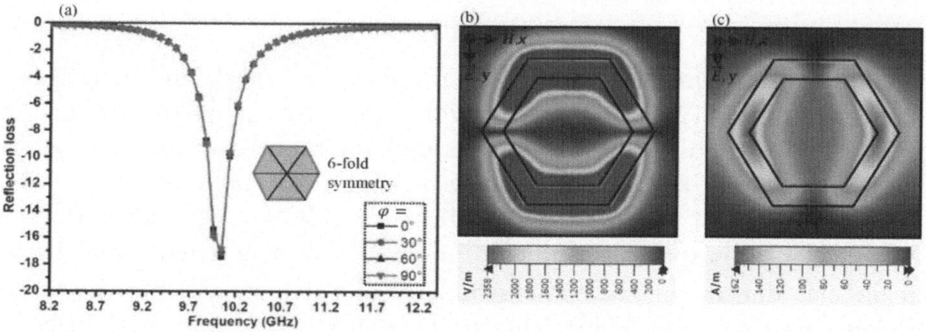

Figure 3: (a) Simulated Reflection Loss Curves at Different Polarization Angles. Distribution of (a) E-field and (b) H-field at 10.05 GHz

II. Fabrication and Measurement

The absorber is fabricated using structural parameters from Table 1 in the following steps-1) urethane rubber is moulded into sheet of 0.7 mm using 2:1 (wt. ratio) mixture of viscous urethane and curing agent, 2) hexagonal aluminium rings with thickness 0.05 are etched on commercial adhesive tape and later pasted on the substrate and 3) aluminium tape of thickness 0.05 mm is adhered to the back. The fabricated MMA sample is shown in Figure 4(a) and 4(b). Measurement of absorption performance is carried out at X-band using waveguide technique with Agilent Vector Network Analyzer (VNA) E8362 as in Figure 4(c). The absorber showed reflection loss of -16.98 dB

(~98% absorption) at 10.13 GHz with a -10 dB bandwidth of 0.5 GHz, Figure 4(d). The experimental result is in good agreement with the simulation result; however, edge diffraction and fabrication tolerance may have led to minor differences between the experimental and simulation results.

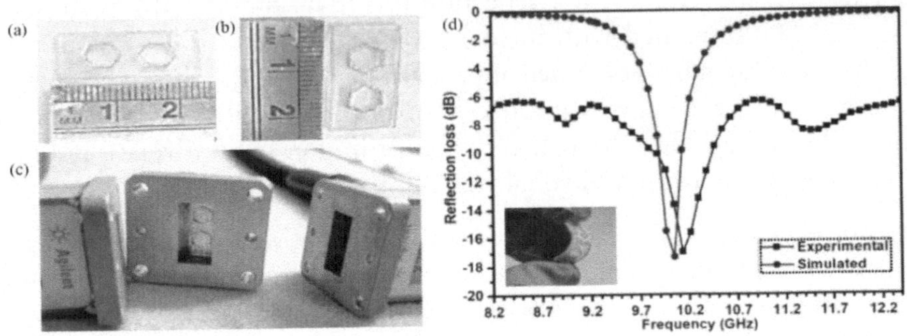

Figure 4: (a), (b) Fabricated MMA without Aluminium Back. (c) Fabricated MMA with Aluminium Back Mounted Inside the Waveguide Setup. (d) Experimental vs. Simulated Reflection Loss.

III. Conclusion

A conformal metamaterial microwave absorber is proposed and investigated for X-band (8.2–12.4 GHz) applications using hexagonal ring resonators made of aluminium pasted on flexible urethane rubber substrate backed by aluminium tape. The absorber showed -16.98 dB reflection loss (98% absorption) at 10.13 GHz with -10 dB bandwidth of 0.5 GHz (narrow) at X-band. The absorption is found to be insensitive to different polarization angles (φ) due to the structural symmetry of the unit cell. Moreover, the designed absorber is of lightweight, flexible and relatively thin with thickness ~0.027λ, hence can be used as absorbing tape.

References

[1] Gándara, M.F., Aluminium: the metal of choice, *Mater. Tehnol*, 47, 261–265 (2013).

[2] Reghunadhan, A. and Thomas, S., Polyurethanes: structure, properties, synthesis, characterization, and applications, in *Polyurethane polymers: Elsevier*, 1–16 (2017).

[3] Landy, N.I., Sajuyigbe, S., Mock, J.J., Smith, D.R. and Padilla, W.J., Perfect metamaterial absorber, *Physical review letters*, 100, 207402 (2008).

[4] Ni, B., Chen, X., Huang L., Ding, J., Li G., and Lu, W., A dual-band polarization insensitive metamaterial absorber with split ring resonator, *Optical and Quantum Electronics*, 45, 747-753 (2013).

[5] Li, H., L. Yuan H., Zhou, B., Shen, X.P., Cheng, Q. and Cui, T.J., Ultrathin multiband gigahertz metamaterial absorbers, *Journal of Applied Physics,* 110, 014909 (2011).

[6] Hossain, M.J., Faruque, M.R.I., Alam, M., Ahamed, E. and Islam, M.T., New Compact Perfect Metamaterial Absorber for Dual Band Applications, in 10th International Conference on Robotics, Vision, *Signal Processing and Power Applications: Enabling Research and Innovation Towards Sustainability, Springer,* 381–386 (2019).

[7] Li, L.-Y., Wang, J., Du, H.-L., Wang, J.-F. and Qu, S.-B., Achieving a multi-band metamaterial perfect absorber via a hexagonal ring dielectric resonator, *Chinese Physics B,* 24, 064201 (2015).

[8] Borah, D. and Bhattacharyya, N.S., Design, fabrication and characterization of flexible and ultrathin microwave metamaterial absorber, in *2017 International Conference on Innovations in Electronics, Signal Processing and Communication (IESC),* IEEE, 190–193 (2017).

Miniaturized Hexagonal DS-CSRR Based Microwave Sensor for High Sensing Applications

Yatish Beria*, Akash Buragohain, Bijay Bikash Chamuah and Gouree Shankar Das

Microwave Research Laboratory, Department of Physics, Dibrugarh University, Dibrugarh-786004, Assam, India

*E-mail: yatishberia786@gmail.com

ABSTRACT: The characterization of unknown liquid sample has been carried out in this work on the basis of their dielectric properties under the response of an applied electric field [1]. This work primarily considers the use of Hexagonal Double Split Complementary Split Ring Resonator (DS-CSRR) based microwave sensor resonating in the licensed free ISM band for the measurement of complex permittivity of liquid samples as a function of frequency. The simulation of the CSRR design has been carried out on a FR4 board with copper cladding on both sides using the High Frequency Structure Simulator (HFSS) software. The value of the resonant frequencies when LUT is introduced into the resonator is taken as a measure to fit polynomial equation for the determination of the real part of permittivity. The magnitude of change in the transmission coefficient w.r.t air is taken as the measure to develop polynomial fit equation for the determination of the imaginary part of permittivity. Here 1-butanol is selected as an unknown sample, for which the complex permittivity is evaluated using the fitted equations. The high sensitivity of the sensor ensures its usefulness in a wide collection of high sensing applications.

Keywords: Material Characterization, Microwave based Sensor, Hexagonal DS-CSRR.

I. Introduction

To incorporate a material into its purposeful application, it is required to know its physical and chemical properties precisely. One of the important parameters for characterizing materials is on the basis of its response under an applied electromagnetic field [1]. The response of a material under test to an applied electromagnetic field primarily depends on its permittivity. To assess the permittivity of a material, various methods have been incorporated, out of which microwave resonance methods are found to be more suitable due to their low-cost and real time monitoring capabilities [2]. The resonance methods using di-electric Split Ring Resonator (SRR) have been used extensively for the measurement of complex permittivity of liquid samples [3]. In this method, the Liquid Under Test (LUT) is allowed to interact with the resonator unit and the shift in frequency due to perturbation of field lines is taken as a measure to interpret the complex permittivity of liquid samples. But due to higher frequency shift and low bandwidth of insertion loss Complementary Split Ring Resonator (CSRR) is preferred over SRR [4]. In this work, we have proposed a novel Hexagonal Double Split SRR (DS-CSRR) resonator unit for the measurement of complex permittivity of liquid samples. High Frequency Structure Simulator (HFSS) has been used to design and optimize the proposed sensor.

The limitations of most of the recently published works is the selection of inconsistent frequency and the use of a very few numbers of samples for developing fit equations. To overcome those shortcomings, we have designed the sensor to operate in the licensed free ISM band (around 2.4 GHz) and used a collection of liquids covering a wide range of values of permittivity.

II. Principle and Design

The underlying method used in this work is the variation of the geometric field configuration when a sample is brought near the sensor to interact with the stored electric and magnetic energy of the resonator sensor. This perturbation in turn results in the shift in resonant frequency as well as change in magnitude of the transmission coefficient, S_{21}, by which the complex permittivity of unknown liquid samples can be determined.

The bottom and top view of the design of the DS-CSRR sensor is shown in Figure 1. As per literature, the rings engraved from the copper clad in CSRR sensor acquires the behaviour of capacitance while the metallic gaps linking the engraved rings acts as inductance. Therefore, the structure equivalently forms an LC shunt circuit which can be made to resonate based on Babinet's principle [5]. The substrate loss and metallization loss of the sensor can be equivalently represented by a resistance R. The lumped equivalent model of the proposed sensor is shown in Figure 1.

The sensor is designed using a low cost FR4 substrate with copper cladding on both sides of 0.035 mm. The loss tangent and the relative permittivity of the substrate is 0.03 and 4.4 respectively. The overall dimension of the substrate used in this work is 20 × 20 mm^2. The hexagonal complementary rings are engraved in the bottom plane of the substrate with the outermost ring having an outer radius of 6.46 mm. The width of each ring is equal to S =

0.36 mm and the gap between two rings is $D = 0.29$ mm and the inductance gap is $W = 0.5$ mm. The microstrip transmission line (MTL) is etched in the top plane with a width of 1 mm. To hold the material under test, a plastic holder made up of an Acrylonitrile-Butadiene-Styrene which can hold about 100 μl LUT is used.

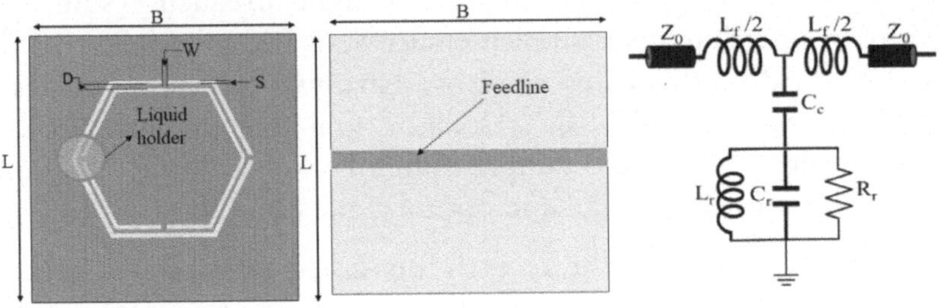

Figure 1: Bottom View, Top View and Equivalent Circuit Model of the Sensor

III. Results and Analysis

The transmission coefficient (S_{21}) is primarily used to quantify the characteristics of the sensor. The notch of the S_{21} coefficient (which gives the resonant frequency) as well as the deep magnitude ($|S_{21}|$ in dB) are used as the variable parameters for analysis. In the HFSS software, first the sensor without any sample is simulated with a frequency sweep from 0.5 GHz to 3 GHz. The notch frequency and the deep magnitude are recorded which are used as reference values. Eventually, seven liquid samples having dielectric constants values ranging from (1-111) are filled in the ABS holder for the simulation study. The corresponding data for the liquids are noted. The results of the S_{21} (in dB) vs frequency have been plotted in Figure 3. The sensor's performance can be quantified by its normalized sensitivity and is

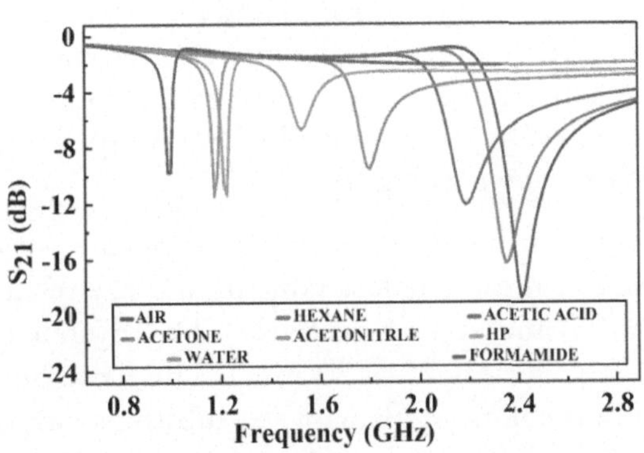

Figure 2: Frequency vs S_{21} Plot for LUTs

used to compare with other recent works [6]. The found normalized sensitivity for the Hexagonal DS-CSRR is about 1.2%. The Shifts in respective resonances and the change in the resonance deep with respect to the reference value are used for generating mathematical fit equations for real and imaginary parts of permittivity respectively.

The polynomial fit equations for real and imaginary parts are given in equation (1) and (2) respectively.

$$\varepsilon' = 60.851f^2 - 281.62f + 328.13 \quad \dots (1)$$

$$\varepsilon'' = 0.382\Delta|S_{21}|^2 - 3.823\Delta|S_{21}| + 3.7533 \quad \dots (2)$$

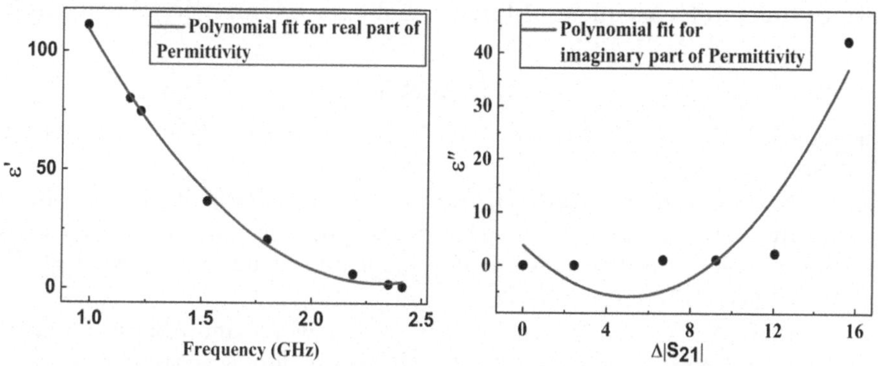

Figure 2: Fitting Curves for Real and Imaginary Part of Permittivity

The polynomial fitting curves for the respective real and imaginary part of permittivity are shown in Figure 4. Now, in order to detect the real and imaginary part of any unknown sample, the values of resonant frequency f and the magnitude of change in transmission coefficent $\Delta|S_{21}|$ of the respective sample are inserted in equation 1 and 2. In this work, we have selected 1-butanol as the unknown sample and the values are listed in Table 1.

Table 1: Complex Permittivity of 1-butanol

Unknown Sample	*Standard Value*		*Experimental Value*		*% Error*	
	ε'	ε''	ε'	ε''	ε'	ε''
1-butanol	17.8	9.764	16.964	10.209	4.69	4.55

4 IV. Conclusion

This work presents a Hexagonal shaped sensor as a planar resonator sensor for the measurement of complex permittivity of liquid samples. The shift in frequency and the change in the resonance deep is taken as the measure for evaluating the real and imaginary parts of complex permittivity respectively. Hence, we have developed mathematical fit curves for both real and imaginary parts separately in this work and 1-butanol is taken as an unknown sample. Also, the sensing ability of the sensor is found to be very high in comparison to some other works which ensures the efficiency of the sensor for detecting the permittivity of chemical samples.

ACKNOWLEDGEMENT

The authors sincerely acknowledge the DST FIST (level-I) Project No (SR/FSI/PSI-217/2016) funding by Department of Science & Technology, Govt. of India to Department of Physics, Dibrugarh University, Assam, India.

REFERENCES

[1] Hippel, V.A. (1954), Dielectrics and Waves, New York, NY: John Wiley and Sons.

[2] Jha, S., Narsaiah, K., Basediya, A., Sharma, R., Jaiswal, P., Kumar, R. and Bhardwaj, R. (2011). Measurement techniques and application of electrical properties for non-destructive quality evaluation of foods—A review, *J. Food Sci. Technol.,* 48(4), pp. 387–411.

[3] Ebrahimi, A., Withayachumnankul, W., Al-Sarawi, S. and Abbott, D. (2014). High-Sensitivity Metamaterial-Inspired Sensor for Microfluidic Dielectric Characterization. *IEEE Sensors J.,* 14(5), pp. 1345–1351.

[4] Ebrahimi, A., Scott, J. and Ghorbani, K. (2019). Ultrahigh-Sensitivity Microwave Sensor for Microfluidic Complex Permittivity Measurement. *IEEE Trans. Microw. Theory Techn.,* 67(10), pp. 4269–4277.

[5] Falcone, F., Lopetegi, T., Laso, M., Baena, J., Bonache, J., Beruete, M., Marqu´es, R., Mart´ın, F. and Sorolla, M. (2004). Babinet Principle Applied to the Design of Metasurfaces and Metamaterials. *Phys, Rev. Lett.,* 93(19).

[6] Buragohain, A., Mostako, A.T.T. and Das, G.S., Low-Cost CSRR Based Sensor for Determination of Dielectric Constant of Liquid Samples. *IEEE Sens. J.*, 21, 27450–27457 (2021).

Section-VI

Interdisciplinary Research and Atmospheric Physics

A Study of Spatio-Temporal Rainfall Variability and Extreme Precipitation Trends over Cachar District of Assam

Abhay Bhattarai* and Atri Deshamukhya

Department of Physics, Assam University, Silchar-788011, India

*E-mail: abhaybhattarai@gmail.com

ABSTRACT: Cachar district is in the southernmost part of the Assam. The Barak is the main river of the district, with a vast network of river tributaries which flows from Dima Hasao district, Manipur, and Mizoram. In the recent period it is experiencing a good number of extreme weather events. Understanding the changes and mechanisms leading to extreme rainfall events is crucial in effective water management for decision making and policy planning. This work is a study of the variability and trends of the extreme precipitation events using high resolution gridded ($0.25^{\circ} \times 0.25^{\circ}$) dataset obtained from Indian Meteorological Department (IMD). For the study 12 extreme precipitation indices are estimated from the gridded rainfall datasets for the selected 25 grids of the study region. A significant decreasing trend of annual and seasonal precipitation is observed with a confidence interval of 95% in most of the selected grids. Furthermore, upwards significant trends in extreme precipitation indices have been observed, though a few of the indices show significant decreasing trend.

Keywords: Extreme Rainfall Events, Precipitation.

I. Introduction

The north-eastern part of India receives heavier rainfall than other parts of the subcontinent. High intensity rainfall for short duration and longer dry spell are the major problems due to impact of climate change. Study of long term pattern of hydrological and meteorological variables are essential to understand the behavior of global as well as regional climate. Water resources management in the north-eastern part of India must include the impacts of climate change. Barak and the adjacent Himalayan River basins are the most vulnerable to this change because of their diversity of climate and topographic [1-2].

The Barak River is flowing parallel to the river Brahmaputra in northeast India. Its climatic scenario is quite different from that of the Brahmaputra

River basin. The Barak basin receives mostly orographic and cyclonic precipitation with about 300 cm of average annual rainfall. With an area of 41,000 km^2, the Barak River basin is considered as one of the large basins in India. The Cachar district almost fall in the mid of the basin, as the urban part of the district is largely associated with the Barak River and mostly remain vulnerable to flood. Several studies have already been done on impact of climate change on water resources of various parts of the North-eastern region of India [1]. Deka *et al.* (2013) [3] reported a significant declining trend of precipitation in monsoon and post-monsoon in the Barak basin. Similarly, Deb and Sil (2019) [2] demonstrated a decreasing trend of annual and seasonal precipitation. Hinge *et al.* 2021 [4] has recently done a district level assessment of changes in extreme rainfall indices in Barak and other basins, where he found that significant increase in most indices. An increase in the number of days in R25 and R20 was noted in the district of Karimganj, Hailakandi, which may worsen the situation in these flood-prone areas. Despite the few studies conducted over the whole Barak region, most of them have given preference to the monsoon period only.

In this context, the main aim of the present study is to understand the spatial and temporal distribution of long-term historical precipitation and extreme rainfall indices over the Cachar district at a regional scale. The pre monsoon period is also explored in the study as in recent period a heavy rainfall and flood events are observed in the study area, like the flood of 2022 in Silchar.

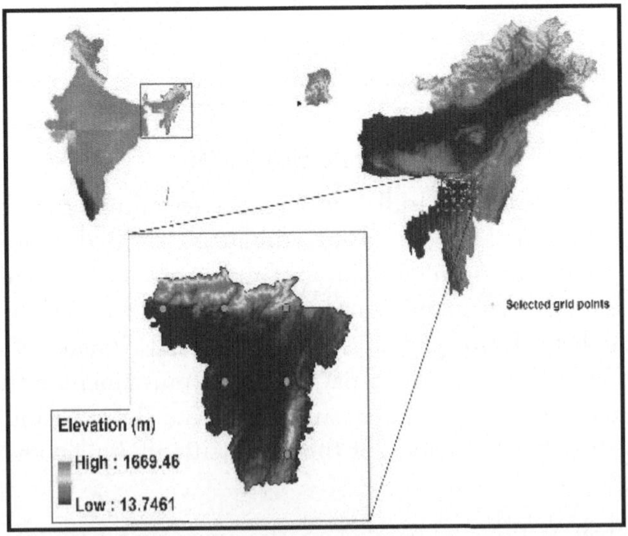

Figure 1: Map of Cachar District along with a over View of Northeast India and Whole India

II. Data & Methodology

A. Rainfall Dataset

A gauge-based, high resolution (0.25° × 0.25°) daily gridded rainfall dataset, prepared by the India Meteorological Department (IMD) is used. This precipitation product was developed using approximately 6955 reasonably good working condition rain gauge stations across India, which was followed by quality checking, handling lots of missing data, and unexpected outliers in rainfall values.

B. Indices

A list of about 50 rainfall and temperature internationally agreed extreme indices can be found on the world meteorological website (www.wmo.int). As the aim of the study is to analyze the spatial and temporal distribution of the extreme rainfall events, hence 12 indices have been chosen from this list which are commonly reported in the literatures and related to rainfall extreme.

Table 1: List of Extreme Precipitation Indices Used in the Study along with Description and Units

Indices	*Name (Unit)*	*Indices*	*Name (Unit)*
R99p	Extremely wet days precipitation (mm)	R10	Number of moderate precipitation days (day)
R95p	Very wet days precipitation (mm)	R20	Number of heavy precipitation days (day)
R1day	Maximum 1-day precipitation amount (mm)	R50	Number of very heavy precipitation days (day)
R5day	Maximum 5-day precipitation amount (mm)	SDII	Simple daily intensity index (mm/day)
CWD	Consecutive wet days (day)	R99pTOT	Extremely wet Days contribution (%)
CDD	Consecutive dry days (day)	R95pTOT	Very wet days contribution (%)

C. Trend Analysis

In the present study, the Modified Mann–Kendall (MMK), a non-parametric test [5], is applied to determine the monotonic trend of the given time-series data. It is a widely used test for the detection of a statistically significant trend in rainfall and temperature. The main advantages of using the MMK test instead of others are as follows: first, it is assumption independent; i.e., there is no need for the assumption of normal distribution of the data, and second, it is not affected by the missing values.

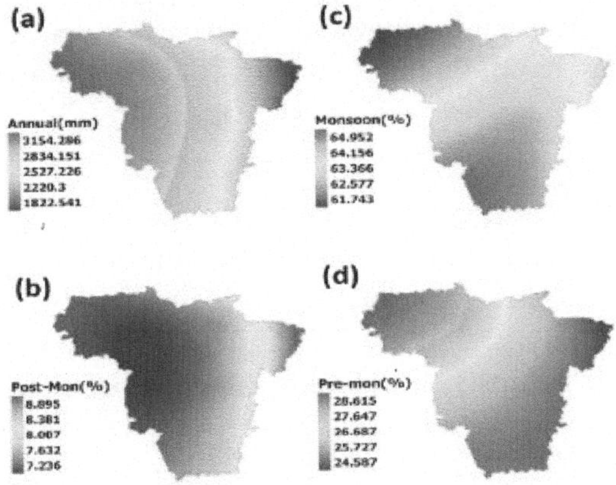

Figure 2: (a) Spatial Distribution of Mean Annual Rainfall. Contribution of (b) Monsoon (c) Pre-monsoon (d) Post-Monsoon Rainfall

III. RESULTS & DISCUSSION

The spatial distribution of the IMD-derived gridded annual mean rainfall along with seasonal mean over the region during 1901–2021 is shown in Figure 2. These figures clearly indicate that the monsoon season have a direct impact on total annual rainfall and the pre-monsoon season add some to it. It is clearly observed that there is a decline in precipitation from the northwest to the south of the region in term of annual period while in Monsoon season it observed that the declination starts from the west to the east. Other seasons also flows a similar trend. The annual mean rainfall varies from 3110 to 1477 mm per year in the overall study area. The maximum rainfall observed in the Cachar district reaches up to 3310/year in the northwest side and minimum is observed over the northeast side of 1803.6 mm/year. In similar fashion the monsoon season maximum observed rainfall is 1948 mm/year and minimum are 1119.2 mm/year. The pre-monsoon season contribute a significant amount of rainfall to the total annual rainfall in the Cachar District. It experiences a maximum rainfall of 886 mm/year and a minimum of 437 mm/year in the pre-monsoon season.

The different indices which are used to characterize the extreme rainfall in the basin are discussed in the Table 1. The trends and changes in the frequency of heavy and very heavy rainy days are generally explained by extreme precipitation indices. Figure 3 and 4 gives the spatial distribution of the different extreme precipitation indices over the Cachar district for the monsoon and pre-monsoon period of 1901–2021 respectively. The magnitude of indices varies in the same way of the spatial distribution of mean annual rainfall, with an increasing gradient from east to northwest (R99p, R95p, R1 day, R5 day, R10, R20, R50, SDII) except for CDDs, CWDs, R99pTOT and R95pTOT which represent a different spatialization.

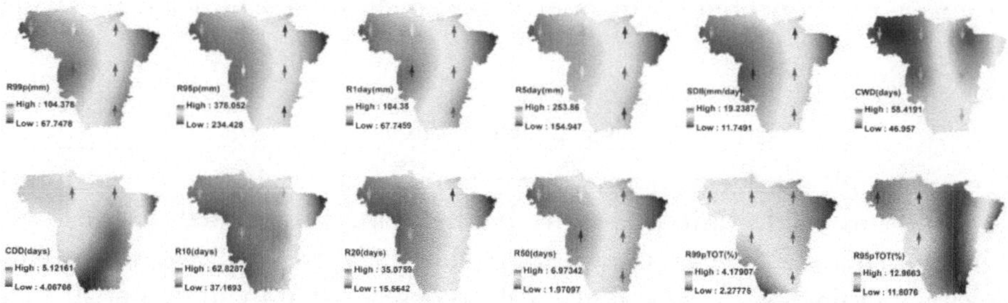

Figure 3: Spatial variability of extreme precipitation indices over the Cachar District during 1901–2021 for the monsoon period.

Significant Increasing trend
Significant Decreasing trend
Non- significant Increasing trend
Non-significant Decreasing trend

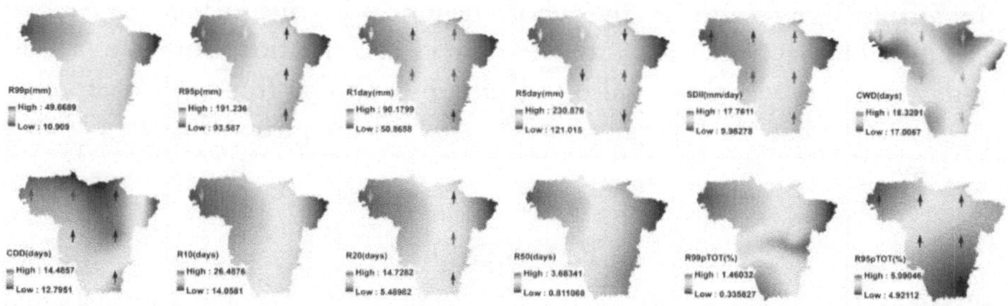

Figure 4: Spatial variability of extreme precipitation indices over the Cachar District during 1901–2021 for the pre-monsoon period.

Significant Increasing trend
Significant Decreasing trend
Non- significant Increasing trend
Non-significant Decreasing trend

In most of region during the monsoon period there is a significant increase trend for the R99p, R95p, R1day and R5day in nearly half of the district mostly in the eastern side. This is also true for the SDII value, which indicate the rainfall intensity in increasing in most of the region expect for the north western side where significant decrease in monsoon rain intensity is observed. Significant decreasing trend is observed for the CWD, R10 is the whole region. Whereas R20 also show significant decreasing trend in north western side. But R50 is showing significant increasing trend in most of the regions. It is interesting to note that the contribution of the 99-percentile rainfall is increasing significantly in whole region. This is true for the 95-percentile value also in most of the region.

For the pre-monsoon season also, the trends are like that of the monsoon season, but not trend is observed for the 99-percentile rainfall. The maximum one-day precipitation is increasing significantly in most of the region. The SDII value for the pre-monsoon is showing increasing trend, where as in most of the region it is significant. The CWD is significantly decreasing in whole region. Non-significant increase in the CDD is observed in whole region. The very wet days contribution to rainfall is also showing significant increasing trend.

IV. Conclusion

Planning and management of water resource at a local scale is very much important to tackle various weather natural hazard associated with extreme rainfall events in this changing world. In this study, we have analysed extreme precipitation indices in the regional scale to observe their trend in spatial aspect of the region. As in the recent period we have observed frequent flooding in the pre-monsoon season also. So, we have introduced this season in our analysis and found that the for both the season the Simple daily intensity index is increasing significantly in the whole region indicating the increase of the extreme rainfall events. And also, the rainfall contribution from the extreme wet day and very wet days is also indicating increasing trend. So, over all we can conclude that the extreme rainfall and its frequency in the region is increasing.

References

[1] Hinge, G., Surampalli, R.Y. and Goyal, M.K., Prediction of soil organic carbon stock using digital mapping approach in humid India. *Environ Earth Sci.*, 77(5), 172 (2018a).

[2] Deb, S. and Sil, B.S., Climate change study for the meteorological variables in the Barak River basin in North-East India. *Urban Clim 30*, 100530 (2019).

[3] Deka, R.L., Mahanta, C., Pathak, H., Nath, K.K. and Das, S., Trends and fluctuations of rainfall regime in the Brahmaputra and Barak basins of Assam, India. *Theoret Appl Climatol,* 114(1–2), 61–71 (2013).

[4] Hinge, Gilbert, Mazumdar, Mriganka, Deb, Saikat and Kalita, Mintu. District-level assessment of changes in extreme rainfall indices in Barak and other basins in Indian Himalayan states: risks and opportunities. *Modeling Earth Systems and Environment,* 8. 1–11.10.1007/s40808-021-01152-1 (2022).

[5] Hamed, K.H. and Rao, A.R., A modified Mann-Kendall trend test for auto correlated data. *J. Hydrol.* 204(1–4), 182–196 (1998).

Investigation on Group Velocity Dispersion and Leakage Loss in Air-Core Photonic Band-Gap Fibers

P. Sarma* and D. Phukan

Department of Physics, Royal Global University, Guwahati-781035, Assam, India

*E-mail: sarmapranami37@gmail.com

ABSTRACT: The wavelength dependence of group velocity dispersion (GVD) and leakage loss are investigated in air-core Photonic band-gap fibre (PBGF) by simulation using the software Finite Difference Time Domain (FDTD). It is one of the most popular simulation techniques for solving electromagnetic equations. It is seen that with the increase of the hole diameter to pitch ratio in air-core PBGF, the leakage losses decrease drastically and we require at least 18 rings of arrays of air holes in the cladding region so that the confinement loss can be reduced to a level of 0.1 dB/km in 1.5 μm wavelength range while using the diameter to pitch ratio 0.8.

Keywords: Group Velocity Dispersion, Leakage Loss, Air-Core Photonic Band-Gap Fiber.

I. Introduction

Group velocity dispersion (GVD) is the change of group velocity with frequency when a signal propagates through a medium[1]. Photonic band-gap fibres (PBGFs) have drawn great interest because of their unique band-gap guiding mechanism. The PBGF with cladding composed of air holes has been widely used in practical applications. It has a wide band-gap guiding covering the visible and near infra-red range of the spectrum constituting relatively low-loss and high confinement of light in the core. Since PBGFs work via diffraction rather than total internal reflection, they have the possibilities of high power delivery with low nonlinearity and extremely low loss transmission. Because of this working principle the nonlinearities decreases more than 1000 times as compared to that of a conventional fiber. Moreover, it has the ability to maintain pulse duration and peak power. There are lots of limitations like dispersion, attenuation, nonlinearities etc. which are relieved in a PBGF [2, 3].

PBGFs with air core/hollow core have many possibilities of nonlinearity in high power delivery and very low loss transmission. They also have possibility of enhanced Raman scattering, soliton transmission and so on. Many researchers have studied about chromatic dispersion and leakage losses [4]. Here, we have studied the characteristics of GVD in air-core PBGFs considering a finite number of air hole rings.

II. Methodology and Optical Simulation

Practically it is not possible to fabricate PBGF with infinite numbers of air-hole rings in the cladding. Therefore leakage loss occurs. The value of leakage loss (L_c) is deduced from the equation [5],

$$L_c = 8.86\alpha \qquad \ldots (1)$$

in decibels per meter, where, α is the attenuation constant.

Again, the waveguide GVD, D_w is calculated by the equation,

$$D_w = -\frac{\lambda}{c}\frac{d^2 n_{eff}}{d\lambda^2} \qquad \ldots (2)$$

Where, λ is the wavelength, c is the velocity of light in vacuum, $n_{eff} = \frac{\beta}{k_0}$, is the effective index, β is the phase constant, k_0 is the free space wave number, d is the hole diameter.

First we have considered an air-core PBGF with the background refractive index 1.5 considering 'd' as the hole diameter and 'Δ' as the center to center distance between the holes which is also known as the hole pitch. Since the number of air-holes is finite in the cladding, so the leakage losses will occur. Considering Δ = 2 μm, $\frac{d}{\Delta} = 0.8$ and the number of air-hole rings as a parameter, we have found the leakage loss which is wavelength dependence for an air-core PBGF with a finite number of air-holes as shown in figure 1. The leakage loss increases as it approaches the bandedge and around the center of PBGF it becomes minimum.

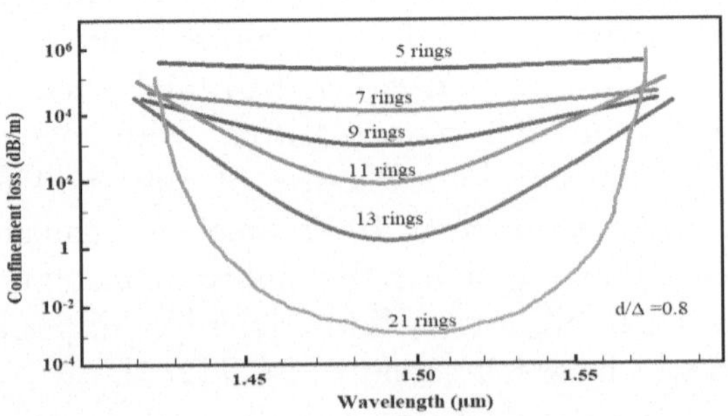

Figure 1: Confinement Loss as a Function of Wavelength for an Air-core PBGF with a Finite Number of Air Holes Considering the Hole Pitch Δ = 2.28 μm

The GVD is strongly wavelength dependent [3]. Considering the number of air-hole rings as a parameter and Δ = 2.28 μm, $\frac{d}{\Delta} = 0.8$, we have shown the GVD of this air-core PBGF, for fundamental mode as shown in Figure 2. The GVD goes from negative to positive values as it goes from shorter wavelength to longer wavelength. Moreover, GVD decreases rapidly near the lower bandedge and increases near the upper bandedge. We have seen that near the upper and lower bandedge, the GVD is influenced by the change in the number of air-hole rings, but is not affected around the center of PBG.

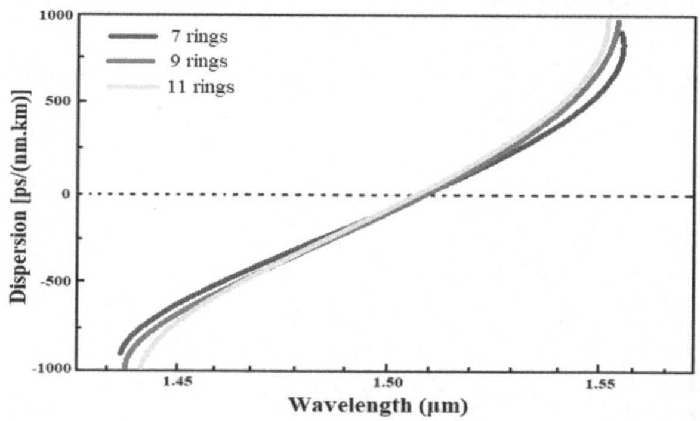

Figure 2: Waveguide Group Velocity Dispersion of an Air-core PBGF with a Finite Number of Hole Considering Hole Pitch Δ = 2.28 μm and $\frac{d}{\Delta} = 0.8$

In Figure 3 the leakage loss is shown as a function of number of rings, where λ = 1.50 μm, Δ = 2.28 μm and d/Δ = 0.8. It is found that with the increasing number of rings the leakage loss deceases steadily. However, the rate of reduction in losses is not rapid. At least 18 rings of arrays of air holes in the cladding region are necessary in order to realize the confinement loss of 0.1 dB/kmin 1.5 μm wavelength range with d/Δ = 0.8.

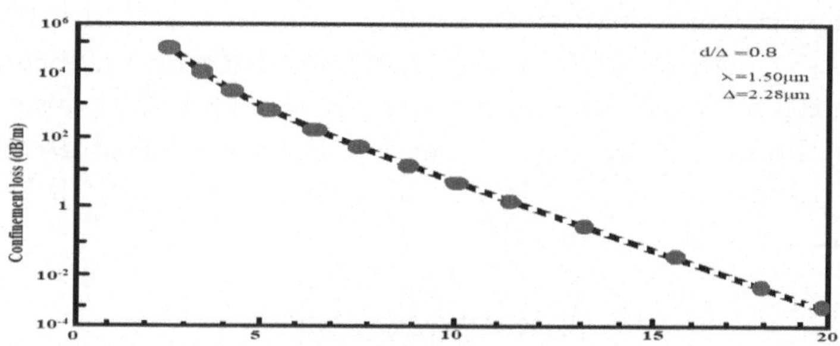

Figure 3: Confinement Loss as a Function of the Number of Rings Considering Δ = 2.28 μm, d/Δ = 0.8 and λ = 1.50 μm

Next, we have considered the GVD in air-core PBGFs and leakage loss which is wavelength dependent of air-core diameter. Taking the hole diameter to pitch ratio $\frac{d}{\Delta}$ as a parameter, the wavelength dependence of the leakage loss for air-core PBGF is shown in Figure 4 considering 11 rings of air-hole in the cladding of the PBGF. It is seen that with the increase of the hole diameter to pitch ratio in air-core PBGF, the leakage losses decrease drastically. From these results, we can say that the value of d/Δ should be large for reduction of leakage losses with a finite number of air-holes in the cladding region of an air-core PBGF.

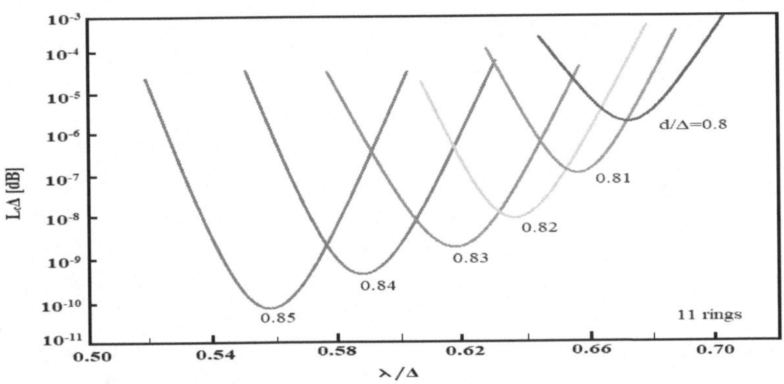

Figure 4: Normalized Leakage Loss as a Function of Normalized Wavelength in an Air-core PBGF with Eleven Rings of Arrays of Air Holes

III. CONCLUSION

The wavelength dependence of group velocity dispersion (GVD) and leakage loss are investigated in air-core Photonic band-gap fibre (PBGF) by simulation. It was shown that even if using large air holes of the diameter to pitch ratio of 0.8, we required at least eighteen rings of arrays of air holes in the cladding region to reduce the leakage losses to a level of 0.1 dB/km in 1.5 μm wavelength range. Moreover, it is seen that with the increase of the hole diameter to pitch ratio in air-core PBGF, the leakage losses decrease drastically. From these results, we can say that the value of d/Δ should be large for reduction of leakage losses with a finite number of air-holes in the cladding region of an air-core PBGF.

ACKNOWLEDGEMENT

Authors acknowledge the Royal Global University for providing the required software for this study.

References

[1] Qian, X., Wang, X., Lu, X., Zhang, T. and Fan, W., Suppression of Frequency Modulation to Amplitude Modulation Conversion with Modified Group Velocity Dispersion Compensation Device in the Front End of High-Power Lasers, *Applied Sciences*, 10.3390/app12020884, Vol. 12(2), pp. 884 (2022).

[2] Mangan, B.J., Farr, L., Langfort, A., Roberts, P.J., Williams, D.P., Couny, F., Lawman, M., Mason, M., Coupland, S., Flea, R., Sabert, H., Birks, T.A., Knight, J.C. and Russell, P.St.J, Low loss hollow core photonic band-gap fiber, post deadline paper PDP24, OFC'04. 2004).

[3] Li, W., Matniyaz, T., Gafsi, S., Dong, M.T.K., Hawkins, T.W., Parsons, J., Gu, G. and Dong, L., 151Wmonolithic diffraction-limited Yb-doped photonic bandgap fiber laser at 978 nm, *Opt. Exp.* 27 (18), 24972–24977 (2019).

[4] Lobanov, V.E., Kondratiev, N.M., Shitikov, A.E., Galiev, R.R. and Bilenko, I.A., Generation and dynamics of solitonic pulses due to pump amplitude modulation at normal group velocity dispersion, *Phy. Rev.* A 100, (1), 013807 (2019).

[5] Ouzounov, D.G., Ahmad, F.R., Muller, D., Venkataraman, N., Gakkagher, M.T., Thomas, M.G., Silcox, J., Koch, K.W. and Gaeta, A.L., Generation of megawatt optical solitons in hollow-core photonic bang-gap fibers, *Science* 301, 1702–1704 (2013).

Experimental Investigation of Double Layer Near a Fireball Boundary

Paragjyoti Sut[1,3*], Prarthana Gogoi[1,3], N.C. Adhikary[1] and H. Bailung[2,3]

[1]Physical Sciences Division, Institute of Advanced Study in Science & Technology (IASST), Paschim Boragaon, Guwahati-781035, Assam, India

[2]Department of Physics, Bodoland University, Kokrajhar-783370, Assam, India

[3]Academy of Scientific and Innovative Research (AcSIR), Ghaziabad-201002, India

*E-mail: sutparagjyoti@gmail.com

ABSTRACT: Investigating Double Layers (DLs) in laboratory plasma systems is crucial to comprehend the underlying physical processes involved. One of the special ways to produce DL in the laboratory is near a fireball boundary. In our experiment, a planar electrode with a diameter of about 1.5 cm is immersed in the plasma and biased positively by (50–90) V, which is much higher than the plasma potential, to create a fireball within the plasma. An emissive probe is used to measure the axial potential profile from the bulk plasma towards the electrode. The existence of a DL at the border between the fireball and the background plasma is clearly shown by the measured potential structures, and the DLs are presented for different electrode biasing conditions.

Keywords: Fireball, Double Layer, Emissive Probe.

I. INTRODUCTION

The DL is a nonlinear phenomenon within a plasma where the quasineutrality condition is violated. It consists of two oppositely charged layers separated by a small distance in the order of Debye length and the potential differences between two layers in the order of electron temperature. An electric field exists between two layers responsible for particle acceleration [1]. DL has been an interesting topic in astrophysics as it is responsible for particle acceleration in space plasma [2]. Numerous theoretical studies have been done so far to understand the DL characteristics. On the other hand, DLs have been produced in the laboratory, and studies are being carried out to explore basic physical understandings[3].

One of the complex phenomena within a plasma is a fireball. It is a typical additional discharge in front of an immersed electrode when it is biased far above the background plasma potential. The additional discharge is generally quasi-spherical in size, but in the presence of a magnetic field and depending on the electrode geometry, it may have different shapes[4].As the fireball-containing electrode is an anode, the fireball is sometimes termed an 'anode spot.' A couple of theoretical works have been done so far which explained the formation mechanism of the fireball in front of a biased electrode [5], [6]. Besides, fireball is a new research topic in plasma physics, as different kinds of oscillations are associated with it[7], [8].Moreover, recent research has shown that fireball can be used to form nanodot pattern and surface modification. So, it can be considered a new tool for material science [9], [10].

This chapter will discuss the formation of the fireball and the axial potential profiles showing a DL between the bulk plasma and the fireball.

II. Experimental Device and Fireball Production

The experiment is performed in a double plasma device, and details of the device can be found in the reference [11]. A schematic of the device is shown in Figure 1.

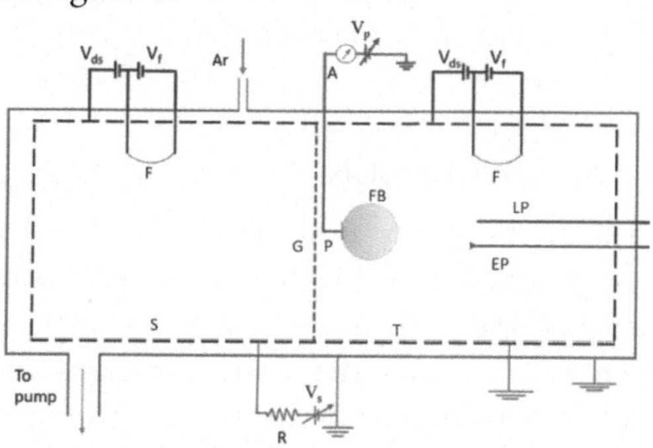

Figure 1: Schematic of the Double Plasma Device. S – Source anode, T - Target anode, V_{ds} – Discharge Voltage, V_f - Filament Voltage, G- Grid, F- Filament, P – electrode, V_s - Source Bias, R- Resistor, V_p = Plate Bias, LP-Langmuir Probe, EP- Emissive Probe, FB- Fireball, A-Ammeter

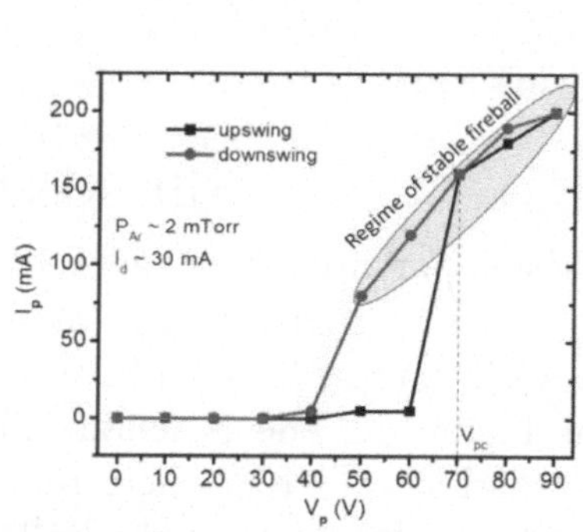

Figure 2: I-V characteristic exhibiting hysteresis for a 1.5 cm diameter SS plate where plasma conditions are I_d ~30 mA and P_{Ar}~ 2 mTorr. The gradually increasing and subsequently decreasing electrode bias (V_p) and corresponding current values (I_p) are shown as black and red circles respectively.

The chamber is evacuated up to ~10^{-6} Torr using a turbo pump backed by a rotary pump, and then argon gas is introduced up to ~2×10^{-3} Torr. Plasma is produced at pressure P_{Ar}~ 2 mtorr and discharge current at I_d~ 30 mA. With increasing steps, a positive bias is applied to the electrode 'P' (a 1.5 cm diameter stainless steel plate) and the fireball is initiated at a critical bias V_{pc}.

While decrementing plate bias, a stable fireballis sustained within a 50–90 V range and the I-V characteristic follows a hysteresis as depicted in Figure 2.

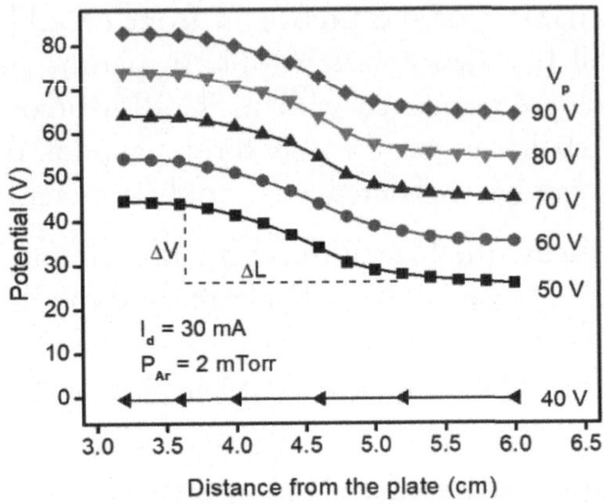

Figure 3: Axial Profiles of the Measured Plasma Potential for Various Plate Bias Voltages V_P = 90 V to 40 V Are shown. The P_{Ar} and I_d are Maintained at 2 mTorr and 30 mA Respectively.

III. Results and Discussion

Plasma potential values are measured from x = 6 cm (bulk plasma) to x = 3.2 cm (within fireball). The monotonic potential transition is observed from x = 5.2 cm to 3.6 cm, where the plate position is considered at x = 0, as seen in Figure 3. The potential enhancement towards the plate indicates the formation of a DL where the fireball is considered as the higher potential region with respect to the bulk plasma. At 40 V, there is no fireball and we got a flat potential profile. The potential difference between the two regions is called DL potential (ΔV); and the separation between them is known as DL thickness (ΔL). For this case, ΔV values fall within 16–17 V, which is nearer to the first ionization energy of the argon gas. Moreover, the thickness of the DL is within 1.5–1.6 cm. The DL potential and DL thickness are normalized by the electron temperature (T_e) and Debye length (λ_D), respectively. This experiment normalized potential $\Delta V/\kappa T_e$ ~3.2, which signifies the weak DL. Again, the normalized thickness is $\Delta L/\lambda_D$~30. The electric field due to these DLs is ~ 11 V/cm.

IV. Conclusion

This chapter introduces the plasma fireball and discusses laboratory production of it in a dc discharge plasma. The emissive probe measurements confirm a DL between the fireball and the bulk plasma. The normalized DL potential value indicates that the strength of the DL is weak.

ACKNOWLEDGMENT

The authors acknowledge IASST, Guwahati for providing the lab facilities to conduct the experiment.

REFERENCES

[1] Block, L.P., A double layer review, *Astrophys. Space Sci.*, Vol. 55, No. 1, pp. 59–83 (1978), doi: 10.1007/BF00642580.

[2] Raadu, M.A., The physics of double layers and their role in astrophysics, *Phys. Rep.*, Vol. 178, No. 2, pp. 25–97, (1989), doi: 10.1016/0370-1573(89)90109-9.

[3] Charles, C., A review of recent laboratory double layer experiments, *Plasma Sources Sci. Technol.*, Vol. 16, No. 4, (2007), doi: 10.1088/0963-0252/16/4/R01.

[4] Stenzel, R., Ionita, C., and Schrittwieser, R., Plasma fireballs, *IEEE Trans. Plasma Sci.*, Vol. 36, No. 4 PART 1, pp. 1000–1001, (2008), doi: 10.1109/TPS.2008.924498.

[5] Baalrud, S.D., Longmier, B. and Hershkowitz, N., Equilibrium states of anodic double layers, *Plasma Sources Sci. Technol.,* Vol. 18, No. 3, (2009), doi: 10.1088/0963-0252/18/3/035002.

[6] Scheiner, B., Barnat, E.V., Baalrud, S.D., Hopkins, M.M. and Yee, B.T., Theory and simulation of anode spots in low pressure plasmas, *Phys. Plasmas,* Vol. 24, No. 11, pp. 1–12, (2017), doi: 10.1063/1.4999477.

[7] Irimiciuc, S.A., Konrad-soare, C.T., Dimitriu, D.G., Ionita, C. and Schrittwieser, R.W., On the interactions between three fireballs in low-temperature plasma, *Plasma Sources Sci. Technol.,* Volume 31, No. 8, (2022), [Online]. Available: https://iopscience.iop.org/article/10.1088/1361-6595/ac8447/pdf.

[8] Scheiner, B., Beving, L. and Baalrud, S.D., On the hysteresis in fireball formation and extinction, *Phys. Plasmas,* Vol. 26, No. 1, (2019), doi: 10.1063/1.5080945.

[9] Chauhan, S., Barman, T., Bhatnagar, M., Ranjan, M., and Mukherjee, S., Plasma fireball:A unique tool to fabricate patterned nanodots, *Rev. Sci. Instrum.*, Vol. 88, No. 6, (2017), doi: 10.1063/1.4989701.

[10] Sudheer et a*l., Plasma fireball-mediated ion implantation for nonvolatile memory application, Appl. Surf. Sci.*, Vol. 607, No. September 2022, p. 154999, (2023), doi: 10.1016/j.apsusc.2022.154999.

[11] Pathak, P., Sharma, S. K., Nakamura, Y. and Bailung, H., Observation of second order ion acoustic Peregrine breather in multicomponent plasma with negative ions, *Phys. Plasmas,* Vol. 23, No. 2, (2016), doi: 10.1063/1.4941968.

Molecular Docking Studies of Azoles-Anti Fungal Drugs via Computational Method

Nidhi Awasthi[1*], Arvind Kumar Tiwari[1], Jitendra Kumar[2] and Anamika Shukla[2]

[1]Department of Physics, B.S.N.V.P.G. College, Charbagh, Lucknow-226001, India

[2]Department of Physics, Babasaheb Bhimrao Ambedkar University, Lucknow-226025, India

*E-mail: nidhimsc51@gmail.com

ABSTRACT: MLKs are one type of proteins identified as signalling molecules in the nervous system. They are also play an important role in the cell cycle. Some in vivo studied also suggested that MLKs are not only expressed in cells of the immune system or nervous system, Due to its versatility this protein could be use in skin treatment like-fungal infection. In the present study three antifungal drugs Itraconazole, Metronidazole, Nirvanol is docked with MLK protein with PDB ID 3 dtc. The docking results shows good relibility and these results can be used for further drug design studies.

Keywords: Anti-Fungal Drugs, Autodock, Chimera, PDB etc.

I. Introduction

MLKs are one type of proteins identified as signalling molecules in the nervous system. They are also play an important role in the cell cycle. Some in vivo studied also suggested that MLKs are not only expressed in cells of the immune system or nervous system, but also may be implicated selectively in the signalling pathway that follows on toll-like receptor ligation in innate sentinel cells, such as the dendritic cell [1]. Due to its versatility this protein could be use in skin treatment like-fungal infection. For the treatment of fungal infection antifungal drugs are used. These drugs are generally known as Azoles. There are some anti-fungle drugs which can be used as anti-TB drugs.

So for this purpose, firstly we have to find the best bind position of few anti-fungal drugs with protein MLK. The PDB form of mentioned protein is used downloaded from RCSB PDB site with PDB ID 3DTC [2]. Here Itraconazol, Metronidazole, Nirvanol [3] anti-fungal drugs are used for docking. Here Autodock [4], computational metho are used for docking.

Finally, current calculations show good results as best binding position, charge distribution and hydrogen bonding of drug with protein which we discuss later.

II. Computational Method

To study the interactions between the protein and selected anti-fungal drugs, the following theoretical steps were followed.

A. Optimization

We modelled and optimized structure of drugs using Gaussian 09 software package [5]. These drugs were obtained from the literature survey [6]. Optimization of compounds are done at B3LYP level with basis set 6-31G. There is no imaginary frequency is found.

B. Molecular Docking

For molecular docking, a segment of protein of PDB ID 3DTC was downloaded from the RCSB PDB website [34, 35]. The nonstandard ligands, receptors and water molecules were removed from the PDB ID 3DTC using the Chimera software [7]. The molecular docking process is done using the program set AUTODOCK [4]. After the Optimization drugs are made to dock with the protein 3DTC, providing the most preferred binding position of drugs within protein. Both ligand and macromolecule were prepared in the form of PDBQ and PDBQT files for grid and dock calculations. Grid boxes with dimensions 126 × 126 × 126 were prepared for each couple of protein and ligand. Docking calculations were done using 10 LGA runs with various default settings for docking run. The docked pose having minimum binding affinity was extracted and combined with the macromolecule for analysis and further study.

C. Results and Discussion

The molecular docking calculations were performed after the optimization of drugs. The docking results are tabulated in Table 1 and Table 2. The Table 1 included binding energy, reference RMSD (root mean square distance), partition function (Q), free energy (A), internal energy (U), entropy (S), while Table 2 included, estimated free energy of binding and estimated inhibition constant (Ki). All present energies are in Kcal/mol and reference RMSD is in Å.

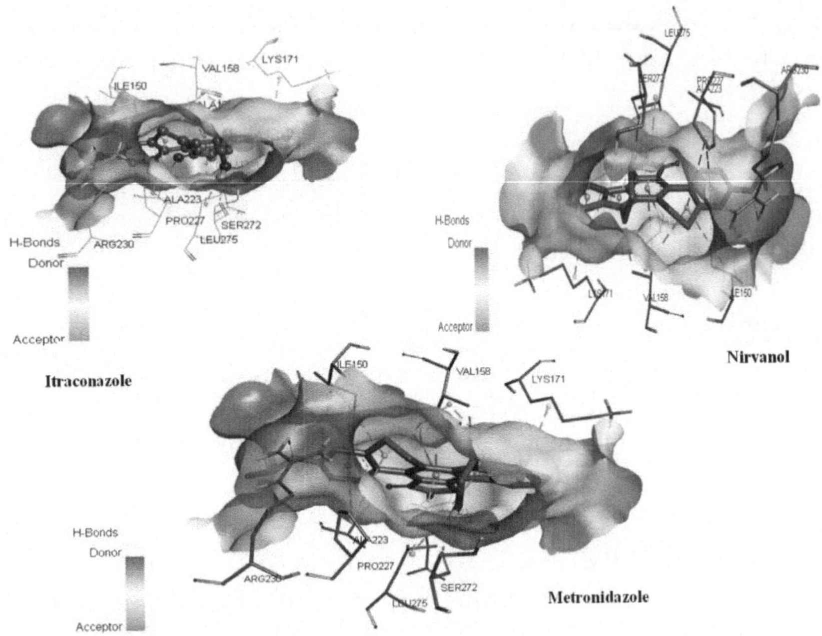

Figure 1: Figure Representing the H-bond Donor and Acceptor Regions in Best Docked Posed Complexes 3D

Table 1: Binding Energy, Reference RMSD, Partition Function (Q), Free Energy (A), Internal Energy (U), Entropy (S) of Complexes

S. No.	*Name of Complex*	*Binding Energy*	*Reference RMSD (Å)*	*Partition Function (Q)*	*Free Energy (A)*	*Internal Energy (U)*	*Entropy (S)*
1.	3DTC + Intraconozole	-5.66	35.16	10.07	-1368.5	-4.26	4.58
2.	3DTC + Metronidazole	-6.19	44.97	10.10	-1370.3	-6.10	4.58
3.	3DTC + Nirvanol	-4.74	47.74	10.08	-1368.8	-4.60	4.58

Table 2: Estimated Value of Binding Free Energy and Inhibition Constant of Docking Complexes

S. No.	*Name of Complex*	*Estimated Free Energy of Binding (Kcal/mole)*	*Estimated Inhibition Constant (Ki) (μM)*
1.	Itraconazole	-5.66	71.33
2.	Metronidazole	-6.19	29.22
3.	Nirvanol	-4.74	333.99

The hydrogen bond in 3D of each drugs with protein is shown in Figure 1. Dipole-dipole interactions, π-π stacking, and hydrogen bonds between the protein and ligand were responsible for the stability of the docked poses.

III. Conclusion

This paper studied the interaction of anti-fungal drug with protein via molecular docking. The Molecular Docking study carried out to achieve the protein binding affinities of the anti-fungal drugs. Itraconazole, Metronidazole, Nirvanol bind in the minor groove with protein, i.e., they act as a minor groove binder. The present study gives detailed insight on the interaction of protein with various anti-fungal drugs. It will be helpful in further study of various anti-fungal drugs which are helpful in fungal infection as well as TB treatment.

References

[1] Matthew, E.H., Jane, R., Benjamin, M.C. and David, R.K, Mixed lineage kinases (MLKs): A role in dendritic cells, inflammation and immunity? *International journal of Experimental pathology*, 88, 111–126 (2007).

[2] Akbarzadeh, S., Ji, H. and Frecklington, D., Mixed lineage kinase 2 interacts with clathrin and influences clathrin-coated vesicle trafficking. *J. Biol. Chem.*, 277, 36280–36287 (2002).

[3] Lesley, K., Rao, M.D., Alicia, M., Flaker, A.S., Christina, C., Friedel, B.S., Evan, D. and Kharasch, M.D., Rolc of Cytochrome P4502B6 Polymorphisms in Ketamine Metabolism and Clearance, *Ketamine Metabolism and Clearance,* 6, 1103–1112 (2016).

[4] AutoDock Vina Documentation (2022).

[5] Roy, D., Todd, A.K. and John, M.M., Semichem Inc., S.M. Gauss View, Version 6.1 (2016).

[6] Jiachen, W., Divya, C., Shana, R.M. and Kyle Hadden, M., Truncated Itraconazole Analogues Exhibiting Potent Anti-Hedgehog Activity and Improved Drug-like Properties, *Medicinal Chemistry Letters,* 10, 1290–1295 (2019).

[7] Pettersen, E.F., Goddard, T.D., Huang, C.C., Couch, G.S., Greenblatt, D.M., Meng, E.C., Ferrin, T.E. UCSF Chimera—A Visualization System for Exploratory Research and Analysis. *J. Comput. Chem.*, 25, 1605–1612 (2004).

Spectroscopic Study of Benzyl Acetone-Deuterated Chloroform Binary Liquid Mixtures: Raman and DFT Study

Th. Gomti Devi[1*] and Ranjan Changmai[2]

[1]Department of Physics, Manipur University, Canchipur-795003, Manipur, India

[2]Dibru College, Dibrugarh, Assam, India

*E-mail: tgd@manipuruniv.ac.in

ABSTRACT: Benzylacetone (BA) is a liquid having wide applications in cosmetic and food industries due to its pleasant flowery smell. Chloroform-d (CLFd) is a solvent used in manufacturing of chemicals, drugs, paints and other industrial purposes. The study of BA-CLFd binary liquid mixtures shall insights the interacting environment and molecular properties of the mixtures which will help in finding their potential applications in industries. In this paper, Raman technique has been used to study the local order and interacting environment of C-H stretching vibration of Benzyl Acetone (BA) in liquid mixtures using deuterated Chloroform (CLFd) solvent. Theoretical calculations of the molecules and in their associated states are performed by using Density functional theory (DFT) method with basis set b3lyp/6-31+G(d,p). Theoretical findings and experimental data are found in good agreement. The molecular parameters such as HOMO-LUMO, MEP analysis, vibrational and NBO analysis were also carried out which support in identifying the complex dynamic system.

Keywords: Raman, DFT, Vibrational, BA, CLFd.

I. Introduction

Raman spectroscopy is an important technique to study vibrations and reorientation of molecules in liquids [1-9]. It has been known that there are three possible phenomena taking place if an excited molecule collides with a ground state molecule of the same liquid, namely population relaxation, resonant energy transfer and pure dephasing. Out of these three phenomena, the pure dephasing process has dominated over all phenomena in vibrational studies [6-9]. In binary mixtures in liquids, there are various types of inter and intra-molecular interactions taking place between molecules namely hydrogen bonding, self-association of molecules and other interactions such as dipole–dipole, dipole-induced dipole interactions, induced dipole-induced dipole interactions respectively [7-9].

The Benzyl acetone (BA) molecule has been chosen for the present study. BA ($C_{10}H_{12}O$) is widely used in making artificial perfume as it produces sweet fruit flavors. Its modifier form benzyl acetate is used to produce different organic compounds. This is the first time that Raman band shape analysis has been performed at this molecule. An attempt has been made to study about the properties, optimize geometrical structure and microenvironment of the molecule in liquid phase. Here, density functional theory (DFT) calculation has been carried out to identify its geometry optimization, vibrational spectra, NBO analysis and Molecular electrostatic potential (MEP) analysis in predicting a complete interacting environment of the molecule in liquid mixtures. In our present work, the C-H stretching vibration of solute BA using chemical and deuterated solvents such as, Chloroform-d (CLFd) has been chosen for the study.

II. MATERIALS AND METHODS

A. Experimental

The chemicals were procured from Sigma Aldrich Company, USA. The chemicals are of high purity, spectroscopic grade and used directly without any further process. Horiba XPloral Micro-Raman system has been used to record Raman spectra of Benzyl Acetone and the 785 nm diode laser was used as the excitation source. The samples were recorded in neat liquid Benzyl Acetone (BA) and BA+CLFd liquid mixtures.

B. Computational Methods

The theoretical calculation is performed by using Gaussian 09W package with the basis B3LYP/6-31G+(d, p). The theoretical frequencies are not getting close to the experimental frequencies due to the negligence of electron-correlation effects and anharmonicity in vibration. This error can be corrected by multiplying the data with scaling factor. Here, the scaling factor of 0.9648 has been used for the present basis set [10-11].

The vibrational assignments of BA-CLFd interacting states are completed by using Gauss view interface which visualizes the vibration of atoms [12].

III. RESULTS AND DISCUSSION

A. Theoretical Works

The optimized geometrical structure of BA-CLFd solvent is calculated by using the basis set B3LYP/6-31+G (d,p). Figures 1a and 1b show the optimize structure of BA and BA-CLFd interacting state and the bond length of the interacting states with CLFd solvent has been shown in Table1. In BA-CLFd interacting system, the interaction may be of the type $C=O\cdots D-C$ having bond distance 2.114Å and the distance between Cl and deuterated hydrogen is found at 3.3461 Å respectively (Figure 1b).

Table 1: Bond Length of BA-CLFd Binary Mixtures

Mixture	*Bond*	*Bond Length (Å)*
Monomer BA (1a)	R (20C-23H)	1.0913
BA+CLFd(1b)	R (20C-22H) R (19O-25H) R (22H-28Cl)	1.0914 2.114 3.346

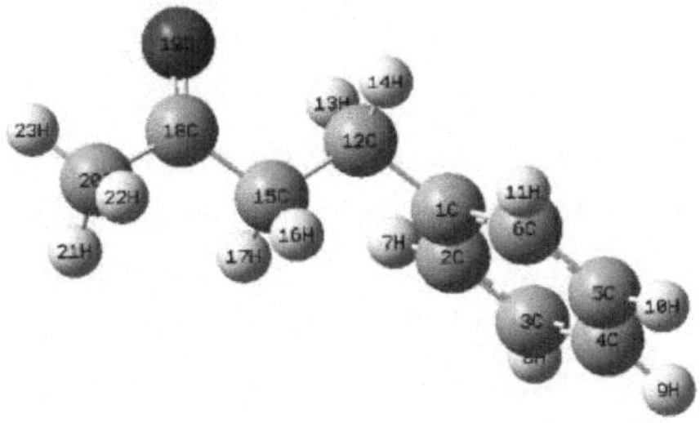

Figure 1a: Optimize Structure of Benzyl Acetone (BA)

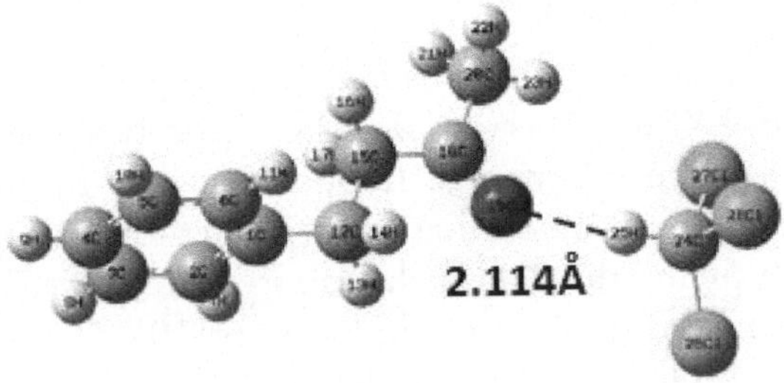

Figure 1b: Optimize Structure of BA+CLFd

B. Analysis of Molecular Electrostatic Surface Potential

The molecular electrostatic surface potential (MEP) predicts the behavior and reactivity of certain molecules. MEP is very helpful to interpret electrophilic and nucleophilic reactions qualitatively on the study of intermolecular interactions in binary mixtures. The MEP surface of BA-CLFD interacting state with solvent are presented in Figure 2.

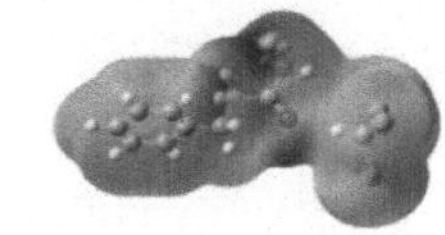

Figure 2

The red color indicates the region of most negative electrostatic potential, whereas the blue one indicates positive electrostatic potential. The green color represents the neutral site. The most favourable interacting sites with carbonyl bond have been shown clearly with red and blue colors in the figures.

C. Natural Bond Orbital (NBO) Analysis

The interaction between Lewis structure and non-Lewis structure provides the delocalization due to inter and intra-molecular interaction. The second order Fock matrix has been used to evaluate the donor–acceptor interaction in this analysis [13].

Table 2 indicates the donor-acceptor interactions of electrons with their stabilization energy for for BA-CLFd state. The interaction between lone pair electron O19 and its neighbor σ^*(C15-C18) anti-bonding orbital is found very strong having the stabilization energy 20.85 kJ/mol. Charge transfer from n1(O19) of BA to σ*(C24-H25) of CLFd have stabilization energy of 5.02 kJ/mol.

Table 2

Donor(i)	*Acceptor(j)*	*Energy(i) (a.u)*	*Energy(j) (a.u)*	*E(2) (kJ/mol)*	*E(j)-E(i) (kcal/mol)*	*F_{ij} (a.u)*
σ(C20-H21)	σ*(C18-O19)	-0.53117	0.57003	3.55	1.10	0.056
σ(C20-H22)	π*(C18-O19)	-0.52802	-0.00854	6.54	0.52	0.053
n2(O19)	σ*(C15-C18)	-0.29325	0.37080	20.85	0.66	0.106
From BA to CLFd						
σ(C18-O19)	σ*(C24-H25)	-1.09887	-0.59883	0.11	1.54	0.012
n1(O19)	σ*(C24-H25)	-0.71263	0.44068	5.02	1.15	0.068
From CLFd to BA						
σ(C24-H25)	σ*(C18-O19)	0.59883	0.57003	0.05	1.17	0.007
n2(Cl 27)	σ*(C20-H23)	0.37126	0.43837	0.48	0.81	0.018

However, charge transfer from CLFd to BA is found less having stabilization energy of 0.05 kJ/mol and 0.48 kJ/mol.

D. Molecular Orbital Analysis

The HOMO represents the capacity to donate an electron, LUMO represents the capability to obtain an electron. The HOMO–LUMO energy gap represents a marker point which manifests the chemical stability of a molecule.

Generally, the molecules having large HOMO-LUMO gaps are not reactive and stable, while those molecules with small gaps are reactive and unstable. In the present study, the BA+CLFd mixture provides large HOMO-LUMO energy gap 5.66 eV which shows a stable and unreactive one.

E. Experimental Results and Discussion

Raman spectra of C-H stretching vibration of BA are recorded for neat solute and BA-CLFd binary mixtures. Figure 3 shows the experimental and theoretical full Raman spectra of BA and BA-CLFd mixtures.

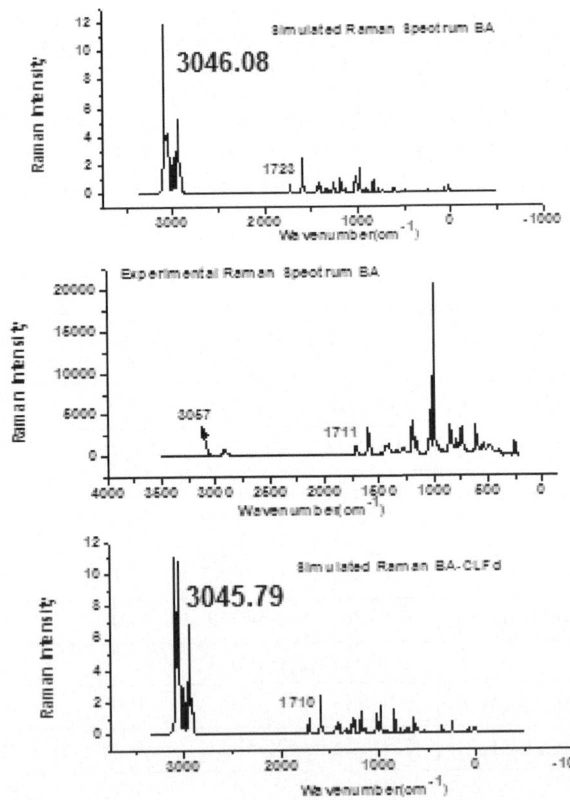

Figure 3: Experimental and Theoretical Raman Spectra showing C-H Stretching Vibration of BA and BA-CLFd Mixtures

The vibrational assignment (selected peak frequencies) of BA has been done using VEDA program and compare with experimental values (Table 3). The theoretical peak frequencies and experimental peak frequencies are found almost comparable.

Table 3: Vibrational Assignment (selected peak frequencies) of BA and BA + CLFd using VEDA and Compare with Experimental Value

Mode	*DFT (BA) (Wave number)*	*Raman (BA) (Experimental)*	*DFT (BA+CLFd state)*	*BA+CLFd (Experimental)*	*Vibrational Assignments*
1.	3056.30				υ(C-H) (ring)93
2.	3046.08	3057	3045.79	3051	υ(C^{20}-H^{23})88
3.	2933.86	2921			υs($C^{20}H_3$)84
4.	2914.95				υs($C^{15}H_2$)94
5.	1723.33	1711	1704.41		υ(C^{18} = O^{19})90
6.	1594.55	1600			υ(C = C) (ring)72
7.	1573.03	1581			υ(C = C) (ring)67

The calculated value of monomer BA is at 3046.08 cm^{-1}, however the peak frequency in the interacting state with CLFd solvent is found at 3045.79 cm^{-1} which are slightly lesser than the monomer state (Figure 3). The hydrogen bond formed in BA-CLFd state of the type C = O---D-C may lead to increase the bond length C-H. Furthermore, the bond lengths of C-H stretching vibration of BA and its interacting states are also found longer than that of monomer bond length (Table 1). The longer, the bond length, the decrease is the force constant which in turn, leads to red shift of the peak frequency.

IV. Conclusions

The band shape analysis of C-H stretching vibration of BA molecule has been carried out in BA monomer and BA-CLFd mixtures. The formation of hydrogen bond between solute-solvent system has shown major role in altering the Raman bandshape. The BA-CLFd mixtures are found stable compound. The experimental vibrational data are compared with the theoretical data and found in good correlation with the experimental results.

REFERENCES

[1] Th. G Devi and Upadhayay, G., Solvent dependent Raman band shape analysis on C = O containing molecules: Vibrational relaxation study, *Spectrochim. Acta*, 91A, 106 (2012).

[2] Upadhyay, G. and Th. G. Devi, Raman spectroscopic study of polar aprotic molecule and its molecular associations with chemical and isotopic solvents: Comparative study with quantum-chemical calculations, *J. Mol. Liqs.*, 97, 263(2014).

[3] Upadhyay, G., Th. G. Devi, Singh, R.K., Singh, A. and P.R. Alapati, Solvent dependent frequency shift and Raman noncoincidence effect of S = O stretching mode of Dimethyl sulfoxide in liquid binary mixtures, *Spectrochim. Acta*, A109, 239 (2013).

[4] Th. G. Devi and Kumar, K., Raman bandshape analysis of o-chlorobenzaldehyde: microviscosity-dependent study, *J. Raman Spectrosc.*, 35, 835 (2004).

[5] Chetry, N., and Th. Gomti Devi, Intermolecular interaction study of L-Threonine in polar aprotic solvent: Experimental and theoretical study, *J. Raman Spectrosc.*, 30, 547 (1999).

[6] Th. Gomti Devi, Vibrational relaxation study of C¼O stretching mode of liquid N, N-dimethylformamide: Comparative study with theoretical calculation, *Vib. Spectrosc*, 75, 65 (2014).

[7] Th.G. Devi, Das, A. and Kumar, K., Raman anisotropic bandwidth study of C = O Stretching vibration of methyl isobutyl ketone: role of van der Waals' volume of the interacting systems, *Spectrochim. Acta*, 60A, 211 (2004).

[8] Th. G. Devi and Kumar, K., Anisotropy shift and Raman bandwidth studies in carbonyl containing molecule o-chlorobenzaldehyde: Role of repulsive forces, *Spectrochim. Acta* 62A, 972 (2005).

[9] Th. G. Devi, Study of vibrational relaxation and molecular reorientation in liquids: A Raman spectroscopic study, *J. Raman Spectrosc.*, 41, 1261 (2010).

[10] Merrick, Jeffrey P., Moran, D. and Radom, L., An evaluation of harmonic vibrational frequency Scale factors, *J. Phys. Chem. A*, 111, 11683.

[11] Frisch, M.J., Trucks, G.W., Schlegel, H.B., Scuseria, G.E., Robb, M.A. and Cheeseman, J.R., *et al., Gaussian, Inc., Wallingford CT*, 2010.

[12] Murray, J.S., Sen, K., Molecular electrostatic potentials, Concepts and Applications, *Elsevier, Amsterdam*, 1996.

[13] James, C., A., A. Raj, Reghunathan, R., V.S. Jayakumar, Joe, I.H., Structural conformation and vibrational spectroscopic studies of 2, 6-bis(p-N, N-dimethyl benzylidene) cyclohexanone using density functional theory, *J. Raman Spectrosc.* 37, 1381 (2006).

Author Index